Low-Temperature Energy S
Applications of Renewable

Low-Temperature Energy Systems with Applications of Renewable Energy

Andriy Redko

Oleksandr Redko

Ronald DiPippo

ELSEVIER

ACADEMIC PRESS

An imprint of Elsevier

Academic Press is an imprint of Elsevier
125 London Wall, London EC2Y 5AS, United Kingdom
525 B Street, Suite 1650, San Diego, CA 92101, United States
50 Hampshire Street, 5th Floor, Cambridge, MA 02139, United States
The Boulevard, Langford Lane, Kidlington, Oxford OX5 1GB, United Kingdom

Library of Congress Cataloging-in-Publication Data
A catalog record for this book is available from the Library of Congress

British Library Cataloguing-in-Publication Data
A catalogue record for this book is available from the British Library

ISBN: 978-0-12-816249-1

For information on all Academic Press publications visit our website at https://www.elsevier.com/books-and-journals

Publisher: Joe Hayton
Acquisition Editor: Lisa Reading
Editorial Project Manager: Michelle W. Fisher
Production Project Manager: R. Vijay Bharath
Cover Design: Greg Harris
Cover Artwork: Andriy Redko and Oleksandr Redko

Typeset by TNQ Technologies

For my wife Nataliia and my son Hrihoriy:
Andriy Redko

For my wife Rayisa:
Oleksandr Redko

For my wife Joan:
Ronald DiPippo

Contents

About the authors

Andriy Redko is Professor of the Department of Heat and Gas Supply, Ventilation and the Use of Waste Heat Recovery (GSP and UWHR) of Kharkov National University of Construction and Architecture (KNUCA), Master of Construction. He received bachelor's and master's degrees from KNUCA; Degree Candidate of Technical Sciences (PHD) and degree D.S. received in the Donbas National Academy of Civil Engineering and Architecture (DNACEA). He has been teaching at KNUCA since 2006. Author of more than 100 publications and has 21 patents. Independently published a monograph on geothermal heat supply (2010) and co-authored the textbook *"Low Potential Energy"* (2016). For five years, he received a scholarship of the Supreme Council of Ukraine for young scientists for their work on improving geothermal heat supply systems. He supervises the work of graduate students in the preparation of a thesis for the PHD degree. He consults and manages projects to improve the efficiency of heat supply systems, heat generation and utilization systems of waste heat (WH) in industrial enterprises. His research interests include the modeling and optimization of thermal circuits, heat pump installations in heating and air conditioning systems of buildings and structures, the transformation of WH using steam turbine technologies. The results of research are regularly reported and published at the main specialized scientific conferences. Currently, he teaches two lecture courses at the Department of GSP and UWHR at KNUCA: Technical Thermodynamics and Gas Supply for Industrial Enterprises and Rational Use of Natural Gas.

Oleksandr Redko is Professor, Head of the Department of Heat and Gas Supply, Ventilation and the Use of Waste Heat Recovery (GSP and UWHR) of the Kharkov National University of Construction and Architecture (KNUCA), Thermal Physics Engineer, Odessa Technological Institute of Refrigeration Industry. The degree of Candidate of Science (PHD) was obtained at the Institute of Natural Gases (Moscow), the degree of Doctor of Technical Sciences (D.S.) at the Kharkov Polytechnic Institute. In KNUCA he has been teaching for over 40 years; for 30 years he has headed the department. He is the author of more than 250 articles in various scientific journals, 35 patents and five books on heat transfer and combustion of solid fuels in a fluidized bed, low-temperature processes of energy conversion and building thermal physics. He supervised the work of many postgraduate students in writing dissertations PHD and D.S. He is a member of the Academy of Construction of Ukraine. He is a member of the editorial boards of many scientific and technical collections and journals. For many years he has advised and managed projects on the use of waste heat utilization systems of industrial enterprises. Projects were implemented in the systems of municipal heat supply and heat generation. After retirement in 2008, he continues to work.

Ronald DiPippo is Chancellor Professor Emeritus, Mechanical Engineering, University of Massachusetts Dartmouth (UMD). He earned his ScB, ScM and PhD in engineering at Brown University. He taught courses as a visiting lecturer/researcher at Brown U., Cornell U., MIT, U. Nevada Reno, and U. Akureyri (Iceland). He has delivered invited keynote lectures in El Salvador, Italy and Japan, is a long-time Associate Editor for *Geothermics*, and a reviewer for many journals. He has over 130 articles in a various journals, written or edited seven books on geothermal power generation, contributed to several science/engineering handbooks, and served as a geothermal advisor to several countries, including Costa Rica, El Salvador, Guatemala and Kenya. The Third Edition of his book, *Geothermal Power Plants*, has been translated into Chinese. He has also conducted research into the thermophysical properties of fluids and ground-source heat pumps. He has provided consulting services to many companies and national labs in the U.S. At UMD, he taught for 37 years, was chair of the ME department for a total of 14 years, president of the Faculty Senate for four years, and Associate Dean of Engineering for three years before retirement in 2004.

Preface

The three co-authors, Andriy Redko, Oleksandr Redko and Ronald DiPippo, first began collaborating in July 2017, working on multistage, low-temperature power cycles. Around the same time, the Redkos proposed that we write a book dealing with low potential energy systems. After some discussion regarding the contents, all three agreed to tackle this project. We then contacted Elsevier to gauge their interest in publishing such a volume, and they expressed their willingness to review a proposal. Our proposal was accepted, and serious writing began in early 2018. The book title was established as "Low-temperature energy systems with applications of renewable energy."

A great deal of the material seen in the book was originally written in Ukrainian by the Redkos. Translation into English was done in Ukraine, and a final editing was carried out by DiPippo with the goal of making the presentation of the material as concise, clear, and as easy to understand as possible. DiPippo also participated in the writing of each section, wrote new material for this volume, and contributed one chapter by himself, Chapter 6.

The text and examples of using low-temperature energy systems are based mainly on the results of the authors' scientific and technological research. This book can be used in engineering and technology educational programs for a deeper understanding of the processes involved in converting low-temperature energy into useful applications. The authors will consider their work successful if this book is found to be useful in educational and training courses for students at technical universities and for technologists in industry.

Motivation for this book stems from the recognition that the world needs to examine its use of energy in light of a growing population and the finite nature of most of the sources of energy in use. Renewables such as solar, wind, biomass, and geothermal are being developed and deployed everywhere, as feasible, but much of the thermal energy derived from conventional energy sources such as coal, oil, natural gas, and nuclear is wasted. On a broad average, about two-thirds of the thermal energy from these fuels ends up being discharged to the environment in the form of medium- to low-temperature heat. This book describes many ways that such waste heat can be at least partially captured and put to use, thus saving a measure of primary energy and mitigating the environmental impact of the energy usage.

One promising technology in the realm of energy saving systems is heat pumps. In many countries, heat pumps are becoming the preferred means of providing indoor comfort conditions as well as industrial uses involving low-grade waste heat.

Chapter 1 focuses on this technology. We begin with a snapshot of world energy usage, with an eye to renewable energy sources. A brief summary is given of pertinent energy policies around the world, along with procedures needed to create and assess energy efficient systems. We then trace the history of heat pumps and refrigerators and describe these machines both qualitatively and quantitatively. Thermodynamically ideal machines that set the standard for real-world systems are analyzed. Practical heat pump and refrigeration systems are described in technical detail and analyzed using the principles of thermodynamics. Both vapor-compression and absorption refrigeration systems are discussed. The process of selecting appropriate working fluids is described with special emphasis on the environmental impacts and international protocols. The chapter concludes with a discussion of modes of operation of heat pumps and cold storage systems.

In Chapter 2 we review the main systems using heat pumps in heat supply systems for residences and public buildings using various low-grade heat sources. Heat supply sources include: ambient air, water, and soil. Limitations on the use of these sources are discussed. Heat pump performance is derived and presented as functions of technical parameters. This is done by means of analytical and practical methods, together with numerical examples of typical calculations. The results show optimum operating points where energy losses are minimized. Finally, we summarize the status of worldwide heat pump usage.

Next, in Chapter 3, comes a wide range of heat pump heating applications from individual private residences to modern skyscrapers to airports and other public facilities. Although mainly technical topics are presented, the economics of various systems is covered in the last section where equivalent coal consumption is used as the basis for comparison. Heat pumps are shown to be very effective in controlling the conditions in pools, particularly indoor units where humidity can pose challenges for designers. Water-loop systems are becoming popular in many places owing to their inherent flexibility to allow heating and cooling simultaneously in large buildings with variable thermal loads. In the case of hot-water heating systems, detailed assessments are carried out and the results presented for a conventional water boiler and heat pumps with three different means of powering the heat pump compressor, namely, electric, diesel, or gas turbine. A heat pump with a gas turbine compressor drive is shown to be the most efficient based on several thermodynamic and practical measures.

As we have discussed in the first three chapters, heat pumps have become well known for providing cost-effective heating and cooling in domestic and commercial buildings. Moreover, for countries with a large agricultural sector, a large share of energy is consumed in crop drying. Chapter 4 explains in detail how heat pumps can be used in the drying industry. One section is dedicated to presenting a compendium of general arrangements that have applications in various aspects of drying. We cover eight different cases in a survey of applications, followed by sections covering two particular applications, namely, grain and wood drying. The first one is important in preparing foodstuffs for humans and animals, while the second one plays a vital role in the construction and furniture industries. The thermodynamic advantages of heat pumps over conventional drying methods are developed. The main obstacles limiting the use of heat pumps in industry are high initial capital cost which might lead to uneconomic returns on investment, and the perceived risk on the part of industry investors who may view heat pump technology as new and not sufficiently proven in commerce, with limited examples of successful applications. This book may help change these largely uninformed perceptions.

In Chapter 5 we change direction and look at geothermal energy as a source of heat. Geothermal energy, a renewable resource if properly managed, can be used either for the generation of electricity or for a great variety of direct heat applications. One of the authors (DiPippo) has published extensively on the first use, while here we look in detail at numerous examples of direct heat applications from around the world. Considerable space is devoted to renewable municipal district heating systems, agricultural uses, recreational applications, and geothermal aquaculture systems, including ones associated with geothermal electric generation stations. The chapter concludes with a theoretical presentation of 1- and 2-stage geothermal heat pumps used for heating purposes, along with several examples showing how the systems perform under a variety of conditions.

Given that geothermal energy has a dual-purpose nature, Chapter 6 deals with combining those two purposes, namely, geothermal combined heat and power (CHP) systems. CHP is a highly efficient means of capturing the available energy, or exergy, in the geothermal resource. The basic configurations and fundamental thermodynamics of CHP systems are presented. The spectrum of direct heat uses is shown as a function of temperature, but detailed analyses of these cases are not covered here as they are presented in other chapters. Technical and environmental considerations determine the best working fluids for CHP plants. Optimum systems are found that share the heat and power loads appropriately to achieve the best thermodynamic and economic performance. Both energy and exergy bases are used to assess system performance. Lastly, two cases studies are described: (1) a municipal district heating system in Japan where a power plant supplied waste, hot brine to heat water for a nearby town, and (2) a campus system in the United States that provides both electricity and heat for all of the buildings including some snow-melting for walkways.

The production of fuels from biological organisms, biofuels, is discussed in Chapter 7 both from a technical and economic viewpoint. A brief summary of the status of biofuel installations around the world provides the context for this chapter, and gives the reader a glimpse at the state of the art of biofuel facilities. The technological processes involved in the transformation of organic waste into useable biofuels are described in detail. The theory is presented along with some typical commercial systems that are available in the marketplace. We show how heat pumps and heat recuperators can be used to improve the efficiency of biogas reactors, and present the results of calculations carried out to illustrate to performance of these systems. All of these topics are important in light of the possibilities presented by waste agricultural products that may be used to generate fuels and reduce the demand for conventional, non-renewable sources.

In Chapter 8 we discuss various hybrid systems involving renewable energy sources. We begin with solar power systems including large-scale photovoltaic power stations. Unique systems such as floating photovoltaic systems and co-located solar-hydroelectric hybrid plants are described. A section covers solar thermal heating systems, followed by a detailed presentation of solar-assisted heat pumps for several applications in which solar energy is used as a topping or bottoming source of heat. Geothermal and solar energy may be used together in a synergistic manner for both district heating and electrical power generation. These applications are presented in technical detail for various arrangements. Geothermal energy can be combined with non-renewable energy resources, and such systems are also described in this chapter. The interesting topic of hybrid geothermal-biomass power systems shows effective ways to combine these two renewable energy sources provided certain conditions are present. Lastly, the chapter deals with the general analysis of optimum operating conditions for hybrid fuel-geothermal heat pumps. The method is illustrated with a case study of a small municipal heat pump system.

The massive amounts of waste industrial heat are examined in Chapter 9. We look at a variety of industrial processes and plants that release useable amounts of waste heat in several forms and at different temperatures. Methods for capturing at least some of that energy are presented. We consider applications where the waste heat recovery system is co-located with the source, such as electricity generation, as well as remote cases, such as district heating systems. In the latter case, heat transfer fluids are needed and several groups of fluids are discussed. Rankine cycles are the primary means of converting waste heat to electricity; systems using both steam and organic working fluids are presented. Natural gas pipelines offer opportunities for waste heat and energy recovery; these are described in detail with practical examples drawn from operating plants. Particular attention is given to the highly energy intensive industries of cement and glass making. The novel supercritical carbon dioxide Brayton cycle is presented as an effective means of capturing waste heat from moderate temperature sources. The chapter concludes with a discussion of waste-heat driven absorption chillers and heat pumps.

Throughout the book we provide nearly 150 review questions and over 30 quantitative exercises for the reader to test his/her understanding of the material. Ample references are cited to allow the reader to carry out an in-depth study of the topics presented.

It is the authors hope that this book will draw much-needed attention to low-temperature energy systems, and show how they can be exploited in ways that will contribute to the energy needs of humankind in a sustainable and environmentally friendly manner.

Acknowledgments

The authors express our gratitude to several people from Elsevier for their encouragement and support from the beginning of this project. Lisa Reading, as Senior Acquisitions Editor, handled all the protocols and paperwork to bring this book from a proposal into being. Michelle Fisher, Editorial Project Manager, took over the day-to-day correspondence and helped us overcome problems along the way. She also coordinated the front cover design and rear cover blurb. Praveen Kumaraswamy, Senior Copyrights Coordinator, initially assisted in acquiring permissions to use materials sourced from others. Later, that task was taken up by Kavitha Balasundaram, Copyrights Coordinator. Vijay Bharath Rajan, Production Manager, ably conducted the production phase of turning our manuscript into a real book.

The Redkos thank our translator, Ms. Nina Tychina, who turned the Ukrainian text into English, and our graduate student, Ms. Anastasiia Davidenko, for her computer typing and graphics that produced many of the drawings. And lastly, DiPippo thanks Andriy and Oleksandr Redko for their close cooperation and patience, as we worked through seemingly endless iterations of passages of text, diagrams, and illustrations, communicating only by e-mail, across seven time zones.

Principles and operation of refrigeration and heat pump systems

1

1.1 Trends in usage of low-temperature technologies

Modern generation of electricity and world energy consumption as of 2016 is provided mainly by fossil energy resources [1]. In many countries of the world there is a diversified use of energy resources. Long-term energy security, meaning "the uninterrupted availability of energy resources at an affordable price" [2], requires the timely supply of energy resources to meet the requirements of the economy and population, taking into account the challenges of sustainable development.

In countries around the world, the substitution of organic fuels is developing in different directions. In European countries, heat pump technology and organic Rankine cycle (ORC) technologies in combination with renewable energy sources (RES) are widely used. Solar energy is developing in Asia (China, Japan), Africa and Australia. Using solar and wind energy, Saudi Arabia plans to become an exporter of electricity. It is predicted that by 2050 global investments in RES will have prevailed in the use of wind (34%), hydro (30%) and solar (18%) energy [3]. Figures 1.1 and 1.2 show statistical data on the state and prospects of the development of global electricity, including RES [4].

The United States (US) plans to generate electricity from three sources: natural gas, solar energy and wind energy. The share of RES in the US of installed capacity was 8.5% in 2017 and is 12% in 2019 [5]. The Parliament of the European Union (EU) adopted a resolution in 2016 on a strategy of heating and cooling, according to which the technologies of direct combustion of fuel for heating purposes are admitted to be technically unpromising. In 2010 Germany, which imports about 71% of its hydrocarbons, adopted a state program aimed at increasing the share of renewable energy by 2050 from 10 to 60%. In Germany in 2015, the share of power generation by RES was 30% [6].

Trends in the development of world power engineering show that the generation of energy using RES will be decisive in the next two decades. At the same time, there is a significant increase in energy consumption in the world due to the economic development and population growth of the planet [3,7,8]. Despite the decrease in energy intensity, the world demand for energy resources will have grown by 30% by 2030 compared to the level of 2014. The main consumer of energy will be industry, where demand will increase by 40%. In second place in energy consumption will be

Low-Temperature Energy Systems with Applications of Renewable Energy
https://doi.org/10.1016/B978-0-12-816249-1.00001-7

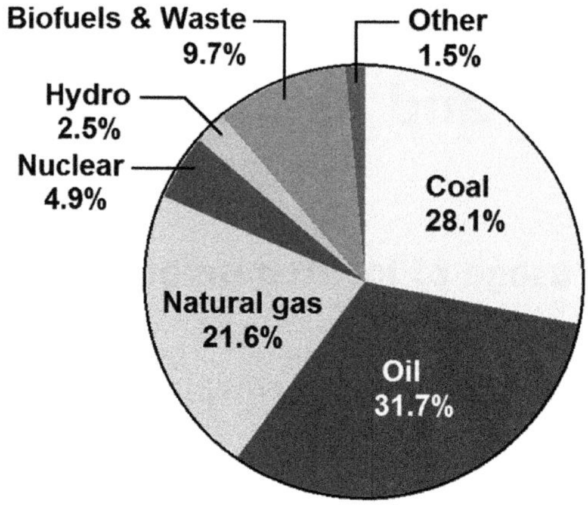

Fig. 1.1 World total primary energy supply (TPES) by fuel in 2015 [4].

the transport sector, followed by commercial and residential buildings [9]. The share of RES in the world fuel and energy balance in 2014 was 18.4%, in 2020 it will be 26% [10].

Assuming that energy consumption increases and energy conservation intensifies, heat pump technologies and ORC technologies in combination with RES will find more applications. Low-temperature power engineering systems are used to produce refrigeration (or "cold"), heat, and electricity, with natural, industrial, and domestic sources providing the heat. The temperature of low-to-medium-grade heat sources varies from 0 °C to 350 °C. In the context of increasing energy efficiency of various industries and processes, there is a need for careful energy audit of heat-consuming technological processes to determine the possible use of reserves to reduce energy

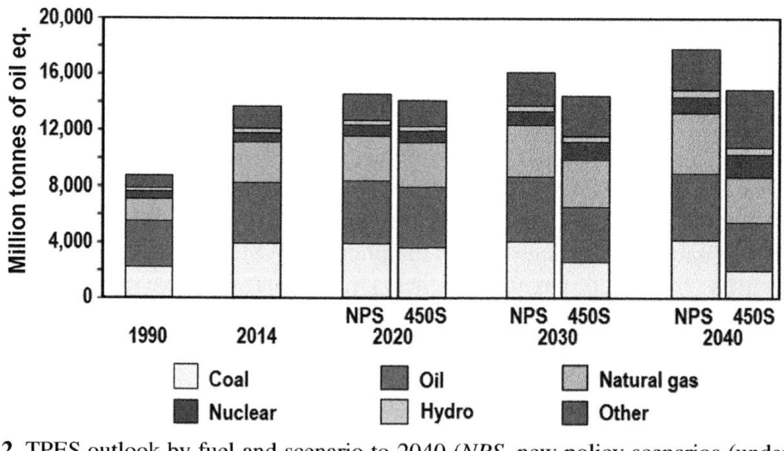

Fig. 1.2 TPES outlook by fuel and scenario to 2040 (*NPS*, new policy scenarios (under consideration), *450S*, policies needed to limit global average temperature increase to 2 °C) [4].

losses. Low-temperature technologies should be characterized by a high degree of energy efficiency, since with a decrease in the temperature of the source of heat, the cost of equipment for energy conversion systems increases and the attractiveness of innovative projects decreases. Sources with a temperature of up to 100 °C are sources for the application of heat pump systems that provide savings and replacement of organic fuel by 1.5–2.0 times or more. Sources with temperatures above 100 °C can be used as sources for heat-using refrigeration machines (i.e., absorption systems), in heat exchangers in cogeneration and systems, as well as in vapor-powered organic cycles (ORC cycles) for generating electricity and heat. A promising direction is the creation of hybrid low-temperature power plants using RES (solar, hydro, geothermal, wind and bioenergy) as well as waste heat of industrial enterprises and household sources.

The main shortcomings of renewable energy sources are the low energy density of flows and their inconstancy over time, dependence on natural and geographical factors, whereas their advantage is their distributed nature that does not require extensive infrastructure for local consumption, which in combination with high-tech, low-temperature energy systems ensures their more efficient use.

1.2 Worldwide energy saving policies

All industrialized countries have applied energy-saving laws that meet specific local conditions in practice since the 1970s. For example, Sweden is pursuing all energy conservation activities on the basis of the "Energy Saving Code" (1977–80); France – on the basis of the "Law on Energy Savings" (1980); Germany – on the basis of the "Law on Energy Saving" (1979); Japan – based on the "Law on the rational use of energy" (1979) [11]. The main laws governing the industry in USA are National Energy Act (1978), Public Utility Regulatory Policies Act (PURPA), Energy Policy Act (1992), and the Energy Independence and Security Act (2007).

The regulatory framework in the EU countries provides the design of the construction and operation of heating and cooling systems using heat pump units. Currently, the following standards are applied:

1. DIN 8901-2002. Refrigerating systems and heat pumps. Safety and environment requirements and testing.
2. DIN EN 378-1, 2, 3, 4 – 2017. Refrigerating systems and heat pumps. Safety and environmental requirements.
3. DIN EN 13313 – 2011. Refrigerating systems and heat pumps. Competence of personnel.
4. DIN EN 14511-1, 2, 3, 4 – 2013. Air conditioners, liquid chilling packages and heat pumps with electrically driven compressors for space heating and cooling.
5. DIN EN 14825 – 2016. Testing and rating at part load conditions and calculation of seasonal performance.
6. DIN EN 15879 – 2011. Testing and rating of direct exchange ground coupled heat pumps with electrically driven compressors for space heating and/or cooling. Part 1.
7. DIN EN 16905-1 (3-5) – 2017. Gas-fired endothermic engine driven heat pump. Part 1, 3, 4, 5.

8. NF EN 12309-2 − 2000. Gas-fired absorption and sorption air-conditioning and/or heat pump appliances with a net heat input not exceeding 70 kW. Part 1, 2.
9. NF EN 378 − 1, 2 et NF EN 14511-4 OENORM.
10. DIN EN 8960. Refrigerants. Requirements and symbols. 1998.
11. VDI 4640. Technical use of soil.
12. DIN EN 12831. Exact indicators of building heat consumption.
13. DIN V 4701-10. Heating load. Coefficient of expenditure.

1.3 Fundamentals of energy management and audit of refrigeration and heat pump facilities

Power engineering management (PEM) is a main means for reducing energy consumption and increasing the efficiency of energy use at industrial enterprises for the production of artificial cold. PEM is a control system based on the implementation of typical measurements and inspections that provide such a work of an enterprise, during which only the required amount of energy is consumed. PEM is an enterprise management tool that provides ongoing research, and, as a result, knowledge relating to the distribution and level of energy consumption in an enterprise as well as on the optimal use of energy resources both for production, heating, and other non-productive needs as well.

Power engineering audit (PEA) is a technical and economic inspection of energy generation and energy demand systems of an enterprise in order to determine the feasibility of cost economy for the needs of fuel and energy resources, the development of measures that ensure the conservation of energy resources and money, to eliminate unacceptable energy losses by the introduction of more economical schemes and processes.

Tasks of PEA are the following: to identify sources of irrational costs and energy losses; and to develop recommendations and programs on energy saving. The overall PEA strategy includes stages, each of which requires a special approach.

At the first stage, the auditor makes preliminary contact with the management of the enterprise, and also becomes acquainted with the main production processes, and concludes an agreement with the management of the enterprise on further work. After the first contact, it is necessary to identify the point of view of the company's employees on energy saving issues and determine their approaches to this problem, in particular, to find out what energy saving at this enterprise was tried earlier and what plans for energy saving are in the future.

At the second stage, the auditor creates energy consumption maps of the enterprise and verifies the possibilities for significant energy savings. The map of energy consumption is created on the basis of additional measurements at the nodal points of the technological scheme of an enterprise production process with the help of various portable or stationary devices and meters. The map can be based on calculations if nominal power and annual output of engines are known. In order to determine the theoretical potential of energy savings, it is useful to compare key data, for example, energy consumption per ton of products or 1 m^2 of production floor area, with data from special literature, information from similar industries, and other similar materials.

The third stage assesses the energy savings and economic benefits of implementing the various possible measures, and chooses a specific energy saving program for immediate implementation. After that, the key technical and economic data are determined.

The fourth stage involves the implementation of an energy saving program, the launch of the PEM system as well as the continuation of the survey, the study of the results obtained, etc. At this stage, a specific energy saving program is planned and implemented, in which the energy auditor should not participate, since it must be a person independent of suppliers and manufacturer of equipment. However, he may enter into a contract for the implementation of monitoring and consulting functions in the process of implementing the program.

For example, the methodology of conducting a PEA of a refrigeration facility includes the following stages:

- Calculation of energy consumption and its cost;
- Calculation of energy costs in different sectors of production;
- Critical analysis of energy flows;
- Formulation of a project on energy saving measures;
- Assessment of the project on energy saving measures;
- Reporting.

A quantitative view of current energy consumption and its cost outlines the scale of the problem and determines where it is necessary to focus efforts to achieve the best results of energy saving. The most comprehensive understanding of the dynamics of the annual energy consumption provides a schedule for recording monthly energy consumption. When calculating energy consumption and its cost, it is extremely important to determine the level of consumption of individual energy sources and their correlation with the total energy consumption. In the analysis of energy consumption for cold production, the main thing is the estimation of electricity costs.

The analysis of electricity consumption has the following tasks:

- Determining the total power of the supplied feeding and total power of connected loads;
- Determining daily and seasonal load fluctuations;
- Determining the possibility of implementing a load curve;
- Determining the average annual power factor;
- Determining the means of correction of the power factors at the facility;
- Classifying the consumed electric energy by special purpose (electric engines, heating, lighting, specific technological processes, etc.).

As a result of the initial calculations, attention should be paid to the availability of information concerning:

- Total cost of energy resources at the facility (including water supply costs);
- Consumption ratio according to the types of fuel;
- Nature of seasonal fluctuations in fuel consumption;
- Prices.

On the basis of the information obtained and the process flow diagram, the relative sizes of energy flows and losses are estimated and lists of main energy consumers are compiled. In determining the power consumed by end-users, it is advisable to use additional meters or other measuring devices. In case of estimating electric power flows in the absence of stationary meters, portable electric meters should be used, the application of which does not require the breakdown of electrical networks.

In the process of inspecting the production site (a compressor shop, a refrigerator as a whole), it is necessary to study all stages of the technological production process and to calculate the balance of energy flows and materials at each stage. When determining the flow rate of heat and cooling agents, refrigerants, and cooling water in the absence of stationary meters and flow meters, it is expedient to use flow meters with external high-frequency sensors, for which there is no need to be in direct contact with the measured flow of substances. The information relating to the consumed and installed capacity of the equipment (compressors, pumps, fans, heaters) can be obtained from the passport data of the equipment manufacturers, but this information should be used with caution as this equipment could be upgraded or reconstructed during operation. Also, one should take into account the nature of the production process (round the clock, continuous or discrete with a certain period); technicians, technologists, operators and process managers are the ones best able to answer these questions.

After the identification of main energy consumers, it is desirable to carry out an assessment of energy consumed. If consumers are particularly large or complex, then further distribution of energy consumption is required to improve the quality of energy efficiency measures.

1.4 Energy-saving cooling and heat pump systems

1.4.1 Physical principles of heat pumps and cooling systems

Thermodynamic principles of heating and cooling require that work must be performed on a system that transfers heat from a body with a lower temperature to a body with a higher temperature, in accordance with the Second Law of thermodynamics [12,13]. In the engineering device performing this task, the thermodynamic properties of a working medium are changed during the processes of decreasing and increasing the temperature.

Lowering the temperature of a working fluid in refrigeration machines may be effected by using various thermodynamic processes, such as: (1) the Joule-Thompson effect, i.e., a flow throttling valve; (2) Ranque-Hilsch vortex tube, i.e., a means of producing a cold and a hot stream from a compressed gas; (3) an expansion process performing external work; (4) the Peltier effect, i.e., a thermoelectric phenomenon; (5) the magneto-caloric effect, i.e., adiabatic demagnetization; and (6) gas desorption. The first one, namely, a throttle valve, is the simplest and the most commonly used in practical vapor-compression systems.

1.4.2 Qualitative description of vapor-compression heat pump operation

Most people have a heat pump of some sort in their house, even without knowing it. A refrigerator or freezer operates as a heat pump, only with a different focus. The refrigerator uses the "cold" side and the heat pump mainly uses the "hot" side. The heat pump takes thermal energy or heat from the environment, i.e., soil, water or air, "pumps" the working fluid (cooling agent or refrigerant) to a higher temperature, which then delivers heat to the heating system. All this happens in a closed circuit in which the cooling agent circulates; an evaporator, compressor, condenser and expansion valve are the significant components of this circuit (see Fig. 1.3).

The components perform the following tasks:

Evaporator: An evaporator is a heat exchanger in which thermal energy is absorbed by the working fluid, thereby cooling its surroundings. Thus there is a transfer of heat from the surroundings to the working fluid inside the evaporator. A cooling agent flows through the evaporator with low pressure and at near-ambient temperature. As it receives heat from the ambient air, which must be at a higher temperature than the cooling agent, it evaporates. The temperature of the air must be higher than cooling agent otherwise there will be no heat transfer from the air to the agent by the Second Law of thermodynamics.

Compressor: A compressor receives the cooling agent in the vapor state from the evaporator and increases its pressure and, consequently its temperature. For this purpose, the compressor requires power input, usually supplied by a motor driven by electricity from the power supply network. Upon compression, the now hot refrigerant vapor goes from the compressor into the condenser.

Condenser: A condenser is a heat exchanger, like an evaporator, but performing the opposite process. Inside the condenser, the superheated cooling agent releases heat energy to a colder heat carrier, for instance, water in a heating coil. Under the action of temperature difference, heat from the hot refrigerant goes to the heat carrier. As a result, the refrigerant condenses and a heat carrier is heated, while the pressure of the refrigerant remains high.

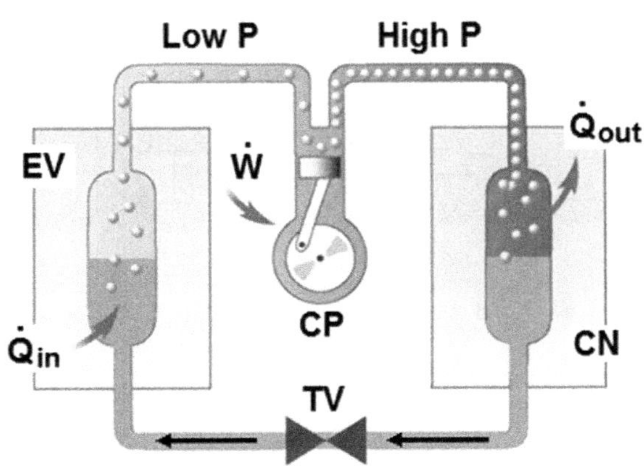

Fig. 1.3 Simplified schematic of heat pump/refrigeration system. *P*, pressure; *EV*, evaporator; *CN*, condenser; *CP*, compressor; *TV*, throttle valve.

Expansion valve: An expansion valve is used to suddenly drop the high pressure created by a compressor, and in so doing reduce the cooling agent temperature to a level lower than the heat source, thus renewing the cycle. This throttling process is the simplest one to achieve the cooling via the Joule-Thomson effect mentioned earlier.

1.4.3 Historical facts on heat pumps

Now that we understand the basic principles of heat pumps/refrigerators, it is interesting to look back at how these now common devices came into being.

The principle of the heat pump follows from the work of Carnot and the description of the Carnot cycle, published in his book "Reflections on the Motive Power of Fire and on Machines Fitted to Develop that Power" in 1824. The practical heat pump system was introduced by William Thomson in 1852. It was called the Heat Multiplier and showed how to use a refrigerator efficiently for heating purposes. In substantiating his suggestion, Thomson pointed out that the limited energy resources would not allow the burning of fuel in heating furnaces forever, and that his heat multiplier would consume less fuel than conventional furnaces. As can be seen from Fig. 1.4, the heat pump proposed by Thomson uses air as a working fluid. Ambient air (1) is sucked into cylinder (2), expanded and as a result cooled off; it then passes through a heat exchanger (3) where it is reheated by external air. After recompression (6) to atmospheric pressure and to a temperature above the environment, air from the cylinder enters the room thereby providing heat.

There is evidence that the first implementation of such a machine was in Switzerland. Thomson said that his heat pump was able to provide the required heat when using only 3% of the energy used for direct heating. Refrigerating machines were already being developed at the end of the nineteenth century, but heat pumps

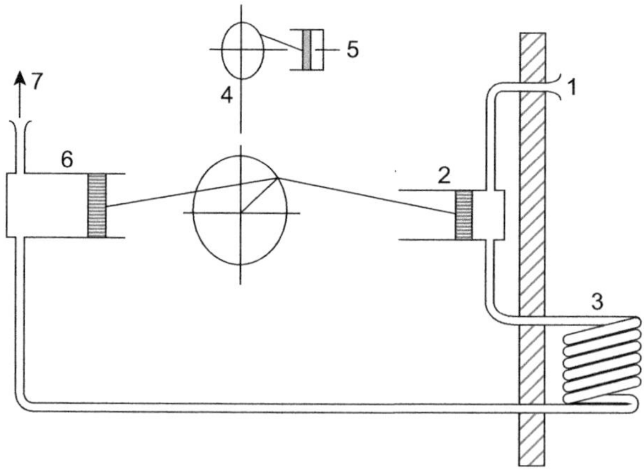

Fig. 1.4 The scheme of Thomson's "heat multiplier": 1 = ambient air; 2 = input cylinder; 3 = heat exchanger; 4 = drive mechanism; 5 = steam engine; 6 = output cylinder; 7 = room being heated.

had a rapid development only in the 1920s and 1930s when the first heat pump unit was set up in England. In 1930 Holden described the testing of a domestic heat pump designed for heating and hot water supply, while using the heat of ambient air. After that, work started in the United States and resulted in demonstration plants, but relatively few projects were brought to this stage, since all of them had only private funding. The first large heat pump plant in Europe was put into operation in Zurich in 1938–9. It used the thermal energy of river water, a rotary compressor, and a refrigerant. It provided heating for the city hall with water at a temperature of 60 °C and at a thermal power of 175 kWt. There was a system of heat accumulation with an electric heater to cover the peak load. In summer months, the installation was operated for cooling [14].

1.5 Efficiency of heat pump and refrigeration systems

This section will develop equations for assessing the performance of heat pumps (HPs) and refrigeration machines (RMs). It is important to first examine ideal thermodynamic systems so that we may compare real systems to these to find where improvements may be made in any particular design [15,16].

1.5.1 Ideal measures of performance

Heat pumps are energy-consuming devices used to create heating or cooling effects. This is the opposite of power plants that use temperature differences to drive cycles that produce power output at the expense of high-temperature energy. Carnot introduced a cycle with the highest thermal efficiency possible given a fixed set of heat reservoirs – the Carnot cycle. That efficiency formula is well-known to all who understand thermodynamics, namely,

$$\eta_C = 1 - \frac{T_L}{T_H} \tag{1.1}$$

where T_L is the absolute temperature of the heat sink and T_H is the absolute temperature of the heat source. Figure 1.5 shows the Carnot cycle in a temperature-entropy (*T-s*) diagram. All processes are reversible in this ideal cycle. Knowing that the heat transfer during a reversible process is equal to the area under the process in a *T-s* diagram, it is easy to derive Eq. (1.1) from its basic definition as the ratio of the net work to the heat input.

Next we will find an ideal measure of performance for a heat pump. Given that a heat pump is built to deliver heat at the expense of work input, the coefficient of performance for a heat pump, COP_{HP}, is defined as the ratio of the heat delivered to the net work input. Figure 1.6 shows the ideal heat pump cycle in a *T-s* diagram, where the cycle is traversed in a counter-clockwise direction, opposite to a power cycle. By analogy with the Carnot efficiency, it is easy to determine that the ideal COP_{HP} is given by

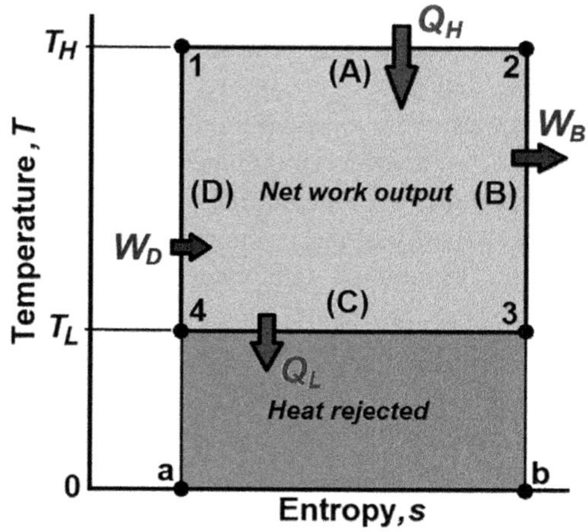

Fig. 1.5 Carnot cycle in *T-s* coordinates.

$$COP_{HP} = \frac{T_H}{T_H - T_L} \tag{1.2}$$

In principle, a heat pump can be viewed as a refrigerator by focusing on the heat removed from the low-temperature heat reservoir as the objective of the device. Thus the system can be used to provide cooling, air conditioning, and ice-making under the proper conditions. In such a case, the ideal coefficient of performance of refrigeration, COP_R, is given by

$$COP_R = \frac{T_L}{T_H - T_L} \tag{1.3}$$

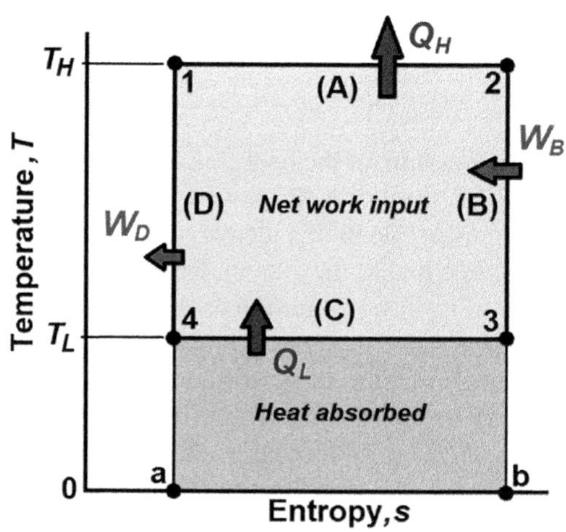

Fig. 1.6 Ideal heat pump cycle in *T-s* coordinates.

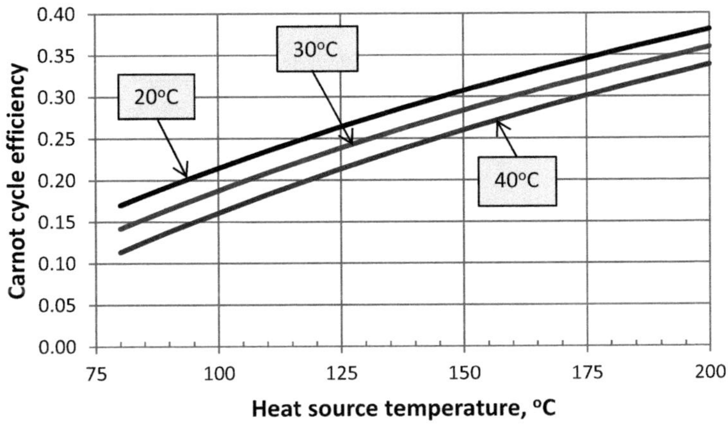

Fig. 1.7 Carnot power cycle efficiency versus heat source temperature for three values of heat sink temperature.

It is useful to plot these three measures of ideal system performance to gain an appreciation of how well real systems perform. Figure 1.7 gives the Carnot cycle efficiency as a function of the temperature of the heat source for three different heat sink temperatures. The lower the sink temperature, the higher the efficiency. For low-temperature sources, the efficiency is less than 30%, even for the lowest sink temperature of 20 °C; for very low temperature sources, the efficiencies are lower than 20%.

Figure 1.8 gives the ideal heat pump COP_{HP} as a function of the heat delivery temperature (or desired room temperature) for three different heat source (or ambient) temperatures. The lower the source temperature, the lower the COP_{HP}. From Eq. (1.3) it is clear that the performance decreases as the temperature difference across the heat pump increases. This is opposite the case for a power producing cycle.

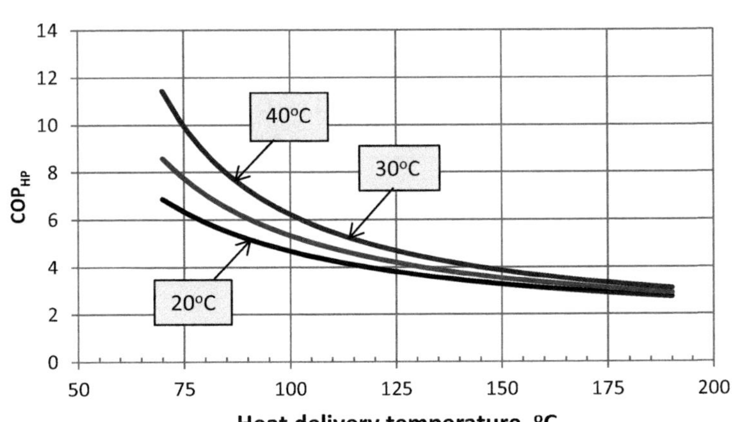

Fig. 1.8 Ideal heat pump cycle COP versus heat delivery temperature for three values of heat source temperature.

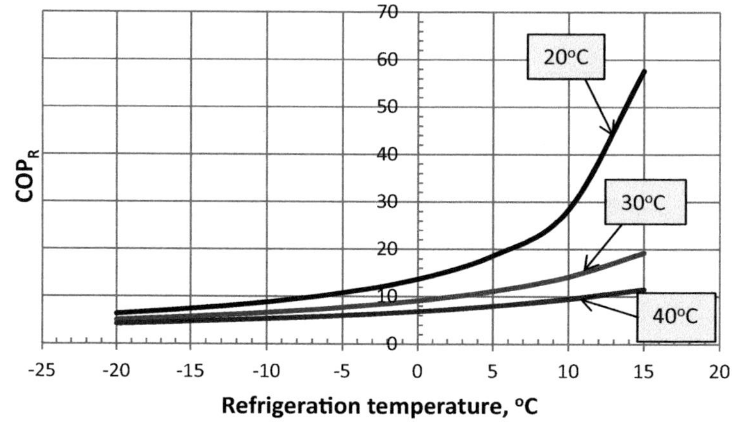

Fig. 1.9 Ideal refrigerator cycle COP versus refrigeration temperature for three values of ambient (or sink) temperature.

Figure 1.9 gives the ideal refrigerator COP_R as a function of the cold space (refrigerator) temperature for three different heat sink (or ambient) temperatures. Contrary to the case of the heat pump, the lower the sink temperature, the higher the COP_R. In fact from Eq. (1.4) it is clear that the performance grows infinitely large as the temperature difference across the heat pump decreases toward zero. Of course, if the temperature difference across the refrigerator approaches zero, there is no need for a refrigerator at all.

Figure 1.10 shows a comparison between semi-ideal refrigeration machines and heat pumps, where the temperature differences needed to allow heat transfer to take place are also shown. Heat transfer in a heat exchanger must always occur from the hotter body to the cooler one. This also illustrates Thomson's multiplier effect since the output heat at the higher temperature is equal to the sum of the heat input at the

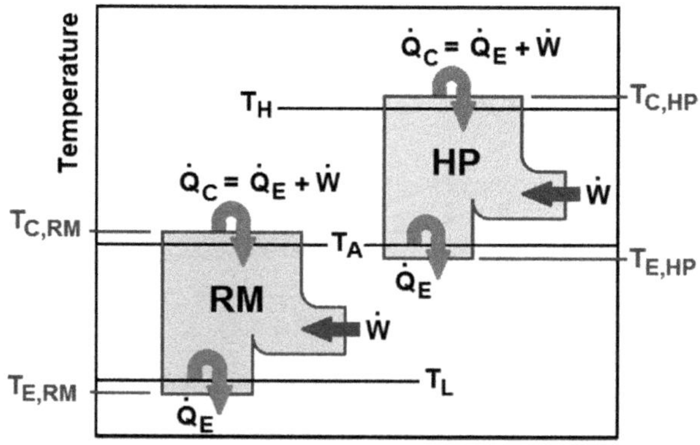

Fig. 1.10 Schematic of heat and work flow for refrigeration machines (RMs) and heat pumps (HPs); see Nomenclature.

lower temperature plus the work input to the cycle. This fact is always true whether the cycles are ideal or real (by conservation of energy), and results in high coefficients of performance for heat pumps.

1.5.2 Thermodynamic efficiency of vapor compression heat pumps

The ideal heat pump derives its high performance from the isothermal heat transfer that occurs at both the high- and low-temperature ends of the cycle. Practical considerations and the natural saturation pressure-temperature relations for real fluids conspire against such ideal heat transfer. Real heat pumps typically employ vapor compression processes to achieve high temperatures needed for the application at hand. This results in non-isothermal heat transfer between the heat pump working fluid and the space to be heated. Isothermal heat transfer is still possible at the low-temperature end.

Figure 1.11 shows in simplified schematic form a heat pump that uses the vapor-compression principle. There are five main components: a compressor (CP), a motor (M), two heat exchangers (one an evaporator, EV, and one a cooler-condenser, CN) and a throttle valve (TV). The motor drives the compressor which raises the pressure of the working fluid from its saturation pressure corresponding to the temperature of the condenser to the saturation pressure corresponding to the high-temperature requirement.

The processes undergone by the heat pump working fluid are shown in Fig. 1.12: (A) *T-s* diagram and (B) pressure-enthalpy (*P-h*) diagram. State d is the dew-point where the compressed vapor first begins to condense upon cooling.

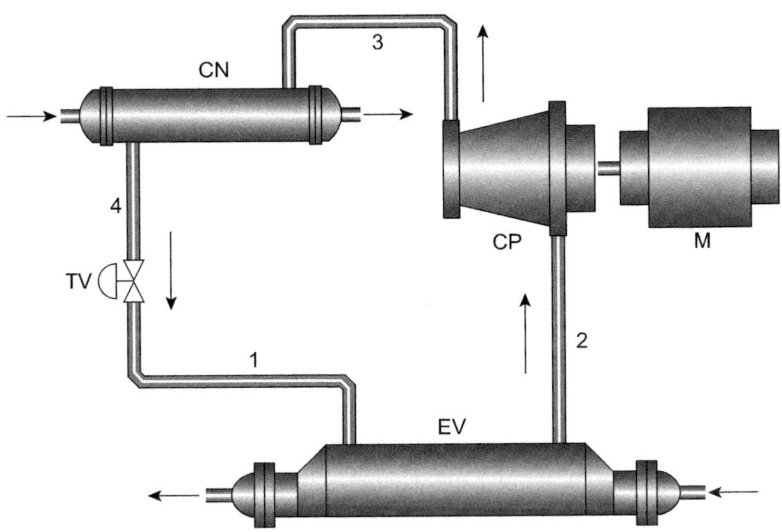

Fig. 1.11 Vapor-compression heat pump schematic.

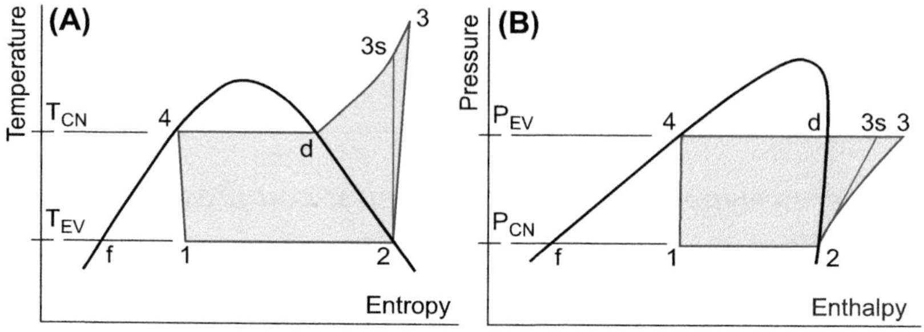

Fig. 1.12 (A) *T-s* and (B) *P-h* diagrams for a vapor-compression heat pump.

The processes may be described as follows:

1-2 – evaporation of heat pump working fluid in thermal contact with the surroundings; heat input to the cycle, $Q_{1,2}$
2-3s – ideal isentropic process of compression
2-3 – actual compression process driven by a motor; work input to the cycle, $W_{2,3}$
3-d – sensible cooling of working fluid in thermal contact with the space being heated; heat output from the cycle, $Q_{3,d}$
d-4 – latent heat of condensation transferred from working fluid to the space being heated; heat output from the cycle, $Q_{d,4}$
4-1 – throttling of the working fluid down to the saturation pressure in the evaporator; no heat or work transfer, i.e., constant enthalpy.

The COP_{HP} is the ratio of the heating effect, $Q_{3,d} + Q_{d,4} = Q_{3,4}$ to the work input, $W_{2,3}$:

$$COP_{HP} = \frac{Q_{3,4}}{W_{2,3}} = \frac{h_4 - h_3}{h_3 - h_2} \tag{1.4}$$

The actual compressor outlet state 3 may be determined from the isentropic outlet state and the compressor isentropic efficiency:

$$\eta_{CP} = \frac{W_{2,3s}}{W_{2,3}} = \frac{h_{3s} - h_2}{h_3 - h_2} \tag{1.5}$$

Thus,

$$h_3 = h_2 + \frac{h_{3s} - h_2}{\eta_{CP}} \tag{1.6}$$

The enthalpy terms can be found from property tables or correlations for the particular working fluid chosen for the application. Digital equations of state, such as Refprop, may be embedded as "add-ins" into the programs used for the computations.

Example 1 – Basic vapor-compression heat pump

With reference to Figs. 1.11 and 1.12, consider a heat pump with the following specifications:

Cycle working fluid, refrigerant R152a
Heating capacity, 350 kWt
Compressor efficiency, 82%
R152a condenses at 100 °C and evaporates at 20 °C.

The NIST software Refprop will be used to determine the required properties of the working fluid R152a. Table 1.1 is a useful aid: **bold values** are specified, *italic values* are found from Refprop.

The enthalpy at state 1 is the same as at state 4 owing to the isenthalpic throttling process. The quality at state 1 is found from the so-called Lever Rule:

$$x_1 = \frac{h_1 - h_f}{h_2 - h_f} = \frac{403.59 - 234.77}{520.09 - 234.77} = 0.5917$$

The enthalpy and temperature at the ideal compressor outlet, state 3s, are determined by the pressure (same as at state 4) and the entropy (same as at state 2). The enthalpy at state 3 is found from Eq. (1.6):

$$h_3 = h_2 + \frac{h_{3s} - h_2}{\eta_{CP}} = 520.09 + \frac{583.80 - 520.09}{0.82} = 597.78$$

Figure 1.13 shows the cycle (to scale) in temperature-entropy coordinates.

The required mass flow rate of R152a is found from the energy equation for the desuperheater-condenser, namely,

Table 1.1 State-point properties for heat pump example.

Point	Pressure, MPa	Temperature, °C	Quality	Entropy, kJ/kg·K	Enthalpy, kJ/kg
f	*0.51291*	**20**	**0**	*1.1219*	*234.77*
1	*0.51291*	**20**	0.5917	*1.6978*	403.59
2	*0.51291*	**20**	**1**	*2.0952*	*520.09*
3s	3.5050	*118.65*	*superheated*	2.0952	*583.80*
3	*3.5050*	*125.75*	*superheated*	*2.1306*	597.78
d	*3.5050*	**100**	**1**	*1.9707*	*536.28*
4	*3.5050*	**100**	**0**	*1.6151*	*403.59*

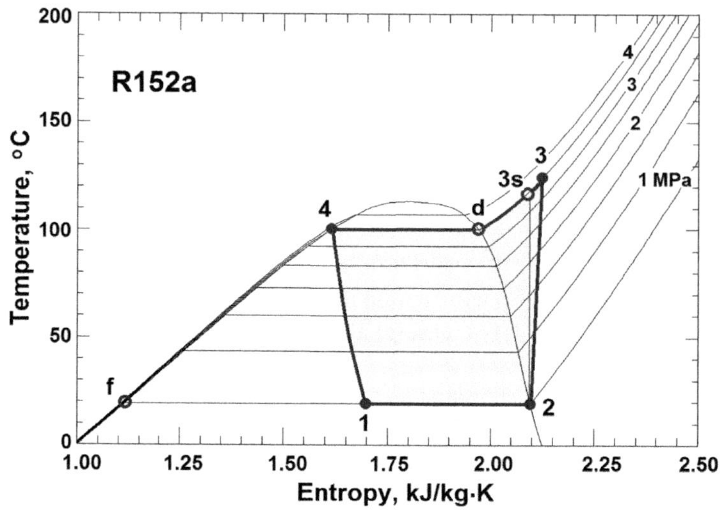

Fig. 1.13 Heat pump processes for Example 1.

$$\dot{Q}_{3,4} = \dot{m} \times (h_3 - h_4)$$

or

$$\dot{m} = \dot{Q}_{3,4} \Big/ (h_3 - h_4) = 350/(597.78 - 403.59) \quad = \ 1.8024 \text{ kg/s}$$

The power required for the compression is found from the energy equation for the compressor:

$$\dot{W}_{2,3} = \dot{m} \times (h_3 - h_2) = 1.8024 \times (597.78 - 520.09) = 140.0 \text{ kW}$$

Lastly, the COP_{HP} is found from Eq. (1.4):

$$COP_{HP} = \frac{Q_{3,4}}{W_{2,3}} = \frac{350}{140.0} = 2.50$$

Using Eq. (1.3) or Fig. 1.8, the ideal heat pump operating over the same temperature limits as the one in this example would have a COP_{HP} of 3.77. This heat pump therefore achieves 66% of the ideal limit. Note that we have not accounted for the efficiency of the electric motor that drives the compressor. Since motors are very efficient, the actual electrical power might be 2–3% higher than calculated in this example.

1.5.3 Thermodynamic efficiency of vapor compression refrigerators

Since the configuration shown in Fig. 1.11 can be viewed as a refrigerator as well as a heat pump, there is no need to devise a new flow diagram for a refrigerator. The state-point diagrams in Fig. 1.12 are also applicable here. The difference lies in the objective of the cycle, namely, to remove heat from the low-temperature reservoir, while discharging heat to the surroundings. Thus, with reference to the definition of terms given above, the coefficient of performance, COP_R, may be written as

$$COP_R = \frac{Q_{1,2}}{W_{2,3}} = \frac{h_2 - h_1}{h_3 - h_2} \tag{1.7}$$

where $Q_{1,2}$ is the cooling effect, i.e., the heat removed from the space that is being cooled. The refrigeration cycle may be seen as thermodynamically the same as the heat pump cycle but with a shift to lower temperatures. For the refrigerator, the heat discharge from the cycle is at or above ambient temperature, whereas for the heat pump, the heat input to the cycle is close to ambient temperature. Otherwise, the working equations to determine the state-point properties are the same.

Example 2 – Basic vapor-compression refrigerator

Let us suppose that Example 1 is now viewed as a refrigerator. This will not be a practical case, but it will illustrate an interesting relationship between the *COP* of heat pumps and refrigerators.

The heat removed from the "cold space" is the objective and the work supplied is the input. Thus the COP_R can be found from

$$COP_R = \frac{Q_{1,2}}{W_{2,3}} = \frac{h_2 - h_1}{h_3 - h_2} = \frac{520.09 - 403.59}{597.78 - 520.09} = 1.50$$

Notice that the heat pump has a COP_{HP} that is greater than the COP_R for the refrigerator by exactly one. Thus,

$$COP_{HP} = COP_R + 1$$

We leave it to the reader to prove that this is a general equation that applies to any cycle operating as either a heat pump or a refrigerator; see Problem 1 at the end of the chapter.

1.5.4 Thermodynamic efficiency of sorption refrigerating machines

Sorption RMs and HPs are divided into two types: (1) adsorption and (2) absorption. Adsorption refrigerators are based on the use of a solid adsorber and a refrigerating working fluid. The refrigerant is adsorbed onto the surface of the solid material. Absorption refrigerators are based on the use of a liquid sorbent and a refrigerating working fluid (or refrigerant), where the two fluids physically mix together. Water-ammonia RMs are commercially available, whereas modern absorption-diffusion RMs and HPs, proposed earlier [17] are being developed.

Absorption refrigeration systems (AR) are a well-known alternative to vapor-compression systems to achieve a cold space, especially where electricity prices are high. Indeed, some AR systems completely eliminate mechanical work input and use gravity to circulate the working fluid. This requires a 3-component working fluid such as water-ammonia-hydrogen, first devised in 1922 by von Platen and Munters in Sweden and described in Section 1.5.5.

For AR systems, the primary motive force is heat, not work, thus eliminating the need for a pump and an electrical supply. The heat can be obtained from steam if a boiler or other supply of steam is available. In many industrial plant applications, excess ("waste") steam is sent to the AR. However, if the steam has to be raised by burning fossil fuel specifically for use in the AR, the question arises as to whether that is the best means to obtain the heat. Renewable sources are to be preferred. One steady renewable source of heat is hot geothermal water found in many fields around the world. Another renewable, but intermittent, source is solar thermal energy collected in reflecting parabolic troughs. One effective use of AR systems is in buildings with high peak loads on the power supply system.

Historically, the first absorption refrigerating machine was created in France by Ferdinand Carré in 1858, using as the working fluid a mixture of water and sulfuric acid. The first two-stage absorption refrigerating machine was created in 1950. A three-stage AR with three condensers and three generators (1985), with double condensers (1993) exceeds the efficiency of the two-stage system by 30–50%. ARs of both direct and indirect heating are in wide use.

AR systems may be used, for example, in ice-making or in air conditioning. Ice-making must achieve sub-freezing temperatures so the working fluid normally is a water-ammonia mixture where ammonia is the refrigerant and water is the carrier. Air conditioning does not require such low temperatures, and a lithium bromide-water mixture may be used where water is the refrigerant and the Li-Br is the carrier. Since the working fluid is a binary-mixture as opposed to a simple pure substance, there is an additional degree of freedom when determining the thermodynamic properties.

Figure 1.14 shows a block diagram for a generic ideal AR system in which the main components, namely, the vapor generator, condenser, absorber and evaporator, are within the boundary of the refrigerator proper. The AR is in thermal contact with three heat reservoirs, i.e., the source of the motive heat at T_H, the created cold space at T_L, and the surroundings at the ambient temperature T_A. Both the condensation and the

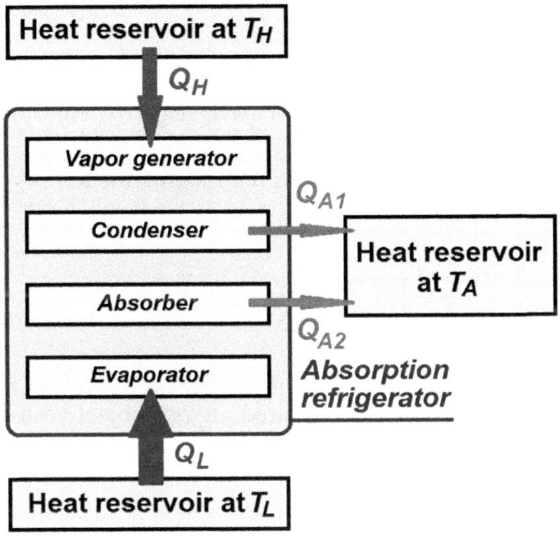

Fig. 1.14 Block diagram for energy flows in an ideal absorption refrigerator.

absorption processes release heat to the ambient sink. For the purposes of the analysis below, it is assumed that reversible heat transfer takes place, allowing the use of the Carnot relations between heat flow and absolute temperature.

The First Law of thermodynamics for the AR gives:

$$Q_H + Q_L = Q_{A1} + Q_{A2} \equiv Q_A \tag{1.8}$$

where all the heat discharged to the surroundings is defined as Q_A.

The Second Law of thermodynamics is written for the overall combined system (assumed to be adiabatic) consisting of the AR and the three heat reservoirs in the form of the Principle of Entropy Increase, i.e., any process in an adiabatic system can only produce an increase in the entropy of the system; for an ideal operation, the entropy change is zero:

$$\Delta S_{T_H} + \Delta S_{T_A} + \Delta S_{T_L} \geq 0 \tag{1.9}$$

Note that the change in entropy for the AR itself is zero because it operates on a cycle.

Each term in Eq. (1.9) can be expressed as the ratio of the heat transfer to the absolute temperature of the associated heat reservoir, as follows:

$$-\frac{Q_H}{T_H} + \frac{Q_A}{T_A} - \frac{Q_L}{T_L} \geq 0 \tag{1.10}$$

$$\frac{Q_A}{T_A} \geq \frac{Q_H}{T_H} + \frac{Q_L}{T_L} \tag{1.11}$$

Notice that the reservoirs that lose heat have negative entropy changes, *et vice versa*.

Substituting Eq. (1.8) into Eq. (1.11) and rearranging the terms yields an important result:

$$\frac{Q_L}{Q_H} \leq \left[\frac{T_L}{T_A - T_L}\right]\left[\frac{T_H - T_A}{T_H}\right] \tag{1.12}$$

This ratio may be used as the absorption refrigerator coefficient of performance COP_{AR} since it gives the ratio of the desired energy objective to the input energy needed to operate the system.

$$COP_{AR} = \frac{Q_L}{Q_H} \leq \left[\frac{T_L}{T_A - T_L}\right]\left[\frac{T_H - T_A}{T_H}\right] \tag{1.13}$$

or

$$COP_{AR,ideal} = \left[\frac{T_L}{T_A - T_L}\right]\left[\frac{T_H - T_A}{T_H}\right] \tag{1.14}$$

When the system is ideal, i.e., perfectly reversible, we find:

$$Q_H = \frac{Q_L}{\left[\frac{T_L}{T_A - T_L}\right]\left[\frac{T_H - T_A}{T_H}\right]} = \frac{Q_L}{COP_{CR} \times \eta_{CPP}} \tag{1.15}$$

where COP_{CR} is the coefficient of performance for a Carnot refrigerator operating between the surroundings and the cold space, and η_{CPP} is the thermal efficiency for a Carnot power plant operating between the high-temperature heat reservoir and the ambient heat sink.

Figure 1.15 shows the ideal COP for an absorption refrigerator as a function of the heat source temperature for several cold space temperatures; ambient temperature was taken to be 25 °C. The best results are obtained for modest cold space temperatures and high heat source temperatures. Thus, we can find the minimum required heat Q_H at a high temperature that must be supplied continuously to the AR to remove any amount of heat Q_L in order to create and maintain the cold space at a given low temperature. Any real refrigerator will require more heat than the ideal case.

1.5.5 Heat-driven, 3-fluid absorption refrigerator

This section presents a unique absorption refrigerator. It is possible to achieve refrigeration with an absorption refrigerator without using any form of energy input except

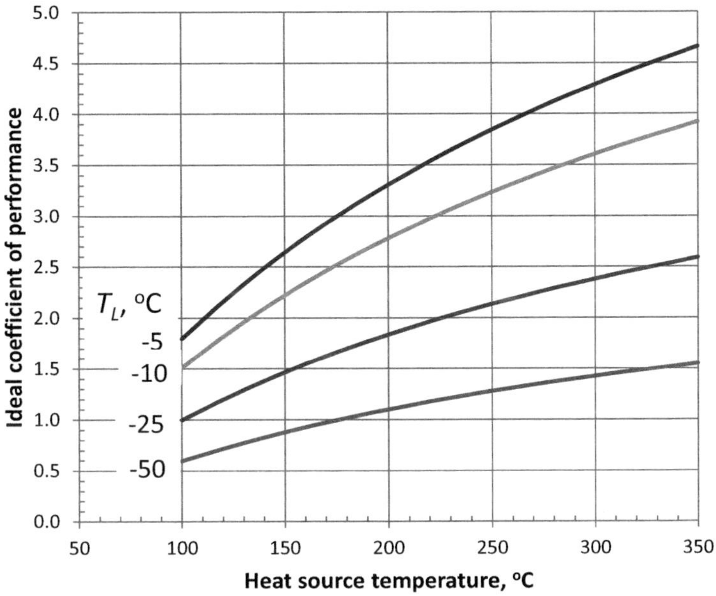

Fig. 1.15 Ideal COP for absorption refrigerators: $T_A = 25$ °C.

heat. This type of system was commercialized in 1933 when many homes were connected to municipal natural gas service. Figure 1.16 shows a simplified schematic of such a machine. There are three working fluids: ammonia (the coolant), water (the absorber) and hydrogen (a catalyst).

The heat input at the generator (3) is supplied by a gas burner. At that place, the working fluid is a liquid mixture of ammonia and water. The ammonia having the lower boiling point boils off and percolates up a tube, lifting the water-ammonia mixture to the rectifier (4) where the hot vapor (nearly pure ammonia) is separated and rises to the condenser (5). There it comes in thermal contact with the ambient temperature room air, releases its latent heat and liquefies. It then passes through a trap (throttle) and into coils surrounding the freezer section of the refrigerator (1). The now cold ammonia receives heat from the freezer space and evaporates. However, before the ammonia enters the freezer it mixes with hydrogen gas rising from the absorber (2), creating a denser fluid than either of the fluids alone. This allows the ammonia and hydrogen mixture to fall under its own weight to the absorber (2) where the ammonia is absorbed into the water that had been separated at the rectifier. Hydrogen cannot absorb into the water (water is saturated with hydrogen) and is released to rise through the absorber coils where it passes through the descending water and eventually rises to the freezer, completing a loop of travel. The hydrogen releases heat at the coils, giving up the heat it had picked up in the freezer.

No mechanical work or electricity is needed to operate this RM, only heat from burning natural or bottled gas, making these units popular for remote residences such as hunting cabins. Of course, if one desires a light bulb inside the refrigerator

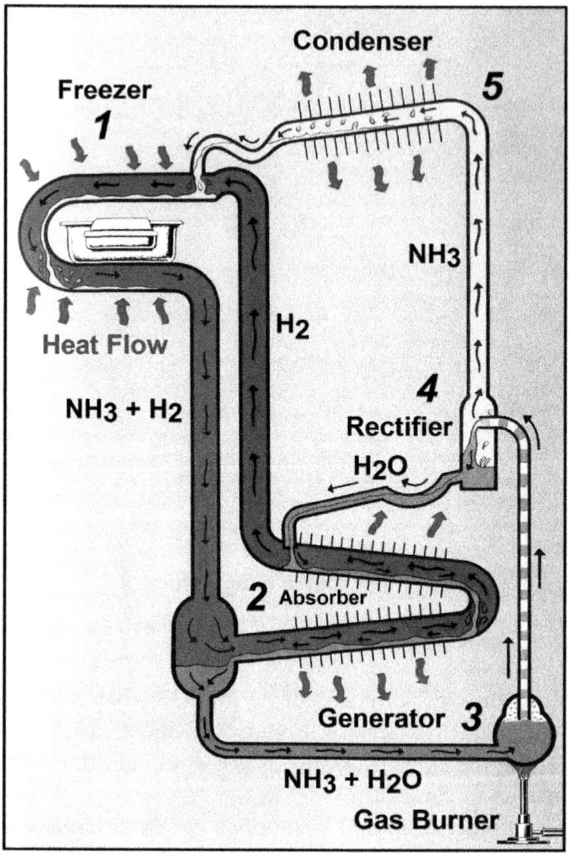

Fig. 1.16 Heat-driven absorption refrigerator, modified from Ref. [18].

for convenience, then electricity must be available. The working fluid circulates under gravity owing to density differences. There are no moving parts and no seals. It is also silent in operation which can be a desirable feature for certain stealthy situations such as in military operations. The last gas-fired refrigerator was manufactured in the US in 1957. The product was banned in 1999 after several deaths caused not by ammonia leakage but by carbon monoxide from faulty combustion of the gas burner. Apparently the refrigeration system itself was still robust and still sealed after decades of use.

One of the writers (RDP) grew up in a home with such a device; it was a puzzle to me as a youngster how it could achieve cold from heat, perhaps spurring my interest in thermal science and engineering.

1.5.6 Exergy assessment of heat pumps

The previous sections considered the performance on the basis of energy. Now the concept of exergy will be applied to heat pumps. The concept of exergy is based on the Second Law of thermodynamics which places limits on the performance of a system. Once the maximum allowable performance is known, actual systems may then be compared to this ideal and deficiencies identified for improvement.

The following equations are used to carry out an exergy analysis of any thermodynamic system. Imagine a general system interacting with its surrounding in several ways: work and heat transfer may occur, and mass may flow in or out of the system at various point on the surface. Each of these carries not only energy but also exergy, e, defined as

$$e_i \equiv h_i - h_0 - T_0(s_i - s_0) \tag{1.16}$$

where h_i and s_i are the enthalpy and entropy of the material flowing at each inlet or outlet, T_0 is the absolute temperature of the surroundings (dead state), and the subscript 0 refers to the dead state conditions. The dead state properties are evaluated at T_0 and P_0. The specific exergy in Eq. (1.16) when multiplied by the mass flow rate yields the rate of exergy or exergetic power

$$\dot{E}_i = \dot{m}_i e_i \tag{1.17}$$

If heat transfer $\dot{Q}$ is involved, the exergy $\dot{E}_Q$ associated with the heat is given by:

$$\dot{E}_Q = \left[1 - \frac{T_0}{T}\right]\dot{Q} \tag{1.18}$$

where the factor in brackets is the Carnot efficiency for an ideal power cycle operating between T and T_0; see Eq. (1.1). Thus the exergy associated with a heat transfer is the maximum amount of work that ideally could be produced from it by a cycle operating as a Carnot cycle.

If work transfer is involved, the exergy associated with that work transfer $\dot{E}_W$ is exactly the amount of work delivered, undiminished by any irreversibility such as friction. Thus,

$$\dot{E}_W = \dot{W} \tag{1.19}$$

The exergy accounting for a heat pump operating as a steady system will involve the exergy associated with each flow stream and each exergy transfer term. Unlike a First Law energy balance in which energy is conserved, an exergy accounting will always show that less exergy leaves the system than enters it. That is, some exergy will always be destroyed. If one imagines a perfect world where all dissipative phenomena could be eliminated and all processes were carried out reversibly, then no exergy would be lost; unfortunately this is not possible in the real world. The objective of an exergy analysis is to identify the lost or destroyed exergy $\Delta\dot{E}_i$ and devise ways to reduce those losses [19,20].

With reference to the heat pump in Fig. 1.6, the following exergy transfers take place: The motor delivers exergy to the compressor which in turn imparts exergy to the working fluid; the working fluid imparts exergy to the heat transfer fluid in the condenser; the working fluid loses exergy via the throttling process; and the working fluid receives exergy from the heat source in the evaporator. The difference between the sum of the input exergy terms and the discharge exergy terms is the exergy loss for each component, as shown below.

$$\text{Compressor}: \quad \Delta\dot{E}_C = \dot{W}_C + \dot{m}e_2 - \dot{m}e_3 \tag{1.20}$$

$$\text{Condenser}: \quad \Delta\dot{E}_{CN} = \dot{m}e_3 - \dot{E}_{Q-CN} - \dot{m}e_4 \tag{1.21}$$

$$\text{Throttle}: \quad \Delta\dot{E}_{TH} = \dot{m}e_4 - \dot{m}e_1 \tag{1.22}$$

$$\text{Evaporator}: \quad \Delta\dot{E}_{EV} = \dot{m}e_1 + \dot{E}_{Q-EV} - \dot{m}e_2 \tag{1.23}$$

$$\text{Heat pump}: \quad \Delta\dot{E}_{HP} = \sum_i \Delta\dot{E}_i = \dot{W}_C + \dot{E}_{Q-EV} - \dot{E}_{Q-CN} \tag{1.24}$$

Example 3 – Exergy analysis of basic vapor-compression heat pump

We will use the Example 1 to illustrate the application of exergy accounting. Figure 1.13 has been modified for this purpose and shown below as Fig. 1.17.

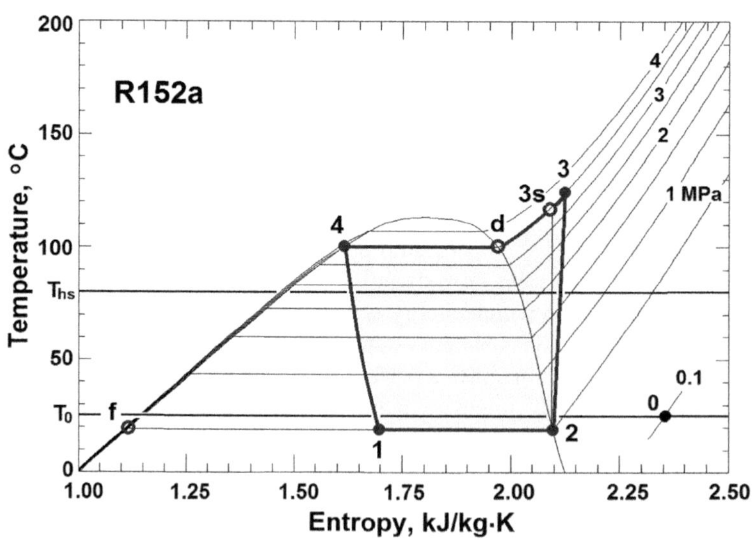

Fig. 1.17 Heat pump processes for Example 3, showing heated-space and dead-state temperatures, T_{hs} and T_0, respectively.

Table 1.1 has also been modified to include data on the dead state and the specific exergy for each state point; it is given below as Table 1.2. The dead state has been chosen as 25 °C and 0.10 MPa, a common choice. Notice that the exergy at states 1 and 2 are both positive although they have a temperature less than the dead-state temperature. All states other than the dead state have positive exergy values because there is a potential for performing work by bringing the state into equilibrium with the dead state regardless of whether the state lies above or below the dead state.

The temperature of the heated space, T_{hs}, receiving the heat discharged at the condenser and the temperature of the low-temperature heat source, T_0, for the evaporator must be specified to allow the exergy terms associated with those heat transfers to be calculated. Let us assume that the condenser heat is delivered to a reservoir at $T_{hs} = 80$ °C (353.15 K) and that the evaporator receives heat from the surroundings at the dead-state temperature, $T_0 = 25$ °C (298.15 K). The exergy accounting for the system and its components may now be carried out.

Compressor: The compressor receives 140.0 kW of mechanical power all of which is exergy. The exergy accounting, Eq. (1.20), yields:

$$\Delta \dot{E}_C = \dot{W}_C + \dot{m}e_2 - \dot{m}e_3 = 140 + 1.8024 \times (57.98 - 125.12) = 18.99 \text{ kW}$$

Condenser: The exergy associated with the heat transfer from the R152a to the heated space can be seen as two distinct processes: process 3-d and process d-4. The exergy for the isothermal condensing portion, process d-4, is easily found from Eq. (1.18):

$$\dot{E}_{Q-CNc} = \left[1 - \frac{T_0}{T_c}\right]\dot{Q}_c = \left[1 - \frac{298.15}{373.15}\right] \times 1.8024 \times (403.59 - 536.28)$$
$$= -48.07 \text{ kW}$$

The negative sign means exergy is being transferred out of the R152a, in the same direction as the heat transfer. The exergy for the variable-temperature desuperheating portion may be found by integrating Eq. (1.18) over the process 3-d. Refprop was used to determine the points along the isobar; a 2nd-order polynomial is an excellent fit to the data that gives T_{ds} as a function of S, and simple integration yields the following result:

$$\dot{E}_{Q-CNds} = \left[1 - \frac{T_0}{T_{ds}}\right]\dot{Q}_{ds} = \int_{S_3}^{S_d}\left[1 - \frac{T_0}{T_{ds}}\right]T_{ds}d\dot{S}$$
$$= \int_{S_3}^{S_d} T_{ds}d\dot{S} - \int_{S_3}^{S_d} T_0 d\dot{S} \cong -61.45 + 47.64$$
$$= -13.81 \text{ kW}$$

Thus, the sum of these two terms gives the magnitude of the total exergy discharged with the condensing heat $\dot{E}_{Q-CN}$, as about 61.88 kW. The exergy destroyed in the condenser, from Eq. (1.21), is:

Table 1.2 State-point properties for heat pump example; R152a.

Point	Pressure, MPa	Temperature, °C	Quality	Entropy, kJ/kg·K	Enthalpy, kJ/kg	Exergy, kJ/kg
f	*0.51291*	**20**	**0**	*1.1219*	*234.77*	62.85
1	*0.51291*	**20**	*0.5917*	*1.6978*	403.59	59.96
2	*0.51291*	**20**	**1**	*2.0952*	*520.09*	57.98
3s	3.5050	118.65	*superheated*	2.0952	*583.80*	121.69
3	*3.5050*	125.75	*superheated*	*2.1306*	597.78	125.12
d	*3.5050*	**100**	**1**	*1.9707*	*536.28*	111.29
4	*3.5050*	**100**	**0**	*1.6151*	*403.59*	84.62
0	**0.100**	**25**	*superheated*	*2.3549*	*539.54*	0

$$\Delta \dot{E}_{CN} = \dot{m}e_3 - \dot{E}_{Q-CN} - \dot{m}e_4 = 1.8024 \times (125.12 - 84.62) - 61.88$$
$$= 11.12 \text{ kW}$$

Notice that the middle term in this equation is numerically negative because Eq. (1.20) assumed that the condenser-heat exergy flowed out.

At the same time, the heat that is received by the reservoir at 80 °C brings an amount of exergy $\dot{E}_{Q-hs}$ where

$$\dot{E}_{Q-hs} = \left[1 - \frac{T_0}{T_{hs}}\right]\dot{Q}_{CN} = \left[1 - \frac{298.15}{353.15}\right] \times 350 = 54.51 \text{ kW}$$

Thus, the exergy destroyed at the condenser caused by heat transfer from the cycle working fluid to the heated space across a large and variable temperature difference amounts to 7.37 kW, i.e., about 12% of the exergy of the condensing heat released, or about 2.1% of the total heat transfer. The efficiency of the exergy transfer is therefore about 88%.

Throttle: Since there is no work or heat transfer at the throttle, the exergy decrease for the working fluid is the only factor. Thus,

$$\Delta \dot{E}_{TH} = \dot{m}e_4 - \dot{m}e_1 = 1.8024 \times (84.62 - 59.96) = 44.45 \text{ kW}$$

This seems quite large but it is the throttling process that creates the low temperature that makes the heat transfer from the surroundings possible.

Evaporator: Here heat is removed from an ambient-temperature space and delivered to the R152a which is maintained at a lower temperature. The method used for the condenser may be applied to the evaporator with the following results:

$$\Delta \dot{E}_{EV} = \dot{m}e_1 + \dot{E}_{Q-EV} - \dot{m}e_2$$

$$\dot{E}_{Q-EV} = \left[1 - \frac{T_0}{T_{EV}}\right]\dot{Q}_{EV} = \left[1 - \frac{298.15}{293.15}\right] \times 1.8024 \times (520.09 - 403.59)$$
$$= -3.581 \text{ kW}$$

This negative result may seem surprising, given that heat is transferred from the surroundings to the R152a. The term is negative because $T_0 > T_{EV}$, meaning that the flow of exergy is actually from the R152a to the surroundings. And since any exergy discharged into the dead state is lost, this term constitutes part of the exergy destroyed in the evaporator. Therefore, we find

$$\Delta \dot{E}_{EV} = \dot{m}e_1 + \dot{E}_{Q-EV} - \dot{m}e_2 = 1.8024 \times (59.96 - 57.98) + 3.581$$
$$= 7.150 \text{ kW}$$

The results of Example 3 are summarized in Table 1.3.

Summarizing: On an overall energy basis, the heat pump COP is 2.5 (i.e., 350/140); that is, 2.5 times more energy was delivered to the heated space in the form of heat than was needed to drive the heat pump. However, on an exergy basis, the COP is just 0.389 (i.e., 54.51/140). That is, of the 140 kW of exergy spent to drive the heat pump, only 54.51 kW of exergy was actually delivered to the heated space, i.e., 38.9%. The high loss in the throttle suggests that improvement may be achieved by staging the throttling into 2- or 3-stages combined with staged compression with interstage cooling. An alternative approach would be to replace the throttle with a total-flow expander capable of producing power that could partially offset the work needed to drive the compressor; this option is left for the reader to explore (see Problem 4 at the end of the chapter). The losses in the condenser might be reduced by lowering the superheat at the compressor outlet (a consequence of staged compression) and lowering the average temperature difference between the R152a and the heating fluid. This last modification would however require an increase in the area for heat transfer and a more costly heat exchanger.

1.6 Working fluids for refrigeration and heat pumps systems

1.6.1 Antifreeze solutions

Antifreeze is a mixture that remains a liquid when used, for example, in the pipeline network of a buried collector coil. The evaporator of the heat pump is a suitable application for antifreeze mixtures. But it is necessary to take into account that the hydraulic evaporator resistance is increased as well. When an antifreeze mixture goes through a pipe, the hydraulic resistance is determined by taking into consideration the viscosity coefficient. Furthermore, it is important to note that the use of antifreeze in some concentrations does not guarantee that it will not freeze at significantly negative ambient temperatures. Thus, the buried pipe coil of the soil heat exchanger should be laid below the frost line.

Here are some antifreeze mixtures that may be considered.

Table 1.3 Exergy destruction in heat pump components, Example 3.

Component	Exergy destroyed, kW	Percent of total, %
Compressor	18.99	23.2
Condenser	11.12	13.6
Throttle	44.45	54.4
Evaporator	7.15	8.75
Total	81.71	100

Mixture of water and ethylene glycol

To obtain such a mixture with a freezing temperature of −13 °C, 75% water and 25% ethylene glycol (by volume) should be mixed. This mixture has an increased viscosity compared with pure water and results in decreasing the thermal power as well as in increasing the hydraulic resistance.

Mixture of water and propylene glycol

To prepare such a mixture it is necessary to mix in a volume ratio of 70% water and 30% propylene glycol. The disadvantage of such a mixture is the increased viscosity, which, as in the previous case, causes an increase in the flow resistance. The drop in thermal power, however, is not as sensitive as in the previous case. The advantage of such a mixture is the lack of chemical aggressiveness of the solution.

Mixture of water and ethanol (alcoholic solution)

To obtain such a mixture with a freezing temperature of −14 °C, 75% water and 25% alcohol are mixed (by volume). The increase in thermal power is insignificant, the advantage being in moderate density. When doing calculations one should take into account that after the mixing of 100 L of water and 100 L of ethanol, the total volume is 180 L. This mixture, as opposed to others, does not require the use of neutral components since the use of particularly pure alcohol does not lead to system aging. In case of using conventional fermented alcohol the reagent life time is about 5 years.

Mixture of water and $CaCl_2$

A mixture of 82% water and 18% calcium chloride salt (by weight) has a freezing temperature of −13 °C. The disadvantage is the drop in thermal power compared to pure water, but it is partly compensated by a higher solution density. Owing to the strong aggressiveness of the solution, however, the use of this solution is not recommended.

1.6.2 Evolution of refrigerants

The world's refrigeration and comfort-control industries depend entirely on the decisions of the global community regarding cooling agents. Several generations of refrigerants have already changed over the past few decades. Some substances, which until recently were considered the most up-to-date and environmentally safe, have now been withdrawn from use. Below we trace the evolution of working fluids for refrigeration machines.

First generation – Everything and anything that worked. Conventional solvents and other volatile liquids were the most commonly used cooling agents for the first hundred years. In fact, the first generation of cooling agents included everything that worked and was available. Almost all the first cooling agents were flammable and toxic, and some were even chemically active. Accidents commonly occurred during the operation of refrigeration equipment. Some companies vigorously pushed propane (R290) as a "safe, odorless cooling agent" over ammonia (R717). Accidents, particularly in domestic situations, often provoked changes in design and new regulations.

Second generation: Safety and durability. The second generation was marked by the transition to fluorine compounds to increase the safety and shelf life of refrigeration systems. Prior to this, early attempts to replace "ice boxes" with refrigerators, in which either methyl formate (R611) or sulfur dioxide (R764) were used as a cooling agent, were unsuccessful; it was impossible to eliminate leakage of these toxic substances. Under the slogan "The refrigeration industry needs a new cooling agent, if it hopes to use it everywhere," a cooling agent R12 was developed by General Motors under the direction of Charles Kettering. Industrial production of R12 began in 1931, and of R11 in 1932. Chlorofluorocarbons (CFCs) and later hydrochlorofluorocarbons (HCFCs) dominated in the second generation of cooling agents, especially since the 1950s, in domestic and industrial air conditioners and heat pumps. Ammonia has been and remains the most popular cooling agent in large industrial systems, especially in the field of production and preservation of products and beverages, and in large ice-houses serving fishing fleets.

Third generation: Protection of the ozone layer (1997). Hypotheses regarding the destruction of the ozone layer and the causes of the formation of "ozone holes" were developed; e.g., the processes of cosmic radiation and the periodic growth of solar activity, the processes of changing atmospheric circulation, etc. According to the hypothesis of the influence of ultraviolet radiation, the breaking of chlorine and bromine bonds of cooling agents makes them highly receptive to absorbing oxygen. Besides the ozone problem, the emission of certain cooling agents has an impact on the "greenhouse effect."

The Montreal Protocol (1987) identified refrigerants that were destroying the ozone layer and those with the potential of destroying it. The Ozone Depletion Potential (ODP) was defined in order to classify refrigerants. As they were the worst ones, R11 and R12 are assigned a value of unity, i.e., ODP-R11 $\equiv$ 1 and ODP-R12 $\equiv$ 1. At present, there are the following synthetic refrigerants containing ozone-depleting substances:

1. Refrigerants of the CFC class. CFC molecules contain chlorine, fluorine and carbon atoms; they are greenhouse gases and ozone depleting with ODP > 0.1;
2. Refrigerants of the HCFC class. They are also greenhouse gases and ozone-depleting gases with lower ODP of $0 < \text{ODP} < 0.1$.

Fourth generation: Counteraction to global warming. The Kyoto Protocol (1997) and "The Paris Summit on Climate" (2015) defined the possibilities of protecting the environment from greenhouse gas emissions. The Kyoto Protocol identified substances that affect the Earth's climate. Greenhouse gases include carbon dioxide, methane, nitrous oxide, sulfur hexafluoride and all synthetic refrigerants. The Global Warming Potential (GWP) was introduced. Carbon dioxide potential is taken as unity, i.e., GWP-CO_2 $\equiv$ 1.

The GWP compares the amount of heat that affects global warming that is absorbed by a certain mass of refrigerant (usually 1 kg) with the amount of heat absorbed by the same mass of CO_2. During the transition period, ozone-friendly refrigerants (i.e., ODP = 0) were created. There exist the following synthetic ozone-safe cooling agents with a high GWP:

- HFC refrigerants containing fluorine, hydrogen and carbon atoms with ODP = 0.
- Refrigerants of the PFC class (perfluorocarbons), that are greenhouse- and ozone-friendly gases containing fluorine and carbon atoms with ODP = 0.

Current generation (after 2010). Since 2010, ozone-depleting refrigerants R21, R22, R123 and R1245b are not allowed. Until 2025 in Europe, the use of refrigerants with GWP < 2500 is permitted; after 2030 it is planned to not use refrigerants with GWP > 150. Refrigerants R134a, R125, R404a, R407c, R507a, R410a, and in the future R32 will not be allowed. The time is coming for hydrofluoroolefins (HFOs), e.g., R1336mzz, R123yf, R1234ze, and R1233zd – all having GWP < 10. Mixtures of refrigerants based on HFO, namely, R448a, R449a, R450a, R513a, and others are being manufactured.

In developed countries, chlorofluorobromohydrocarbons (CFBHC) are no longer used in cooling and air conditioning systems. HFC mixtures, namely, R407c and R410a are considered an acceptable replacement for the ozone-depleting refrigerant R22. ASHRAE 34-2007 allows the use of ozone-safe refrigerants ranging from R429a to R437a, and R510a. The following sources of information may be used for the designation of refrigerants and their possible replacements as environmental restrictions reduce the available working fluids [21–25].

Projecting into the future, it seems clear that the use of synthetic refrigerants will become more and more restricted. Table 1.4 shows the current situation based on the Montreal Protocol and the EU F-Gas 2 Impact. Table 1.5 shows the bans and restrictions facing the refrigerants. All HCFCs are subject to phase-down in the future owing to their medium-level ODP. From Table 1.4, it can be seen there are only seven natural refrigerants and two HFOs that are completely free of restrictions at this time.

The very first synthetic refrigerants developed in the 1930s and deployed thereafter were the best ever conceived from a strictly technical performance standpoint. They were crafted to yield excellent efficiency with outstanding safety characteristics for personnel who had to work with them. The story is told by Charles Kettering [27] about how Thomas Midgley, the engineer who led the development team at Frigidaire, introduced R12 to an audience by inhaling the vapor from a beaker, filling his lungs with R12, and then gently releasing it over a burning candle. It extinguished the flame and he was not harmed by the vapor, thereby demonstrating both non-toxicity and non-flammability simultaneously. It was only decades later that the detrimental impact CFCs on the environment came to light.

Each new refrigerant created in the laboratory has both benefits, mainly environmental, and drawbacks, mainly technological. Certainly they are less effective from an efficiency standpoint, requiring more energy input to operate the RMs for the same cooling effect. Actual operation of RMs and HPs has identified the technical problems related to synthetic working substances. Ozone-safe refrigerants of the HFC and PFC classes are very expensive compared to, say, R22. Higher pressure for the system processes requires an increase in the strength of the heat exchanger materials and metal consumption; the necessity to use expensive high-hygroscopic, polyester, synthetic oils for lubrication; and in the case of refrigerant mixtures, the requirement to completely replace the refrigerant in the RM when there is any leakage, no matter how small, since the preferential leakage of the low-boiling component significantly alters the thermal properties of the working fluid and hence the system performance.

Table 1.4 Classification of selected refrigerants for ODP and GWP [26].

Type	R-number	ODP[a]	ODP level	GWP[b]	GWP level
Natural	717	0	Zero	0	Zero
Natural	744	0	Zero	1	Low
Natural	1270	0	Zero	2	Low
Natural	290	0	Zero	3	Low
Natural	600a	0	Zero	3	Low
Natural	1150	0	Zero	4	Low
HFO	1234yf	0	Zero	4	Low
HFO	1234ze	0	Zero	6	Low
Natural	170	0	Zero	6	Low
HFC	32	0	Zero	675	Medium
HFC	134a	0	Zero	1430	Medium
HFC	407C	0	Zero	1774	Medium
HFC	437A	0	Zero	1805	Medium
HFC	407F	0	Zero	1825	Medium
HFC	442A	0	Zero	1888	Medium
HFC	410A	0	Zero	2088	Medium
HFC	407A	0	Zero	2107	Medium
HFC	427A	0	Zero	2138	Medium
HFC	438A	0	Zero	2265	Medium
HFC	423A	0	Zero	2280	Medium
HFC	417A	0	Zero	2346	Medium
HFC	424A	0	Zero	2440	Medium
HFC	422D	0	Zero	2729	High
HFC	422A	0	Zero	3143	High
HFC	434A	0	Zero	3245	High
HFC	428A	0	Zero	3607	High
HFC	MO89	0	Zero	3805	High
HFC	404A	0	Zero	3922	High
HFC	507A	0	Zero	3985	High
HFC	508B	0	Zero	13396	High
HFC	23	0	Zero	14800	High

Table 1.4 Classification of selected refrigerants for ODP and GWP [26].—cont'd

Type	R-number	ODP[a]	ODP level	GWP[b]	GWP level
HCFC	123	0.060	Medium	77	Low
HCFC	402B	0.030	Medium	2416	Medium
HCFC	401A	0.033	Medium	1182	Medium
HCFC	401B	0.036	Medium	1288	Medium
HCFC	409A	0.046	Medium	1909	Medium
HCFC	22	0.055	Medium	1810	Medium
HCFC	402A	0.019	Medium	2788	High
HCFC	408A	0.024	Medium	3152	High
CFC	502	0.33	High	4657	High
CFC	11	1	High	4750	High
CFC	12	1	High	10900	High

[a] Ozone Depletion Potential, UNEP (2006): R11 = R12 ≡ 1.
[b] Global Warming Potential (100 year), IPCC fourth Assessment Report, 2007: $CO_2 \equiv 1$.

It might be possible that only natural refrigerants will be allowed some time in the near future, essentially returning the world to the starting point of refrigeration; see Fig. 1.18.

Decisions by international commissions will control what engineers may use as working fluids in refrigeration machines and heat pumps. First proposed by Passet in 1979 [29], several studies have been developed on the basis of the integrated concept of sustained development involving ecological, social and economic considerations; see Fig. 1.19. Only where the three aspects are simultaneously satisfied can a solution be considered sustainable and therefore acceptable.

Table 1.5 Restrictions and bans on refrigerants in Table 1.4 [26].

ODP level	Montreal protocol	GWP level	Level	EU F-Gas 2 impact
Zero	No restrictions	<150	Low	No controls
Medium	Subject to consumption phase down	150–2500	Medium	Some supply restrictions and new equipment use bans
High	100% global production & consumption ban	>2500	High	Substantial supply and use restrictions and new equipment bans

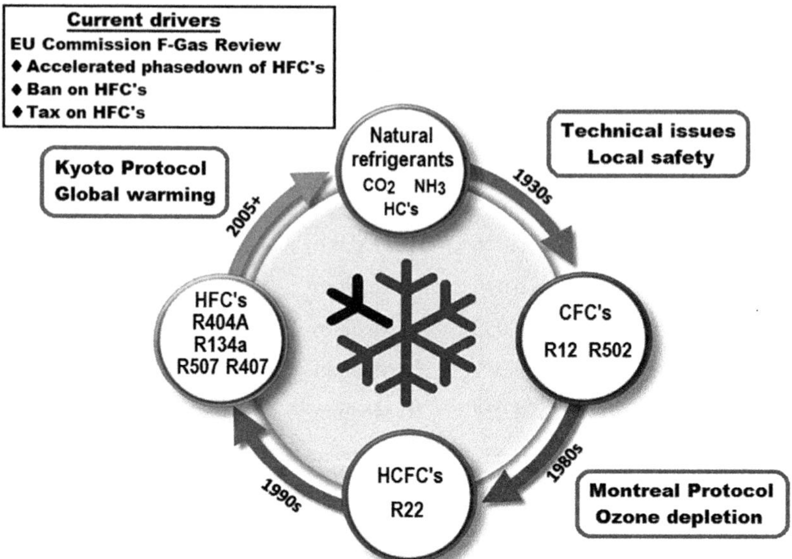

Fig. 1.18 Trends in working fluids for refrigeration units: a closed cycle. Redrawn from [28].

1.6.3 Summary: Basic requirements for working fluids

- Thermodynamic
  - Large heat of vaporization and the largest values of the Clausius number *Cl* which represents the ratio of energy transport associated with the fluid momentum to energy transfer by thermal conduction:

$$Cl = \frac{V^3 L\rho}{\lambda\,\Delta T} = \frac{\rho V^2 \times V}{\lambda\,\Delta T/T}$$

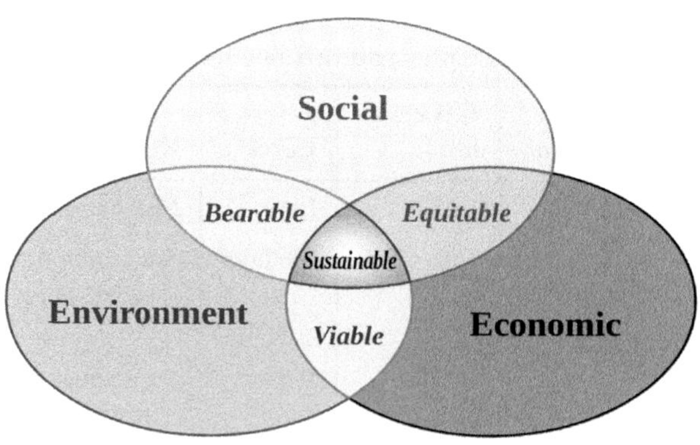

Fig. 1.19 Triune concept of sustained development [30].

where V = velocity, L = characteristic length, ρ = density, λ = thermal conductivity, ΔT = temperature difference.

- Low heat capacity of liquid and high heat capacity of superheated vapor;
- Saturated vapor curve has positive slope in temperature-entropy coordinates;
- Low condensation pressure and low boiling point temperature at atmospheric pressure;
- High thermal conductivity and low viscosity.

- *Operational*
  Thermochemical stability, chemical compatibility with materials, nonflammability, non-toxicity and non-explosion hazard.
- *Economic*
  Availability of commodity production and reasonable prices.
- *Ecological*
  Ozone safety and low potential of global warming.

1.7 Operating modes of heat pumps

1.7.1 Flexible heating and cooling

Heat pumps can provide both heating and cooling. The possibility of heating or cooling using only one device is achieved by "switching" the refrigerant flow path by means of a four-way valve. Figure 1.20 is a simple schematic of how one unit can serve both heating and cooling. Since the compressor CP can only operate in one direction, the flow direction changes in the piping, but remains the same through the compressor whether the unit is in heating or cooling mode.

In heating mode (A), when the rotary 4-way valve is set at "H", the discharge side of the compressor is directed inside the house where the hot vapor condenses (CN) and delivers heat where it is needed. The throttle valve (TV) drops the pressure, thereby lowering the temperature below the outside ambient, and directs the working fluid

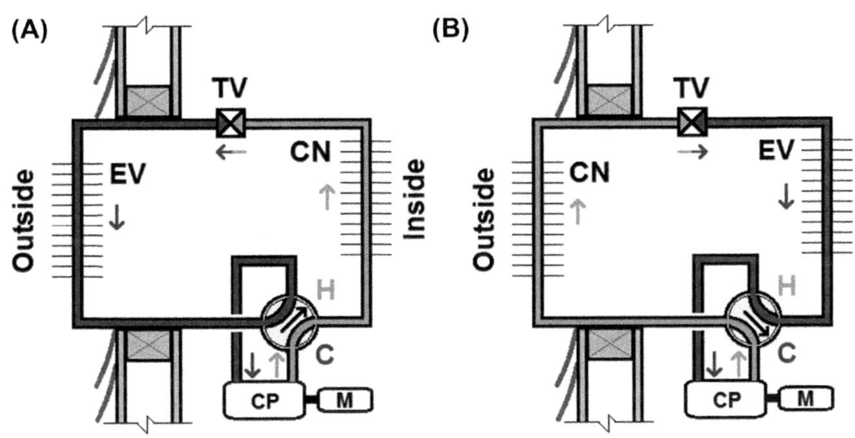

Fig. 1.20 Domestic heat pump schematic: (A) heating mode; (B) cooling mode.

outside. There it picks up heat from the surrounding air and evaporates (EV). The cold refrigerant vapor is shunted into the compressor where it is repressurized, its temperature increases, and the cycle repeats.

In cooling mode (B), with the valve pointed at "C", the discharge side of the compressor is directed outside where the hot vapor is condensed by the surrounding air and dissipates heat to the environment. The throttle valve now sees flow in the opposite direction, but still reduces the pressure thereby lowering the refrigerant temperature and directs it inside the house. There it removes heat and evaporates. The low-pressure, low-temperature refrigerant vapor is shunted into the compressor where it is repressurized, its temperature increases, and the cycle repeats.

In large buildings with many rooms, some of which may have higher heat loads than others, it is possible to use heat pumps to cool some rooms while simultaneously heating others. This is usually the case for large schools or hospitals where the heat or cooling needs depend on the building orientation relative to the sun.

1.7.2 Heating mode design guidelines

The following practical guidelines offer design considerations when operating heat pumps in the heating mode.

- The selection of heating system

  When designing the heating system with heat pumps it is very important to select the type of heating device correctly. Radiant panel heating systems that are placed in the floor or on walls and do not require high temperature of a heat carrier are the most profitable ones. The advantage of these systems is that it is possible to lower the room temperature by 2–3 °C due to heat transfer by means of evaporation. One cannot forget that maximal temperature at the ordinary heat pump outlet is 55–62 °C so that is necessary to take into account when the types and dimensions of heating devices are chosen. Radiator systems require a higher temperature, 90–95 °C, and floor radiation and ceiling water panels, require a temperature of 55–60 °C. Therefore, the choice of heat application of a heat pump depends on the type of heating system.
- Heating by means of panels placed in the floor (the "warm floor system")

  This system is profitable due to the fact that the heat carrier temperature of 35 °C is enough for heating. Its disadvantage is the limitation of maximum power of not more than 96 W per m^2 of heating floor area. That is why the combination of the "warm floor" system with radiator and convector heating devices or fancoils is more widely used. It is advisable to use the "warm floor" system in combination with fancoils. This is because most fancoils operate at a 5 °C temperature difference with the heat carrier. It is particularly advisable to use fancoils in the heating system with an ordinary heat pump that operates for room cooling in summer. In such a case a building can be heated in winter and cooled in summer by means of one system (see the previous section) that reduces greatly capital costs at the building construction stage.
- Heating by means of panels, placed in the floor, with heating radiators

  This variant requires the use of a thermohydraulic regulator (THR) and its own regulating collector to operate properly. In addition to pipe laying, it is necessary to have a circulating pipe, which would provide the cycle with water between a THR radiator and heat pipe. A three-way valve and circulating pump that would supply water into the coil with the temperature of 35 °C should be installed in the "warm floor" system coil as well.

- Heating by means of radiators

  We recommend choosing the temperature difference taking into account a maximum heat carrier temperature of 45/55 °C for such a system. When designing one should calculate nominal readings of heating device power and increase their area by decreasing the temperature gradient from 90/70 °C to 55/45 °C. In short, the power of one rib at the temperature difference of 55/45 °C is approximately 40% as compared with the power at temperature of 90/70 °C. The heat pump can be directly connected to the heating system for a simple single-loop system. If the system has two or more heating loops, a circulating pump should be installed for each one. Precise calculation is an important factor, i.e., an additional thermo-electric heater should consume as little energy as possible.

1.7.3 Monovalent and bivalent operating modes

Heat pumps can operate in various operating modes depending on the heat source that is used by the heat pump, as well as on the designed building heating unit or heating technique employed in the building. Four modes are discussed below.

- Monovalent mode

  At this mode the heat pump covers the total heat demand for heating and hot water supply (HWS). Soil and ground waters are the best heat sources since they are almost independent of ambient temperature and give quite enough heat even at low ambient temperatures. The monovalent mode is recommended for operating drain water heat pumps.

- Monopower engineering mode

  To cover power loading peaks, heat pump units (HPUs) operating in the monopower mode are provided with an additional electric heater that can support heating and, if possible, hot water supply. In this case the additional electric heater also allows increasing the temperature of additional hot water from time to time to prevent bacterium (legionella) formation. When designing the HPU with integrated additional heating of hot water between the heat pump and regulating devices, it is advisable to provide a tank-accumulator and, if necessary, to equip it with a heating device. The tank accumulator incorporates several functions: (a) short-time accumulation and tariff compensation for electric energy; (b) a bivalency switch; and (c) a thermohydraulic regulating device. If the tank is made with an internal float flow dish for heating the hot water, then it can be used for heating such water as well.

  One more advantage of this variant is that the water temperature in the tank is changed depending on ambient air temperature. Such a regulating model provides the balance of heat pump operation in the maximum profitable mode, since the heat coefficient depends on the rate of decreasing water temperature in radiators. In other cases the calculations of the heating system do not differ from other classical systems. The bivalent heat pump plant, which incorporates a second (back-up) heat generator, operates using the same kind of energy (electric power). In the heating mode, the heat pump plant is supplemented by a heat generator operating from electricity. The heat pump plant is selected from the calculation of 70–85% of the maximum heat consumption of the building in accordance with DIN EN 12831.

- Bivalent and parallel mode

  Heating units with a bivalent and parallel operation mode are equipped with both a heat pump and an additional heat generator. For example, a liquid-full boiler was often used side by side with an air-water heat pump in some apartment buildings a few years ago, with the main heat supply being done by means of the heat pump. When the ambient temperature was lower than the set limit value, say, lower than 0 °C, the supplemental heat generator began to operate.

- Bivalent alternative mode

  Heating units with a bivalent alternative operating mode are also equipped with a heat pump and the second heat generator. However, unlike the bivalent parallel mode the heat pump and the second heat generator never operate at the same time. Meeting the annual energy demand is done equally between the heat pump and a traditional heating boiler. If the ambient temperature is higher, for instance, than 3 °C, only the heat pump operates. When the temperatures are much lower, heat supply is completely provided by the heating boiler.
- Circulating pump selection

  Regardless of the mode of operation, circulating pumps need to be carefully specified to avoid system losses and possible failure. To choose circulating pumps correctly the hydraulic calculation of heating system pipe laying should be done. General pressure of a circulating pump includes the sum of resistances of the heating system and heat pump condenser. When an accumulating tank is used, the circulating pump should have the pressure capability to overcome the total loop pressure that includes the resistance of the heat pump condenser, the resistance of the tank, regulating armature, pipelines, etc. Proper circulating pumps should be installed for the heating loops of the heating system ("warm floor", radiators, water supply, etc).

1.8 Accumulation and transport of low-temperature energy

The problem of storing the energy that is received during peak production hours and then used in hours of inadequate production (especially relevant for solar and wind generation) for heating and cooling systems is solved using energy accumulators.

In Germany, there are home photovoltaic systems (PVs) equipped with power storage units, storage tanks used in heating and air conditioning systems to store heat and cold. Heat-storage accumulators are most often used in systems with periodically uneven energy demand, for example, over a 24-h period. Examples of cold storage accumulators include air conditioning systems in industrial shops, and milk and dairy product cooling systems at dairies. The unevenness of thermal loads on the corresponding cooling systems is determined by the change in external heat flows for 24 h; in the case of dairies by the frequency of milk arrivals over 24 h, and the necessity for its cooling and recycling immediately and in the short term.

One could install a refrigeration unit that has a cold productivity being equal to the peak load on the cooling system. But then, during the periods of low load, an excessive amount of thermal energy (heat or cold) will be produced and there will be excessive electricity consumption. It is better that the cooling system includes a cold-storage accumulator as an alternative variant. In this case, the refrigeration unit is selected with a cold yield slightly higher than its average value (per 24 h), but less than the required peak value. Then, for periods of time when the thermal load on the cooling system is less than the cold productivity of the system, the excess of produced cold can be directed to "charge" a cold-storage accumulator. In periods of time when thermal load is greater than the cold productivity of the cooling system, one can use the cold accumulated in the accumulator to meet the demand. When there comes again a period of small heat loads, the accumulator is "recharged". Such a simple, at first glance, solution allows the problems of energy saving, reducing the product cost, etc., to be solved effectively. Such decisions are encouraged by financial circumstances as well; for example, when different tariffs for electricity are set for over a 24-h period, it is profitable to produce (accumulate) cold at night, and use it during the day.

Of course, in the above considerations, it is assumed that the requirements for the modes of refrigeration processing or storage technology of food products, for air parameters in air conditioning systems, for characteristics of boiling processes, condensation of a cooling agent in refrigeration units, etc. are not violated and are fulfilled, even when using cooling systems with cold productivity being less than the peak values. This aspect is very important because it can happen for certain businesses that the cost of losses of food product and the lowering of its quality can exceed the achieved energy saving on the total electricity consumption for the cold production.

Therefore, the design and operation of cooling systems with cold storage accumulators requires careful calculation of the plans of the cooling system operation, the sequence of refrigeration processing, the storage of various types of products (nominal and for every day), heat loads for heat exchange equipment, heat flows from the environment, necessary volumes of the accumulator (mass of an accumulated cooling agent), dynamics of changes in the temperature of the cooling agent in the accumulator and in the system of its circulation, kinetics of phase transformations in the cooling agent and other characteristics of the cooling system.

Reducing the required amount of an accumulator can be achieved by not only cooling a cooling agent, but also through use of the exo- or endo-thermic effects of phase transformations (water-ice) or chemical reactions (aqueous solutions of crystallohydrates). In air-conditioning systems, water-cooling machines that operate with uneven loads are used. The choice and sizing of equipment relative to peak loads leads to an increase in equipment productivity and the provision of comfortable conditions. Cold accumulators and heat accumulators are used at peak loads in heating systems (Fig. 1.21).

Fig. 1.21 Thermal Battery™ cooling system consisting of Trane® air-cooled chillers, Trane controls and Ice Bank® energy storage tanks; photo by Trane, used with permission [31].

The following ways of storing cold thermal energy are known: accumulating ice on the surface of submerged tubular heat exchangers filled with a circulating anti-freeze solution; directly on refrigerant evaporator surfaces; within small containers, usually made of plastic and spherical in shape; or simply a large volume of chilled liquid water. A system for using a cold thermal energy storage unit is shown in Fig. 1.22 [31]. During periods of off-peak electricity pricing (typically overnight), the system operates in the charging mode (A). The chiller is driven to provide 25 °F (−3.9 °C) heat transfer fluid (water-antifreeze mixture) to the ice storage unit. This causes the water in the tank

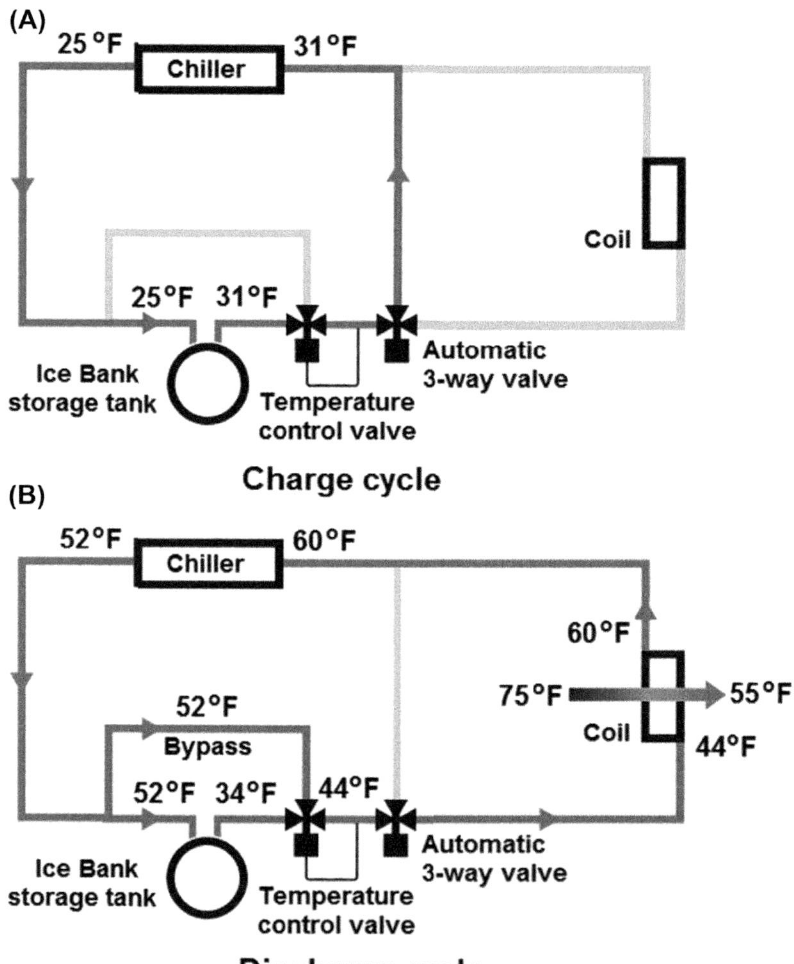

Fig. 1.22 Modes of operation for cooling with a thermal energy storage system. (A) charge cycle; (B) discharge cycle.
Redrawn from [31].

to freeze, releasing the heat of solidification which raises the heat transfer fluid temperature. The loop continues to run until about 90% of the water is frozen solid. During periods of on-peak electricity pricing (typically daytime), the system operates in the discharging mode (B). For the system shown, the fan coil unit needs 44 °F (6.7 °C) cooling fluid to produce 55 °F (12.8 °C) air for the air conditioning system. The chiller is driven to provide 52 °F (11.1 °C) fluid to the ice storage unit where it is cooled to 34 °F (1.1 °C). A bypass line sends some of the 52 °F fluid to mix with the 34 °F fluid, thus making fluid at 44 °F for the coil unit. If low-priced electricity happens to be available during times of demand for cooling, the ice storage section may be completely bypassed, thus saving ice for the high-priced hours.

1.9 Summary

This chapter opens with a snapshot of world energy usage with a special focus on renewable energy sources. A brief summary is given of pertinent energy policies around the world. Given the importance of achieving sustainable energy systems, processes needed to create and assess energy efficient systems are described. Since one of the main foci of this book is the use of heat pumps to enhance energy systems, the history of heat pumps and refrigeration is presented along with qualitative descriptions of these machines. This is followed by quantitative analysis for ideal machines that set the standard for real-world systems. Practical heat pump and refrigeration systems are described in technical detail and analyzed using the principles of thermodynamics. Both vapor-compression and absorption refrigeration systems are discussed. A heat pump uses specialized working fluids to transfer heat between itself and its surroundings, and the selection of the working fluid depends on the constraints imposed by the application and the environment. The working fluid selection process is described with special emphasis on the environmental impacts and international protocols that place limits of the use of certain refrigerants. The chapter concludes with a discussion of modes of operation of heat pumps and cold storage systems.

Nomenclature

Cl	clausius number
COP	coefficient of performance
e	specific exergy
$\dot{E}$	exergy rate
GWP	global warming potential
h	specific enthalpy
L	characteristic length
$\dot{m}$	mass flow rate
P	pressure
Q, $\dot{Q}$	heat, heat rate
s	specific entropy

$\dot{S}$	entropy rate
T	temperature
V	velocity
W, $\dot{W}$	work, power
x	quality

Greek letters

η	efficiency
λ	thermal conductivity
ρ	density

Subscripts

A	ambient
AR	absorption refrigerator
C	carnot
c, CN	condenser
CP	compressor
ds	desuperheater
EV	evaporator
H	high
HP	heat pump
L	low
R	refrigerator
TH	throttle
0	dead state

Review questions

1. What value determines the thermodynamic perfection of the heat pump operation?
2. What data are the starting points for calculating the cycles of heat pump plants?
3. How is the efficiency of the cycle point operation in the heat circuit of a heat pump evaluated?
4. How is the cooling agent for the heat pump selected?
5. How are the cooling agents divided according to ozone safety?
6. What are the criteria for ozone safety and heat exposure?
7. World trends in the application and development of heat pump plants.
8. What are the main principles of legislation on energy saving?
9. What is energy management?
10. Discuss the purpose and main stages of energy audit.
11. Describe the thermodynamic bases of heat pump operation.
12. Describe some sources of low-potential heat.

13. Explain schemes of heat pumps with vapor-compression systems.
14. Sketch diagrams of absorption thermal installations.
15. Give examples of schematic solutions for industrial applications of heat pumps.
16. Describe systems of storage and transportation of heat.
17. Give two examples of systems with a thermal energy accumulator and compare them.
18. What are the advantages of using heat pumps for heating residential and industrial premises?
19. Explain the principle of vapor-compression heat pump operation.
20. Explain the main purpose of each element of a vapor-compression heat pump.
21. Show several operating modes of heat pumps.
22. What heat sources are used for a heat pump?
23. Describe some technological systems of heat pumps for different sources of heat.
24. What heat-carriers are used in the heat supply pipeline with a heat pump?
25. What parameter gives the efficiency of heat pump operation and how is it defined?

Exercises

1. Prove in general that a system operating as a heat pump has a COP_{HP} that is greater than the COP_R as a refrigerator by exactly one, namely, that $COP_{HP} = COP_R + 1$, provided that the unit operates with the same set of heat reservoirs.
2. Consider a basic vapor-compression heat pump (see Figs. 1.11 and 1.12) with the following specifications:
   Cycle working fluid, refrigerant R134a
   Heating capacity, 500 kWt.
   Compressor efficiency, 85%
   R134a evaporates at 5 °C.
   Solve the following parts for two cases of different condensing temperatures; Case (a): 80 °C and Case (b): 60 °C, and discuss the differences.
   Part A. Determine the state-point properties; use Table 1.1 as a template.
   Part B. Calculate the mass flow rate of the R134a, kg/s.
   Part C. Calculate the compressor power if the electric drive motor is 92% efficient, kW.
   Part D. Calculate the heat pump COP_{HP}.
   Part E. Calculate the ideal Carnot heat pump COP_C if it operated over the same temperature limits as the actual heat pump.
3. Refer to Problem 2. Assuming all the heat released from the working fluid in the desuperheater-condenser is delivered to water that circulates between the condenser and the space to be heated, calculate the efficiency of exergy transfer from the heat pump to the water if the water return temperature is 35 °C and the pinch-point temperature difference between the R134a and the water is 5 °C. Do the calculations for both Case (a) and (b). The dead-state temperature is 2 °C (275.15 K) and the pressure is 0.1 MPa.
4. Refer to Problem 2, Case (a). Explore the feasibility of replacing the throttle valve with an expansion machine. The saturated liquid at state 4 will be sent to a device that can handle a mixture of liquid and vapor (such as a helical screw expander) to generate electrical power. That power can be used to offset part of the input needed to drive the compressor.
   Part A. Determine the power developed by the device if the machine has an isentropic efficiency of 60% and its generator is 93% efficient.
   Part B. Calculate the COP_{HP} for this mode of operation.

References

[1] BP statistical review of world energy. London, UK: BP; 2016. http://www.bp.com/statisticalreview.

[2] OECD/IEA. World energy model documentation. International Energy Agency; 2015. www.worldenergyoutlook.org.

[3] OECD/IEA. Energy technology perspectives – Harnessing electricity's potential. International Energy Agency; 2014. http://www.iea.org/publications/freepublications/publication/EnergyTechnologyPerspectives2014.pdf.

[4] Key world energy statistics. IEA, International Energy Agency; 2017. https://www.iea.org/publications/freepublications/publication/KeyWorld2017.pdf.

[5] Electric power monthly with data for December 2017, US Energy Information Agency, Table B.1 major disturbances and unusual occurrences, year-to-date 2017; February 2018. https://www.eia.gov/electricity/monthly/current_month/epm.pdf. https://www.eia.gov/electricity/monthly/epm_table_grapher.php?t=epmt_b_1.

[6] Jahreswirtschaftsbericht 2016: Zukunftsfähigkeit sichern – Die Chancen des digitalen Wandels nutzen. (Annual economic report 2016: securing future viability using the opportunities of digital change) Bundesministerium für Wirtschaft und Energie (Federal Ministry of Economy and Energy), Germany (in German). https://www.bmwi.de/Redaktion/DE/Publikationen/Wirtschaft/jahreswirtschaftsbericht-2016.pdf?__blob=publicationFile&v=18.

[7] Statutory report (in accordance with Norwegian authority requirements). Stavanger, Norway: Statoil; 2014. https://www.statoil.com.

[8] Annual report and form 20-F for the year ended December 31, 2016. Royal Dutch Shell plc, The Hague, The Netherlands. http://reports.shell.com/annual-report/2016/.

[9] OECD regional outlook 2016: productive regions for inclusive societies. Paris: OECD Publishing; 2016. https://doi.org/10.1787/9789264260245-en.

[10] The OECD Post-2015 reflection series, OECD. http://www.oecd.org/dac/post-2015.htm.

[11] Federal power act. 1920. https://en.wikipedia.org/wiki/Federal_Power_Act.

[12] Dossat RJ, Horan TJ. Principles of refrigeration. John Wiley and Sons; 2001.

[13] Moran MJ, Shapiro HN. Fundamentals of engineering thermodynamics. 5th ed. John Wiley & Sons; 2006.

[14] Zogg M. History of heat pumps: Swiss contributions and international milestones. 9th International IEA heat pump conference, Zürich, Switzerland, 20–22 May 2008.

[15] Forsen M. Heat pumps – technology and environmental impact, Part 1. July 2005. http://ec.europa.eu/environment/ecolabel/about_ecolabel/reports/hp_tech_env_impact_aug2005.pdf.

[16] Heating and cooling with a heat pump. EnerGuide, Office of Energy Efficiency, Natural Resources Canada; 2004.

[17] Stierlin H. Beitrag zur Theorie der Absorptionskältemaschinen (in German), (contribution to the theory of absorption refrigeration), Kabeltechnik (cable technology) 16, 1964.

[18] A to zero of refrigeration, Public relations staff, general motors, Detroit, MI, 13th printing; 1964, p. 81.

[19] Bejan A, Tsatsaronis G, Moran M. Thermal design and optimization. John Wiley and Sons Inc.; 1996. p. 113–62.

[20] Kim SM, Oh SD, et al. Exergoeconomic analysis of thermal systems. Energy 1998;23(5): 393–406.

[21] Coulomb D. World tendencies and priorities in the development of low-temperature engineering. Vestnik of International Academy of Refrigeration 2012;(4):3–7.
[22] Calm JM, Domansky PA. Replacement status of refrigerant R22. ASHRAE Journal 2004; 46(8):29–39.
[23] Tsvetkov OB, Laptev YA, et al. Paris idioms and working fluids of refrigeration systems. Refrigeration Equipment and Conditioning 2016;(2).
[24] Tsvetkov OB, Baranenko AV, et al. Ozone layer-safe refrigerants. Refrigeration Equipment and Conditioning 2016;(2).
[25] Designation and safety classification of refrigerants, ANSI/ASHRAE standard 34-2007.
[26] Refrigerants environmental data: ozone depletion and global warming potential. Germany: Linde Gases AG, Gases Division; 2018. Seitnerstrasse 70, 82049 Pullach, www.linde-gas.com/refrigerants.
[27] Kettering CF. Biographical memoir of Thomas Midgley, Jr., 1889-1944. National academy of sciences of the United States of America, biographical memoirs, vol. XXIV. Eleventh Memoire; 1947. p. 359–80. Presented to the Academy, Annual Meeting.
[28] Infographic. Driving natural alternative refrigerant solutions. Fridgehub: SMPR (Simply Marcomms Ltd); 2018. http://news.cision.com/simply-marcomms/r/infographic–driving-natural-alternative-refrigerant-solutions,c9451968.
[29] Passet R. L'Économique et le vivant. Payot; 1979 (in French).
[30] Sustainable development, Wikimedia commons; 2006. https://en.wikipedia.org/wiki/Sustainable_development.
[31] CALMAC®, A Trane Portfolio, "Why Energy Storage," and "Ice Bank Energy Storage Model C tank". http://www.calmac.com/thermal-energy-storage.

Characteristics of low-temperature energy sources for heat pumps

2

2.1 Ambient air usage in space conditioning

Although ambient air is the most available of all possible low-potential heat sources, its temperature potential is not high enough to permit its use as a heat source for heat pumps in cold seasons. The coefficient of performance (COP) is lower than an alternative effective system when the temperature of ambient air is lower than -3 to -5 °C [1–3]. The corresponding comparisons of water-heating heat pumps having conventional boiler houses and condensing boilers (see Chapter 1) are shown in Ref. [4]. It is also shown that, when comparing heat pumps with a conventional boiler house, significantly low limiting temperatures (-10 to -15 °C) are possible only for low-temperature radiant floor heating systems. Critical ambient air temperatures for heat pumps take the values of 0–10 °C for low-temperature and 2–7 °C for conventional radiator heating systems.

An important conclusion from Ref. [5] is that only subtropical climates are considered suitable for effective application of heat pumps using ambient air in heating systems without requiring supplementary energy sources. Various kinds of reversed, split-type heat pumps operating for cooling in summer and for heating in cold seasons are the simplest technical solution in this case.

Moreover, in summer the temperature potential of ambient air is relatively high and it should be used for hot water systems in many cases. For instance, often a boiler house does not operate in summer, leading to problems of hot water supply. These problems can be solved by means of an "air-water" heat pump installed in the facility where hot water is normally prepared (Fig. 2.1). Besides the conventional equipment of the heating plant (items 1–7), the system includes a heat pump 8, the condenser of which heats the water that is then circulated by the pump 10, located between the condenser and accumulating tanks 9.

The heat potential of ambient air can be used more effectively in the heat pump if other heat sources besides air ambient are also used. For example, an ambient air receiver can be made in a heat exchanger (HX) placed on a house roof. The heat from the HX is transferred by means of glycol circulating between the HX and the heat pump evaporator. Heat transfer intensity increases from air to glycol under the effect of wind; furthermore, the HX receives solar radiation energy that makes heat pump operation more effective.

Ambient air heat for heating systems can be used under appropriate conditions by combining it with the heat of ventilating exhaust of the same building. As an example, a positive experience of using such a system was obtained when

Low-Temperature Energy Systems with Applications of Renewable Energy
https://doi.org/10.1016/B978-0-12-816249-1.00002-9

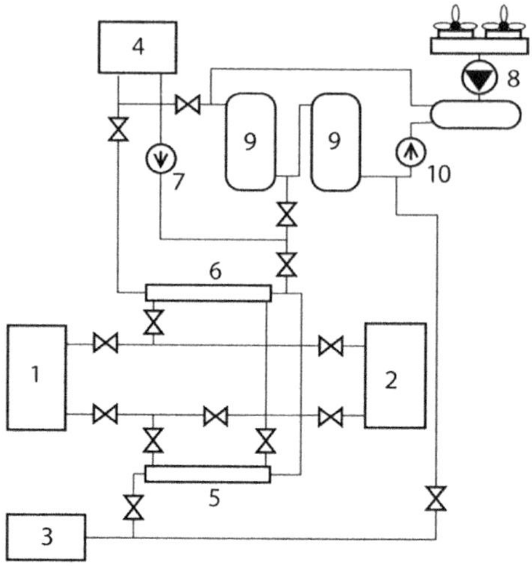

Fig. 2.1 Schematic flow diagram of a heating system with an "air-water" heat pump: *1*, heat network; *2*, heating system; *3*, cold water pipe; *4*, hot water supply system; *5,6*, first and second stage water heaters; *7*, circulating pump; *8*, "air-water" heat pump; *9*, heat storage tanks - hot water accumulators; *10*, condenser feed pump.

implementing a heat-pump heating system for a 4-story office in Kiev, Ukraine. That project uses an off-the-shelf heat pump of the "air-water" type which was placed in a special technical room on the upper floor rather than on the building roof. Fresh air is taken out of the air intake shaft by air handling units, heated inside them, and forced into the heated rooms. Return air goes by means of a ventilation system into the technical room where the heat pump is located. As a result of mixing with ambient air, it is expected that the air temperature at the evaporator input will not be lower than -15 °C during the coldest season. Heating the building rooms is done by means of externally supplied air convectors-conditioners, connected with the heat pump by heat-cold-supply pipelines. In summer the heat pump operates like a refrigerator with condenser air-cooling, but externally supplied air conditioners provide all rooms with conditioned air.

Real possibilities of temperature range expansion using space-conditioning heat pumps come about by using ambient air when water crystallization heat is employed to preheat air; these will be discussed below. The effectiveness and economy of using space conditioning heat pumps with the use of ambient air and the heat of water crystallization increases as a result of using night electric power rates (where available) that are lower than day-time rates, allowing for more cost-effective ice storage for use in air conditioning systems in summer (see Section 2.9).

2.2 Building and construction ventilation air

2.2.1 Residences and multi-story apartment buildings

2.2.1.1 "Air-air" ventilation systems

In places with climates that do not conform to ambient temperatures in the range of 5–7 °C for best performance during heating seasons, it is not feasible to install off-the-shelf ambient-air heat pumps. Nevertheless, even in winter, sources of low-potential heat such as ventilation discharges from industrial and residential establishments fall into the acceptable range.

Private homes (or small houses) and collective multi-story buildings where there are people (e.g., apartment houses, schools, universities, hospitals, medical centers, day-care centers, offices, etc.) all require ventilation at rates from 80% to 100% of room volume per hour. In fact, the rate of ventilation might be as high as 300–400% in business centers.

Ventilation air in modern buildings is ejected or absorbed by ventilators. Such a system of air ventilation permits not only the use of heat pumps, but also heat exchangers so that the fresh air stream can be heated directly by the exhaust stream. If the temperature of exhaust air is up to 20 °C, the heat pump at such temperature will have a COP between 3 and 4. In older houses this system cannot be effectively used unless they have separate inlet and discharge ventilation channels. Heat pumps are used mostly for hot water supply and for heating fresh cold air.

The most economical use of heat pumps is after-air cooling in a heat-exchanger, operating on the principle of thermosiphons, conventional heat-exchangers, or Ljungström regenerative type heaters with metallic heat-accumulating chambers, being rotated in contact with warm and cold air in sequence. Exhaust air is first cooled in the heat exchanger and then in the heat pump evaporator. Fresh air is first heated in the heat exchanger and then directed into the heat pump condenser. It is generally assumed that for individual buildings, the specific energy requirement for ventilation heat transfer is less than 0.3 W/(m^{3}K) which corresponds to a circulation rate of 70–90%. The heat pump should have a defrost system that starts operating when the ambient temperature is approximately 3–5 °C.

The heat pump assembly in private dwellings can be sited under the roof taking into account that the size of the heat pump is small. Noise protection should be provided. For this purpose the heat pump should be placed on additional flooring consisting of double-layered wood with noise protecting isolation between them. The additional floor dimensions must be 2–4 times larger than the heat pump footprint area. The heat pump equipment for the house can be installed in the garage or cellar as well, thereby providing vibration and noise protection. When cooling the air, some condensate is formed owing to the humidity of the air, and it is necessary to collect and remove it from the building or into a special holding tank.

Heat pump heating of multi-story apartment buildings using air requires an additional conventional heat source which must be decentralized. The heat pump can cool and dehumidify the air in summer.

When choosing a heat pump of the "air-air" type, one must pay attention to minimizing electric power supply to ventilators and to overcoming aerodynamic air duct resistance. It is recommended to keep the total electrical cost under 1 W/m/h of air. Air velocity in the heat exchangers should not exceed 2.5 m/s. Velocity control is made easier by installing two compressors.

The heat pump assembly should be located at an area adjacent to a multi-story building or in a cellar, taking measures to lower vibration and noise. The air duct must be connected with flexible heat pump elements. Condensate must go into the sewer system. The outlet of the heated air into the rooms should be done near the outer walls in main rooms or through holes in the floor or making the floor from thermo-porous material which can pass warm air through itself. All air ducts must be effectively insulated. When the air duct goes through a room being not heated, the insulation must be not less than 50 mm thick, and through a room being heated, about 2–3 mm thick. Absorbing air ducts for the collection of the exhaust air must be placed in kitchens, lavatories and bathrooms. They must not be near the doors of rooms that are not heated. Supplementary heaters (electric heaters generally) are placed in each room that is heated. They cannot be located under exhaust air ducts.

2.2.1.2 "Ventilation air-water" heat pump

Air ducts cannot always be installed in existing and new houses which is why water is often used to distribute heat energy in rooms. In buildings having mechanical ventilation, air is taken in from certain rooms such as kitchens and bathrooms and directed to the heat pump evaporator. Water that is heated in the heat pump condenser goes to the heaters and its temperature should not be higher than 50 °C in order to maintain a high COP. Low heat-carrier temperature requires increasing the heat transfer surface area, which is why embedding the heat exchangers inside the floor is the best solution in new buildings. In old ones it is necessary to install additional thermal insulation on the walls and to change windows so that the existing radiators can be saved without increasing their heat exchange surface area.

Heat pump power is chosen depending on ventilation air consumption calculated for ventilation of 70–80% of room volume per hour, and during very cold weather, conventional electric-resistance heaters are turned on in rooms or a centralized boiler is activated. In common multi-story buildings the use of ventilation air is possible if there are ventilation channels, however, in many older buildings such channels are not available.

It is important to note that the use of ventilation air is undesirable in cases where the building is not completely inhabited because the temperature of the ventilation air will be insufficient and will drop below the recommended minimum values, typically 15 °C. In case of cold weather, a supplementary alternative energy source, usually direct electric resistance heating, is used in the same way as in private dwellings. A thermostat must be installed for the electric heaters. Maximum usage of the heat pump is the priority of temperature control in the room. Experiments have shown that the use of a ventilation-air heat pump can decrease electric energy consumption by 40% compared with the direct heating [6].

2.2.2 Underground constructions

Significant potential heat variations, operational irregularities, and the distance from potential heat consumers can prevent the effective use of ventilation air. However, there are no such disadvantages for ventilation exhausts from certain underground objects, for example, subways, metropolitan railways, and coal mines. Due to the high thermal inertia of these objects, the temperature of ventilation exhausts is not lower than 12−16 °C even in the coldest season, which guarantees a high heat pump COP and economic competitiveness. Metropolitan railways and mines use external heat energy for their own needs and thus are the consumers of the used ventilation waste heat.

The effectiveness of using heat pumps in metropolitan railways, for example, has been demonstrated experimentally by designing and constructing a test heat pump system in a railway station in Kiev, Ukraine [7]. Test runs showed that under metropolitan conditions the COP of the off-the-shelf ambient-air heat pump was significantly higher than the certified value of 3.1 and reached 4.9 in the air-heating mode. The main technical characteristics of the base case of the heat supply and heat pump system for the Kiev system are given in Table 2.1.

Furthermore, it should be noted that, since heat pumps can be operated in both heating and cooling modes, the use of heat pumps in heat ventilation metropolitan systems allows combining functions of separate heating, ventilation, and air conditioning systems in one heat pump system. Thus they can cool ventilation air in a warm period which improves both working conditions for staff as well as general technical and economical heat pump performance.

Table 2.1 Technical characteristics of base case and heat pump systems of heat supply.

		Values		
		Base	Heat pump system	
Item	**Units**	**case**	**Calculated**	**Actual**
Calculated heating temperature	°C	21	21	21
Average heat capacity of room heating	kW	26.2	26.2	26.2
Annual heating period	h	4500	4500	4500
Annual heat consumption	GJ	424.4	424.4	424.4
Heat pump chilling capacity	kW	–	21.6	21.6
Electric power for powerhouse hall cooling system and serviced heating rooms	kW	30.4	5.6	5.35
Annual powerhouse hall cooling period	h	6935	6935	6935
Annual electricity consumption	MWh	210.82	38.84	37.10
COP	–	–	5.4/4.7[a]	5.6/4.9[a]

[a] Lower values include electricity used to drive an additional axial ventilator.

The same situation using the heat of ventilating exhaust streams occurs in coal mines. A system that decreases energy costs for heat treating air that goes into the mine during almost the whole year (both for heating in winter and cooling in summer) was proposed in Ref. [7]. Due to higher heat potential of the mine outlet ventilation air relative to the ambient air, the energy efficiency of using off-the-shelf heat pumps under the conditions in question considerably improves upon their certified characteristics.

Therefore it is possible to obtain 100% of prepared air heat energy with electric energy costs for the compressor drive that does not exceed 20–22% in winter and 30–35% in summer. Thus even if there are boilers in the mines that operate using their own solid fuel, the use of heat pump systems can provide not only energy saving but also economic benefits.

2.3 Natural water as a source of energy

2.3.1 Well water

Well water or moist soil can be an ideal heat source for the heat pump, but the decision concerning its use can be made only after hydrogeological research that should answer the following questions:

1) How many wells are necessary for meeting the requirements of the heat pump?
2) What pressure does the pump need to develop to force the water into the water bearing horizon?
3) What are the chemical characteristics of the soil water?

To answer the first question one should have information concerning required water flow rate for the heat pump. The water flow rate G (in kg/h) can be determined by the equation:

$$G = \frac{1.1 \times 3600\, Q_{HP}}{c_p\, \Delta t_w} \tag{2.1}$$

where Q_{HP} is the heat pump thermal power, kW; c_p is water specific heat, kJ/(kg. K); 1.1 is a factor that takes into account a possible decrease in well output; Δt_w is water temperature difference at the evaporator inlet and outlet, generally it is 5 °C.

The required pump pressure depends on soil porosity of the water-bearing bed (see Table 2.2). When soil water penetration is insufficient, owing to low porosity or permeability, the well pump electric power consumption will be comparable with that of the heat pump compressor, rendering the system unfeasible for use as a heat source.

It is important to compare the chemical composition of artesian water with the manufacturer's specification limits for the heat pump components. Unless the water quality meets those criteria, it will not be possible to employ that water source for the heat pump. Most commonly, deep (>700 m) water wells are used according to the method

Table 2.2 Range of values for soil porosity.

Soil	Porosity, %	Soil	Porosity, %
Clay	45–55	Sandstone	5–30
Mule	35–50	Limestone	1–20
Sand	25–40	Shale	0–10
Gravel	25–40	Rock	<1

described in Ref. [5], and depicted in Fig. 2.2. Water is pumped via the water well 4, drilled into the water-bearing bed 3, and delivered to the heat pump evaporator 1.

Water cooled in the heat pump evaporator drains into the same water-bearing stratum through the well 5 and check value 7. In the formation, the water migrates in the direction of the water well 4 following the pressure gradient. On its way from the well 5 to well 4 the water restores its temperature as a result of heat exchange with the warm soil and again is fed into the evaporator, completing a cycle. Pressure that is developed by the pump 6 must be calculated taking into account the hydraulic resistance of the well bore, the pipeline from the well to the heat pump, the heat pump evaporator, and the water-bearing formation.

A significant efficiency parameter of the heat pump unit that uses soil water heat is the distance between the water production well and the disposal well. On the one hand, this distance should not be very large so as to keep the hydraulic resistance of the water-bearing bed manageable. On the other hand, it may be possible that a small distance between wells will not provide sufficient residence time of the water in the soil bed to restore its original temperature. Table 2.3 gives guidance on choosing the optimal distance between wells [5].

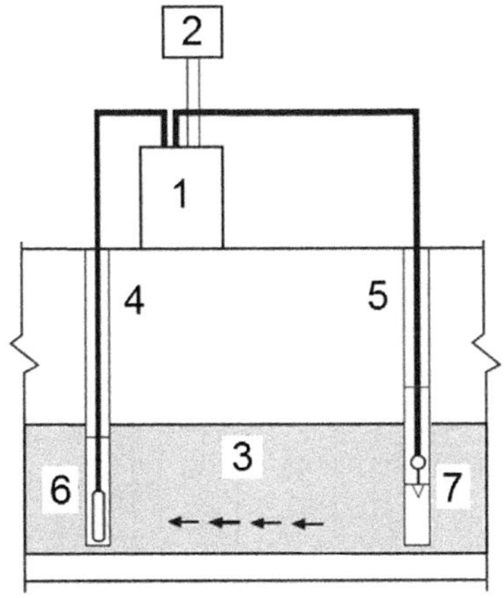

Fig. 2.2 Schematic of using soil water: *1*, heat pump; *2*, heating system; *3*, water-bearing formation; *4*, water production well; *5*, water disposal well; *6*, well pump; *7*, check valve.

Table 2.3 Water consumption (m^3/h) for optimal distance (m) between a producing and disposal well with 20% water-bearing formation porosity.

Water-bearing bed thickness, m	No. of operating days per year	Optimal distance between wells, m								
		2.3	4.6	9.2	13.8	18.4	23	46	115	230
3	100	35	48	69	85	99	111			
	140	39	54	77	94	110	122			
	210	44	61	88	109	127	144			
	280	50	68	100	124	145	125			
6	100	30	41	59	72	84	94			
	140	33	48	67	82	96	108			
	210	39	54	77	95	111	124			
	280	44	61	87	107	126	141			
9	100	26	37	53	63	74	82	117		
	140	30	43	59	73	85	94	136		
	210	36	50	70	85	99	110	157		
	280	41	57	81	97	113	126	178		
12	100	24	33	47	58	67	75	108		
	140	29	38	54	66	77	87	125		
	210	33	46	65	80	92	103	148		
	280	37	54	76	93	106	119	170		
15	100	22	31	44	54	62	68	94	155	217
	140	25	36	50	62	72	80	112	178	251
	210	30	42	60	75	86	94	133	212	301
	280	35	48	70	87	100	108	153	247	350
24	100	17	25	36	45	52	58	81	128	180
	140	21	30	42	52	60	67	94	151	208
	210	25	36	51	63	72	81	115	178	252
	280	29	42	60	74	85	95	136	206	296
30	100	17	23	33	40	47	52	73	116	165
	140	20	27	38	47	54	61	86	136	193
	210	24	32	46	56	65	73	105	162	230
	280	28	37	53	66	76	85	123	187	268

It is shown from Table 2.3 that the distance between the wells must be increased with an increase of water consumption. The table is valid for soils having about 20% porosity. When porosity is 10%, the distance should be increased by 5%, and when it is 30%, it should be decreased by 5%. If natural water movement in the water-bearing bed is detected during the hydrological test, the disposal well must be placed at a distance no less than 5 m from the water well and in the direction of the water movement [5].

The system of using soil-water heat can be simplified [5] if the object with the heat pump is located near a pond and the water-bearing bed is naturally fed with the water out of that pond (Fig. 2.3).

This system differs from the previous one by the fact that the cooled water is drained into the pond rather than the water-bearing bed, the former being hydraulically connected with the latter. In order that water from the pond does not lower the soil-water temperature that is used in the heat pump, the well 4 must be drilled at a distance from the pond shoreline according to Table 2.3 considering the pond to be similar to the receiving well.

If the soil water quality does not meet the requirements of the heat pump manufacturer and it is considered to be chemically aggressive, then in such a case the scheme shown in Fig. 2.4 can be used. The soil water with aggressive properties is cooled in the heat exchanger 9, transferring heat to glycol that circulates in the intermediate circuit between the heat exchanger 9 and heat pump evaporator. Soil water cooled in the heat exchanger is drained through the well 5 into the water-bearing bed or pond close to where the object with the heat pump is located. When soil water with aggressive properties is used, the heat pump efficiency decreases and it is desirable to examine the possibility of using other heat sources, in particular soil heat by means of a soil heat exchanger.

2.3.2 Water of open ponds

When constructing buildings near open ponds, the possibility arises of using the thermal energy of these ponds as sources of low-potential heat in the heat pump. The

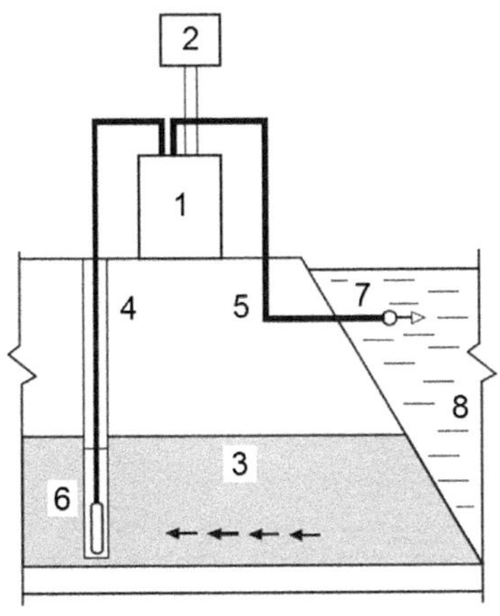

Fig. 2.3 Schematic of using filtrated water from a pond: see Fig. 2.2 for terms; *8*, pond.

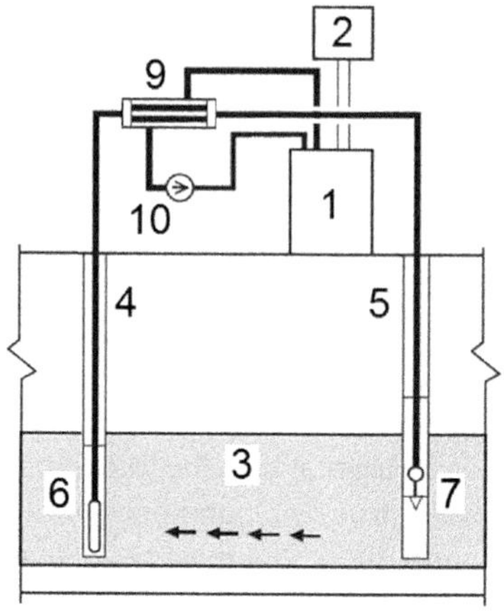

Fig. 2.4 Schematic of using soil water with aggressive properties: see Fig. 2.2 for terms; *9*, heat exchanger; *10*, glycol circulating pump.

feasibility should be estimated accounting for physical characteristics of heat exchangers. When common heat exchangers are used, one can use a scheme with an intermediate circuit [5], shown in Fig. 2.5. In that case, water from the pond is fed with a submerged pump 9 into the heat exchanger 7 and drained back into the pond. In the intermediate circuit filled with glycol, heat is transferred from the heat exchanger into the heat pump evaporator 2.

Fig. 2.6 illustrates a system without the intermediate circuit [5]. In such a case, thermal energy of the water is taken by means of a coil made from cross-linked polyethylene tubing submerged into the pond; the coil may be compact (Fig. 2.6) or large-scale (Fig. 2.7).

To determine the length of the polyethylene pipe, one can use the following equation [5]

$$L = C\,\Delta t^{m} \tag{2.2}$$

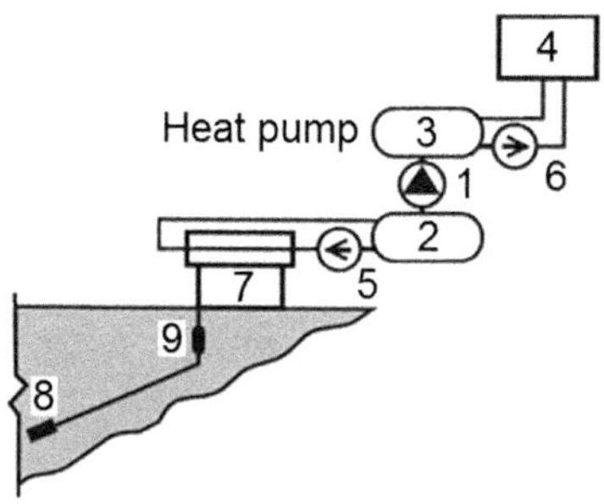

Fig. 2.5 Schematic of using an open pond with an intermediate circuit: *1*, heat pump compressor; *2*, evaporator; *3*, condenser; *4*, heating system; *5,6*, circulating pumps; *7*, heat exchanger; *8*, filter; *9*, submerged pump.

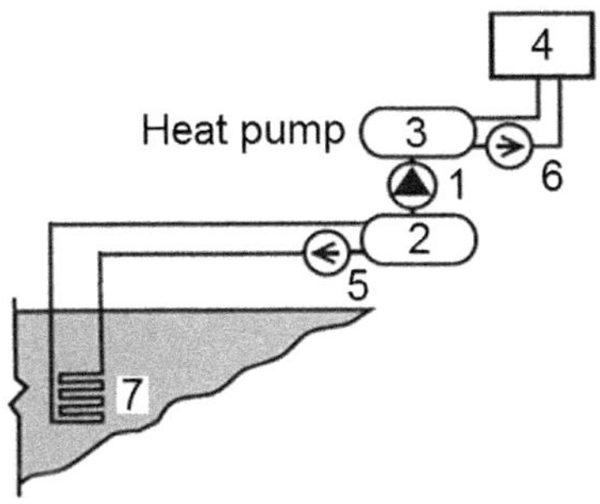

Fig. 2.6 Schematic of using an open pond with a heat exchanger formed in compact bends made from polyethylene pipe: *7*, compact heat exchanger.

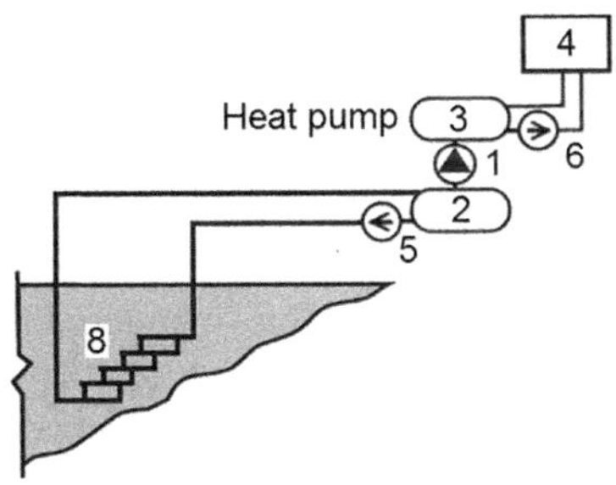

Fig. 2.7 Schematic of using an open pond with a large-size heat exchanger made from polyethylene pipe: *8*, large-size heat exchanger.

where L is the length of the polyethylene pipeline needed to transfer 1 kW of heat flow from the open pond water to the circulating glycol; Δt (°C) is the difference between the water temperature in the pond and the average glycol temperature circulating inside the polyethylene pipe; and C and m are a coefficient and exponential index, respectively: their values are given in Table 2.4.

Table 2.4 Values for C and m in Eq. (2.2) for three pipe diameters [5].

D_{ex}, mm	***C***				***m***			
	Large-scale coil		**Compact coil**		**Large-scale coil**		**Compact coil**	
	Summer	**Winter**	**Summer**	**Winter**	**Summer**	**Winter**	**Summer**	**Winter**
33	82	51	98	63	−0.78	−0.72	−0.75	−0.71
42	56	46	65	56	−0.64	−0.72	−0.79	−0.70
48	69	48	83	61	−0.72	−0.72	−0.70	−0.73

Note: D_{ex} is the external diameter of the polyethylene pipe.

Equation (2.2) is valid if liquid flowing inside the pipe is turbulent, i.e., Reynolds number, $Re > 3000$. The data presented are valid if the polyethylene piping is placed in the ponds with water currents, and can be used both for extracting heat from the pond in winter and discharging heat taken from air conditioning systems into the pond in summer.

2.3.3 Ocean water surface layers

The world's oceans constitute the largest collector of solar radiation wherein a temperature difference is created between warm surface waters and colder deeper layers. This difference exceeds 20 °C over a kilometer of water depth across large expanses of the Pacific, Atlantic and Indian oceans; see Fig. 2.8. Thus there is a continuous supply of ocean thermal energy that can theoretically be transformed into other kinds of energy.

Research has shown the possibility of using such a scheme under conditions of tropical latitudes, where the temperature difference between the surface layers and waters at 1000 m depth is more than 24 °C and changes very little from season to season. Fig. 2.9 shows a typical scheme for an Ocean Thermal Energy Conversion (OTEC) power plant. However, commercial use of these systems has yet to be achieved owing to: (1) the very high cost of capital investments owing to difficult service conditions in the open sea and energy transfer to the mainland; (2) the enormous size of heat exchangers needed to produce meaningful amounts of power; and (3) the necessity of combating biological over-growth of the heat-exchanger surfaces, as a result of which 50% of gross power can be spent merely to overcome fluid friction in pipelines and heat exchangers.

Organic working fluids having a low-temperature boiling point, e.g., refrigerants (ORC) and even water (at a vacuum), may be used for running a power cycle. If water is used, its boiling point should be decreased to the surface water temperatures by

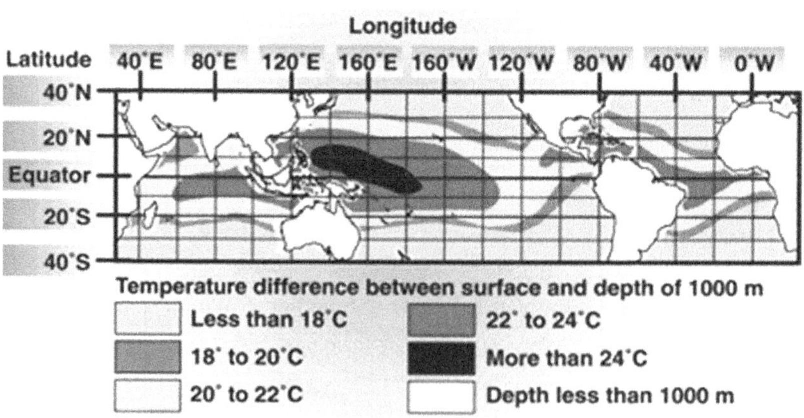

Fig. 2.8 Temperature difference between ocean surface and depth of 1000 m [OTEC, Wikipedia; Public domain].

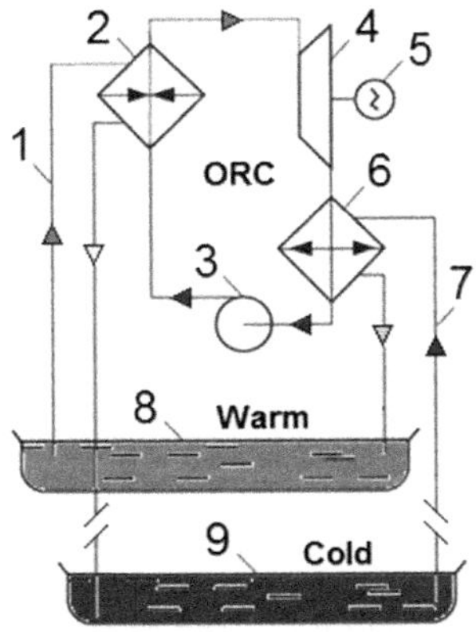

Fig. 2.9 Schematic of converting ocean thermal energy to electrical power using an organic working fluid in a closed cycle: *1*, warm water feed; *2*, evaporator; *3*, circulating pump; *4*, turbine; *5*, generator; *6*, condenser; *7*, cold water feed; *8*, ocean surface; *9*, ocean depths.

creating a partial vacuum, i.e., a vacuum-flash system. Open systems where warm surface waters are used as a working fluid are based upon this principle. Theoretically, energy conversion systems of the OTEC type could be used not only to produce electric energy but also to obtain desalinated water.

The transition from experimental research designs to wide industrial acceptance is challenging because the general biological consequences of raising enormous quantities of water enriched with biogenic compounds up to warm ocean layers are not known. But this scientific and technical idea has attracted the attention of scientists, engineers and funding agencies, such as the U.S. Dept. of Energy, because there are no theoretical thermodynamic obstacles for successful operation of such systems, except for the Carnot cycle efficiency that limits such systems to very low thermal efficiencies. Furthermore, there are many serious practical difficulties to be overcome which are the subject of several ongoing research and demonstrations [see: https://en.wikipedia.org/wiki/Ocean_thermal_energy_conversion].

2.4 Industrial water as an energy source for heat pumps

2.4.1 Cooling water discharge from thermal power stations

Despite the modern tendency for decentralization of energy supply sources, many consumers in large cities remain within range of central power stations. Simultaneous production of electric power and thermal energy has proven advantages regarding efficiency of fuel use. That was the motivation for such heat supply systems which justifies their continued existence; ways to further improve their performance are appropriate. When modernizing them technically, first one must take into consideration the greatest losses of low potential thermal energy in (1) the cooling water system of combined heat and power generating stations (CHPSs), thermal electric stations (TESs) and nuclear power stations (NPSs), as well as large heat losses in (2) the heat distribution piping network [8].

The first loss (1) greatly influences environmental thermal pollution which may contribute to global climate change. If energy consumption increases, this effect may lead to a dangerous situation in the near future according to experts [8]. The use of heat pumps that take advantage of power plant waste heat can mitigate this problem. In traditional heat supply systems, an amount of primary fuel is used to generate the required amount of energy for the heating needs. In heat pumps a considerable part of the useful thermal energy, 70–80%, is produced by low potential environmental thermal energy.

Heat pumps (Fig. 2.10A) apply a reversed thermodynamic cycle using a low-boiling-point working fluid that captures the low potential thermal energy from the water cooling systems of TES, NPS and industrial operations, and increases its potential up to the level required for heat supply systems. In doing so, they use much less primary energy than with direct combustion of fuel. The energy flow diagram for such a scheme is shown in Fig. 2.10B.

Heat pump units can be used instead of cooling towers. Besides using low potential waste heat, this allows the turbine pressure drop to be increased, thereby raising the electric power production, decreasing cooling water energy losses, and reducing pump circulation power, while allowing optimal condenser vacuum and cooling water temperatures to be set.

Efficiency calculations for a heat pump [8] that was installed instead of a cooling tower showed that additional costs were small. Furthermore, there was fuel saving and usage of heat that had been lost before. In summary, fuel saving at the CHPS can be achieved in two ways:

(1) Direct use of industrial cooling water from CHPS as a source of low potential heat for a heat pump; and
(2) Use of recycled power plant cooling water that returns to the CHPS as a low potential heat sink for the condenser, decreasing its temperature by 20–25 °C.

The first way is used when the HPU is placed near the CHPS, and the second one when the HPU is located near the heat users.

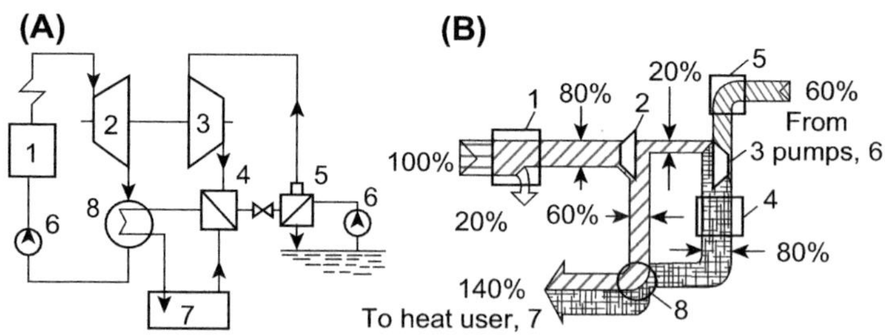

Fig. 2.10 Schematic for (A) heat pump unit (HPU) at CHPS and (B) energy flow diagram showing energy savings: *1*, boiler; *2*, steam turbine; *3*, compressor; *4*, condenser; *5*, evaporator; *6*, pumps; *7*, heat consumer; *8*, water heater/turbine condenser.

2.4.2 Recycled district heating water using extraction steam from thermal power stations

One of the ways of improving a heating system is to use combined systems that include central power station extraction steam (CPSES) and heat pump stations which use that steam [9]. The schematic diagram of such a system is shown in Fig. 2.11.

To supply heat to those buildings that cannot be connected to existing district heating networks, a heat pump station (HPS) with a heat pump unit (HPU) and a peak load boiler is constructed. A portion of the recycled network water is taken by the pump 11 and fed to a HPU evaporator 12 where it is cooled. Network water of a local heat supply system is heated in the HPU condenser 15 and the peak load boiler 6, and is fed to the users in the heating season. To cover the hot water supply load in summer when CPSES network water losses are greatly decreased, the heater 17 is switched on that provides for additional heat loading of CPSES.

The advantages of such a scheme are:

- possibility to include additional users without interfering with the hydraulic operation of the heat network;
- decrease in recycled network water temperature that results in an increase of heat loading of CPSES network heaters and additional production of electric energy;
- decrease in the amount of fuel burned for heat supply of new loads;
- reduction of heat losses when network water is transported in the main pipelines.

The efficiency of the combined scheme was evaluated in Ref. [9] using the example of heat supply of a new city area with a heat load of 10 MWt. The temperatures in supply/return lines of CPSES network are 110/70 °C, and of a local system are 95/70 °C. Network water cooling in the evaporator was taken as 15 °C. Calculations showed that the HPU COP ranged from 5.5 to 8.31 due to high temperature of network water (38–70 °C) and the increase in electric CPSES power is 2.1–2.3 MW. This growth is 2–3 times larger than the HPU compressor power.

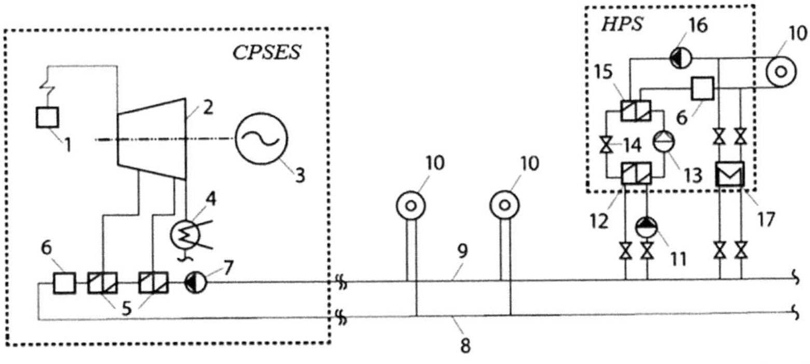

Fig. 2.11 Schematic diagram of a combined system of centralized district heat supply: *1*, boiler; *2*, steam turbine; *3*, electric generator; *4*, condenser; *5*, network heaters; *6*, peak load boiler; *7,16*, network pumps; *8,9*, main lines of feed and recycled network water; *10*, consumers; *11*, pump; *12*, evaporator; *13*, compressor; *14*, valve; *15*, condenser; *17*, hot water supply system heater.

2.4.3 Sewer drains

2.4.3.1 Sewer water

Sewer drains are considered to be an almost ideal source of low potential heat for heat pumps. Utilization of such heat can be much more effective than the use of soil heat due to constant high temperature of waste water during a heating period, high thermal power, and good heat and physical characteristics. The direct use of sewer drain heat by means of heat exchangers without their being located inside the flows or in vessels of liquid drains is of practical interest.

The technical and economic efficiency of such a solution relating to heat pump heat supply for city infrastructure loads was evaluated in Ref. [10]. High-efficiency commercial heat pumps of water-water mode were used for the analyses. Two variants were considered for locating heat exchangers to extract heat from sewer drains (Fig. 2.12), namely:

Variant 1: coiled-tube soil heat exchangers made of polymer pipe directly attached to the lower part of a self-flow concrete sewer;

Variant 2: "pipe-in-pipe" heat exchangers formed with a metal casing around a metal pipe of a 1 m diameter head sewer.

The temperature of fluid in the sewer is taken as 15 °C; the inlet temperature to the sewer heat exchanger is 4 °C and the outlet temperature is 11 °C which is also at the evaporator inlet.

The investigation carried out according to the given conditions concluded that the use of heat pumps in the sewer drain waste heat utilization system is one of the most

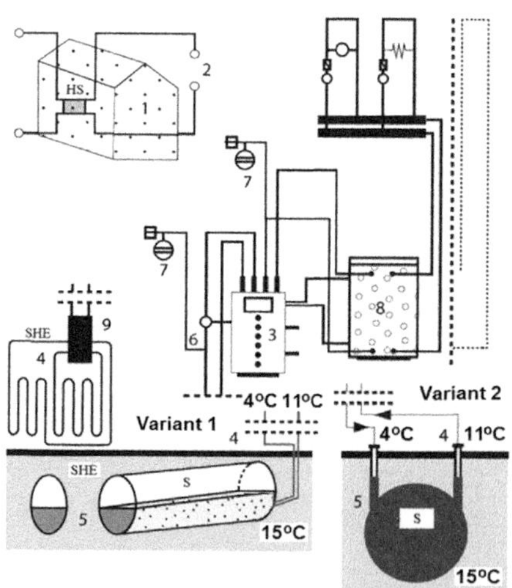

Fig. 2.12 Schematic flow diagram of a heat-pump heat supply system using sewer drain heat: *1*, heat load; *2*, back-up traditional heat supply; *3*, heat pump; *4*, circulating heat exchanger coils; *5*, collector heat exchangers; *6*, circulating pumps; *7*, overflow tanks; *8*, buffer tank of hot water; *9*, distributors; *HS*, heating system; *SHE*, soil heat exchanger; *S*, sewer.

promising methods of energy saving. In addition it is important to note that the efficiency of this system depends greatly on the heat transfer intensity when using the heat of low-potential drain heat. In compact heat exchangers with metal head collectors, the heat transfer intensity is much greater than in coiled-tube soil heat exchangers made of polymer pipes with which headless concrete sewers are equipped.

2.4.3.2 Conventionally pure building sewer drains

Building sewer drains, the temperature of which can reach 32 °C, can be used as lower heat sources for local heat pumps, if there are hot water supply systems. For this purpose special heat exchangers, operating together with a heat pump in a residential building hot-water supply system, may be located in the basement. The heat exchanger design should not obstruct or prevent the natural movement of waste liquid in the sewer pipes. A possible solution may be a "pipe-in-pipe" heat exchanger placed around the casing of the sewer pipe and fed into the formed space among heat-carrier pipes from the heat pump evaporator. Heat extraction efficiency in such a heat exchanger is not high, but may be intensified if use can be made of a separate drainage system of warm, so-called "conventionally pure" waste water from baths, sinks and kitchens, and separately, cold and more polluted water from the toilets.

The hot water supply system is characterized by the fact that most of the thermal energy of the hot water is almost never used. The thermal energy in the sewer drains is not much less than the heat amount that is spent for water preparation in the hot water supply systems. Thus, the thermal energy of conventional pure waters can be used as the heat supply for heat pumps.

At the same time one must note that the wasted thermal energy is not sufficient to provide heating to a building. That is why if the heat pump using ambient air heat is used to heat a building, conventional-pure waste waters can be collected in a special accumulator (either directly or after using the heat pump in the hot water supply system) and crystallization heat of this water can be used for preliminary heating of the ambient air in front of the heat pump in a cold season. The efficiency of this variant of building heat supply greatly increases when lower rates for night electricity is available.

2.4.4 Waste water heat of industrial enterprises

Waste heat from water at industrial enterprises can be of various origins but the most widely used are cooling waters of technological and electrical power generating stations. In closed systems of service water supply, this water goes to cooling towers where the waste heat goes into the surrounding air. The properties of waste water are characterized by relative stability so that the heat of such water may be quite effectively used in heat-supply heat pump systems (Fig. 2.13).

Water heated in a piece of technological equipment 1 is fed by the pump 3 into the intermediate tank 4 through which water is pumped by pumps 5 and 6 into the cooling tower 2 and evaporator 7 of the heat pump. When it is required to provide heat to the user 11, the heat pump is used and the heat from the heat pump condenser 9 starts

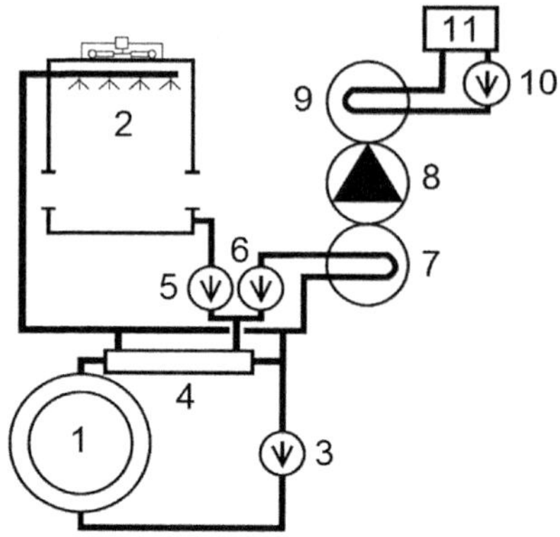

Fig. 2.13 Schematic flow diagram of a heat pump that operates with technological equipment waste heat: *1*, equipment requiring cooling (e.g., an air compressor); *2*, water cooling tower; *3*, circulating pump of water cooling loop; *4*, intermediate tankage; *5*, cooling tower pump; *6*, heat pump evaporator pump; *7,8* and *9*, heat pump evaporator, compressor and condenser, respectively; *10*, circulating condenser pump; *11*, heat supply system.

going into the supply system by means of a circulating pump 10. At the same time the cooling tower loading is reduced and in some cases it may be completely switched off; then the technological heat will be completely used.

As a result of the wide use of refrigerators both in domestic life and industries, there appear new possibilities for recycling heat energy, additional pumps being not always necessary because technologically a refrigerator has all the elements of a heat pump. Some manufacturers provide air and water condensers installed in series for off-the-shelf refrigerators that makes it possible to use some part of waste heat for hot water supply while serving the normal cooling needs inside the refrigerator. The scheme of such a device is shown in Fig. 2.14.

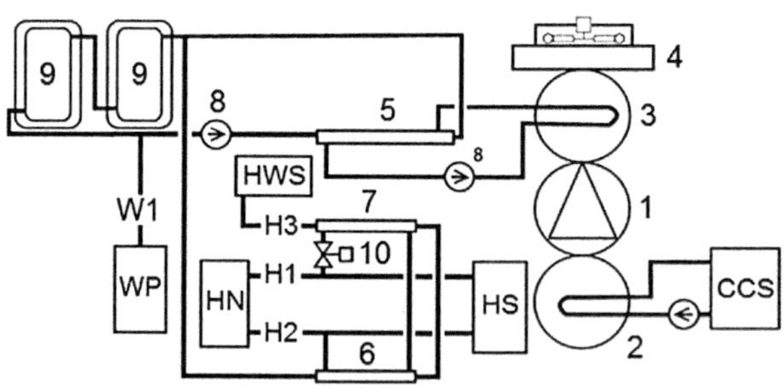

Fig. 2.14 Schematic flow diagram showing how the condenser heat from an air conditioning system's refrigerator is used for heating water in a hotel: *WP*, water pipe; *HN*, heat network; *HS*, heating system; *CCS*, conditioner cold supply; *HWS*, hot water supply system; *1*, refrigerator compressor; *2*, evaporator; *3*, water condenser; *4*, air condenser; *5*, heat exchanger; *6,7*, water heaters of the first and second stage HWS; *8*, circulating pump; *9*, water accumulating tanks; *10*, HWS temperature regulator.

When the refrigerator compressor 1 operates, cold is produced that is used in a cold supply system. At the same time the refrigerator working fluid condensation heat is partly drawn by an intermediate heat carrier that circulates between the water condenser 3 built into the refrigerator and a heat exchanger 5 in which water of a hot water system is warmed up. The working fluid condensation process finishes in the air condenser. Heated water is accumulated in the tanks 9 into which water from the water pipe WP is fed. Water finishes heating up in the water heater 7 by the heat carrier from the heat network HN before it is fed into the hot water supply system. The system requires minimal heat supply from an external source when the temperature of water in the tanks 9 is quite high; when the weather is rather hot, the stored water may provide water supply when it is switched off from the heat network for maintenance procedures.

2.5 Use of soil heat

Unlike soil water that is not always available in necessary amounts for a heat pump, soil itself is everywhere, and its stored thermal energy can almost always be used. Thus, a large number of systems with soil heat collectors have been installed recently, with great capital investments being made for the sake of increasing the COP in the coldest month, January. However, it is not easy to capture the soil heat and the main part of the capital expense goes to constructing soil heat exchangers and connecting them to the HPHS system. Either horizontal or vertical soil heat exchangers may be used to draw the heat from soil.

2.5.1 *Horizontal in-ground heat exchangers*

Horizontal soil heat exchangers can be put into the foundation ditch or a trench around a building. To determine the ditch or trench dimensions and general pipe length one must know the heat pump thermal power Q_{HP} and its calculated COP, φ. [Note: φ will be used in equations for simplicity.] Then the general pipe length L_h can be defined by the equation:

$$L_h = \frac{10^3 Q_{HP}}{q_s}\left[\frac{\varphi - 1}{\varphi}\right] \tag{2.3}$$

where q_s is the specific heat flow through 1 m of the pipe put into the soil, W/m. The q_s parameter depends on a number of factors and its precise value can only be found experimentally. For preliminary calculations q_s is taken to be 25 W/m, but for soils with different heat conductivity this value can vary widely. In addition, the q_s value is characterized by an unsteady heat extraction process and it decreases toward the end of the heating period. Heat transfer with the soil can worsen from year to year especially if there is an unbalance in heat requirements from summer to winter. However, heat transfer will not decline provided there is a balance between heat extraction for heat supply in winter and heat replacement in summer during operation of the air conditioning system.

The analysis that leads to the empirical equation to estimate the heat flow from the soil to a horizontal pipe heat exchanger, given below Eq. (2.4), is developed in Ref. [5]:

$$q_s = (1.4\lambda - 0.5) \cdot \left(\frac{W_c}{W_w} + 1\right) \cdot (t_s - t_c), \tag{2.4}$$

where λ is the soil thermal conductivity, W/m·°C; W_c and W_w are amount of cooling and heating used per year; t_s is soil temperature in the natural state, °C; t_c is the mean fluid temperature that cools the soil, °C. Data on thermal conductivity of some soils are given in Table 2.5.

Horizontal heat exchanger pipes are arranged so that there is no thermal interference with adjacent pipe temperature fields. It is generally assumed that when the distance between pipes is 1 m, the interference between adjacent pipes is negligible. If the pipe spacing is taken to be 1 m, then its area in square meters will be numerically equal to the common pipe length that is determined by Eq. (2.3).

As for the heat exchanger pipe diameter (typ. 27–34 mm), its value does not practically affect the specific heat flow value q_s because only a small range of polymer pipe diameters are deployed. The heat transfer coefficient from the pipe wall to fluid flow is sufficient provided the mode of fluid flow in the pipe is both turbulent and transient. The transient mode is desirable when the heat pump operates periodically as the soil somehow manages to return to its initial temperature state. Fully turbulent flow is obtained in polyethylene pipe at $Re > 3000$. The corresponding minimal fluid flow values are given in Table 2.6.

If fluid flow in the pipe exceeds the minimum, its diameter or discharge capacity of the circulation loop is chosen based on the hydraulic head. When using polyethylene pipe of the widely used diameters, one can refer to the curves shown in Fig. 2.15.

Being aware of the circulation discharge capacity and total fluid flow in the lower loop of the heat pump system, one can determine the number of parallel loops of the soil collector. Total fluid flow is defined by the equation:

$$G_{sc} = \frac{3600 \cdot Q_{HP}}{c_p \cdot \rho_f \cdot \Delta t_{io}} \left(\frac{\varphi - 1}{\varphi}\right) \tag{2.5}$$

where Q_{HP} is the heat pump thermal power, kW; φ is the COP; c_p is the fluid specific heat circulating in the loop, kJ/kg·°C; ρ_f is fluid density, kg/m^3; Δt_{io} is fluid temperature difference at the soil collector input and output that is taken equal to 5°C, according the data in Ref. [5].

Table 2.5 Soil thermal conductivity λ in W/(m·°C).

Dry soil density, t/m^3	Humidity fraction	Horizontal (H) and vertical (V) pipelines					
		Sands		Sandy soils		Loam and clay	
		H	V	H	V	H	V
1.4	0.35	–	–	3.81	3.89	3.32	3.40
	0.30	–	–	3.68	3.76	3.07	3.14
	0.25	4.03	4.12	3.32	3.39	2.82	2.88
	0.20	3.32	3.38	2.82	2.87	2.33	2.37
	0.15	2.93	2.99	2.32	2.37	1.84	1.88
	0.10	2.32	2.36	1.96	2.00	1.48	1.51
	0.05	1.57	1.61	1.35	1.37	0.97	0.99
1.6	0.30	–	–	3.98	4.02	3.56	3.63
	0.25	5.28	5.37	3.80	3.88	3.20	3.26
	0.20	4.53	4.62	3.42	3.49	2.81	2.87
	0.15	3.79	3.86	3.06	3.12	2.32	2.37
	0.10	3.05	3.11	3.41	3.47	1.84	1.87
	0.05	2.21	2.24	1.70	1.74	1.20	1.22
1.8	0.20	5.63	5.72	3.93	4.00	3.32	3.38
	0.15	4.78	4.84	3.54	3.60	2.96	2.99
	0.10	4.34	4.21	3.05	3.11	2.21	2.26
	0.05	3.04	3.09	2.1	2.12	1.47	1.50
2	0.10	5.73	5.82	3.66	3.72	2.69	2.74
	0.05	4.40	4.46	–	–	–	–

Table 2.6 Minimum volumetric fluid flow (m^3/h) for turbulent flow of water and glycol mixtures in pipes of different diameters.

Fluid	Temperature, °C	Inner diameter, mm		
		27	34	39
Ethylene glycol 20%	−1	0.7	0.88	1.02
	+18	0.5	0.64	0.7
Propylene glycol 20%	−1	0.95	1.22	1.38
	+18	0.64	0.82	0.93
Water	+18	0.32	0.39	0.45

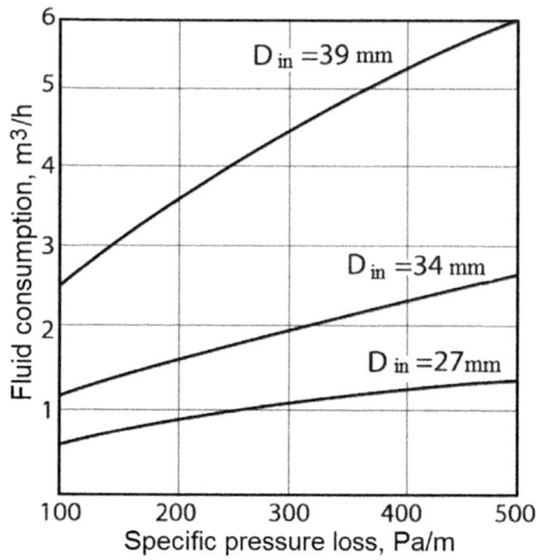

Fig. 2.15 Nomogram for estimating circulation loop volumetric discharge capacity.

Considering that soil can have temperatures lower than 0 °C and the evaporator will always have negative Celsius temperatures in winter, an intermediate heat transfer fluid is used for heat extraction from the soil: anti-freezing liquids, e.g., ethylene glycol and propylene glycol. The first is used more often, and for temperatures of −5, −10 and −15 °C, its required water concentrations are 12, 20 and 27.4%, respectively. Thermal and physical properties of ethylene glycol and propylene glycol water solutions are given in Table 2.7.

Table 2.7 Properties of aqueous ethylene glycol and propylene glycol solutions.

Concentration, ζ, %	Density, kg/m^3	Freezing temperature, t_{fr}, °C	Heat capacity, c, kJ/(kg·°C)
Ethylene glycol			
12.2	1015	−5	3.98
19.8	1020	−10	3.85
27.4	1035	−15	3.73
35.0	1045	−21	3.56
42.6	1055	−29	3.43
Propylene glycol			
20	1021	−7.5	4.06
30	1036	−12.3	3.92
40	1043	−20.5	3.74

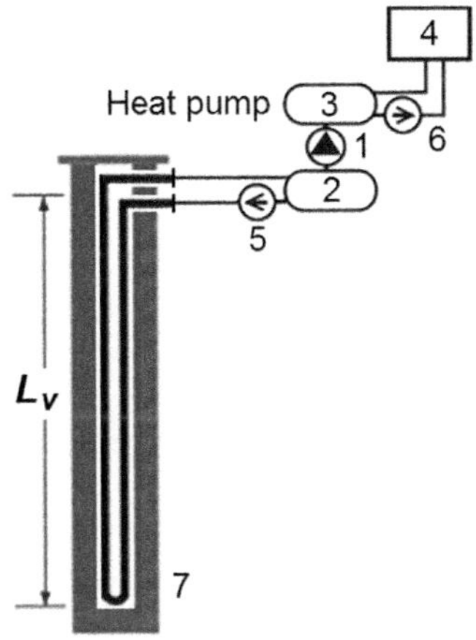

Fig. 2.16 Schematic diagram of a vertical soil heat exchanger in a heat pump: *1*, heat pump compressor; *2*, heat pump evaporator; *3*, heat pump condenser; *4*, heating system; *5*, circulating pump; *6*, heating system pump; *7*, vertical soil heat exchanger.

2.5.2 Vertical in-ground heat exchangers

Fig. 2.16 illustrates a vertical soil heat exchanger. This heat exchanger represents a U-tube polyethylene heat exchanger inserted into a well. The pipeline is connected with the heat pump evaporator via a closed common loop in which a water glycol solution circulates by means of a pump 5.

A general vertical soil heat exchanger length L_v can be estimated by the equation:

$$L_v = \frac{10^3 Q_{HP}}{q_w}\left[\frac{\varphi - 1}{\varphi}\right] \tag{2.6}$$

where q_w is a specific heat flow for a 1 m well, W/m; Q_{HP} is the heat pump capacity, kW; φ is the COP of the heat pump. The reader will notice that this equation is identical in form to Eq. (2.3), except that they are applied to different geometries: Eq. (2.3) to a horizontal pipe and Eq. (2.6) to a vertical pipe.

To determine q_w by analytical methods is very difficult because of unsteady heat transfer in inhomogeneous soils that depends on many site-specific parameters. In such a case, for preliminary calculations the value $q_w = 50$ W/m is assumed, but it should be determined more precisely. Reference [5] gives an approximate equation based on analyses of actual data:

$$q_w = K(1.4\lambda - 0.5)\left(\frac{w_c}{w_w} + 1\right)(t_s - t_c), \tag{2.7}$$

where all symbols are the same as in Eq. (2.4), and K takes the value $K = 1$ if one U-tube pipe is inserted into the well and $K = 1.28$ if there are two pipes. Equation (2.6) is valid if the space inside the well between the pipes and the soil is filled with a specially-prepared heat conducting material, namely, bentonite.

A weighted-average thermal conductivity λ should be used in Eq. (2.7) when the well passes through soil layers with different values of thermal conductivity, and it is estimated by the equation:

$$\lambda = \frac{\sum \lambda_i \delta_i}{\sum \delta_i}. \tag{2.8}$$

where λ_i is the thermal conductivity of i-th soil layer and δ_i is thickness of the i-th soil layer.

Generally for a large project several wells of 100 m depth are used. In order to avoid interference of temperature fields around separate wells, the distance between them is taken to be no less than 6 m. The required number of wells is determined by the total soil heat exchanger length divided by the well depth.

Polyethylene pipes with a diameter in the range of 20—40 mm are used in wells. It is recommended that the pipe diameter be chosen so that the hydraulic head of a well should not exceed 50 kPa. If the hydraulic head of a well is less than 10 kPa, it is recommended that two or more wells be connected in series.

2.6 Optimal usage of low-temperature heat sources

When using low temperature heat sources in HPHS, there is some ambiguity in choosing operating conditions for the heat pump evaporator. In some references one finds recommendations relating to the temperature difference at the heat pump evaporator inlet and outlet, depending on the nature of the lower heat source, but these suggestions are given without justification. We present here a technique for estimating the temperature difference mentioned above that characterizes the nature of various types of low-temperature heat source.

2.6.1 *Specific external energy losses for HPHS using different energy sources*

The temperature level of the heat-transfer fluid that cools the low temperature energy source (e.g., ambient or ventilation air, water, soil) in the evaporator affects the operating conditions of both the heat pump compressor and the mover (or driver) of the surroundings (e.g., a fan or a pump). The problem arises in estimating the optimal level of ambient cooling which minimizes the total specific electric energy loss for the HPHS system. The situation is complicated by the fact that the energy losses change in opposite directions depending on the heat-carrier temperature at the heat pump evaporator outlet.

The efficiency of heat pump systems is usually assumed to be measured by the heat pump COP. But for complex systems, the overall operating efficiency depends not only on the efficiency of the heat pump itself, but also on other system elements.

Thus, it is desirable to use other characteristics to find the efficiency of the entire system. Therefore, in the following analysis the value of specific losses of external energy are determined for the heat supply system, which, in case of energy losses for only the heat pump, is a value different from the heat pump COP.

In general the total specific losses of external energy l_h for the heat pump and ambient coolant driver of a low-temperature heat carrier can be presented in the following way:

$$l_h = \frac{L_{tot}}{Q_h} = \frac{L_c + L_s^{m \cdot e}}{Q_c}, \tag{2.9}$$

where L_c, $L_s^{m \cdot e}$ are energy losses for a heat pump compressor and ambient coolant driver, respectively, kW; Q_c is the heat flow extracted from the heat pump condenser, kW.

Energy losses for the heat pump L_c are defined by the equation:

$$L_c = \frac{Q_{ev}}{\varphi - 1}. \tag{2.10}$$

Heat flow in the heat pump evaporator Q_{ev} is given by:

$$Q_{ev} = V_s \rho_s c_p \left(t_s^{in} - t_s^{out}\right), \tag{2.11}$$

where V_s, ρ_s, and c_p are volume loss, density and isobaric specific heat of the surroundings, respectively; t_s^{in} , t_s^{out} , are surroundings temperatures at the evaporator inlet and outlet, respectively.

Energy losses for ambient coolant driver are:

$$L_s^{m.e} = \frac{V_c \Delta p}{\eta_s^{m.e} \eta_{dr}}, \tag{2.12}$$

where Δp is aerodynamic and hydraulic pressure losses at the heat pump evaporator that depend on the chosen energy source; in the case of using soil heat, $\Delta p = \Delta p_{ev} + \Delta p_{s.h}$, namely, total pressure drop in the lower loop (in the heat pump evaporator and soil heat exchanger); $\eta_s^{m.e}$ and η_{dr} are the efficiency of ambient coolant driver and its motor, respectively. One can assume that at the optimum operating point $\eta_s^{m.e} = 0.8$ for a fan and pump, and the driver motor efficiency, $\eta_{dr} = 0.95$.

Heat flow Q_c in Eq. (2.9) is determined from the HP heat balance equation:

$$Q_c = Q_{ev} + L_c \tag{2.13}$$

The actual COP of the heat pump φ can be presented in the following way:

$$\varphi = \varphi_t \eta_{HP}, \tag{2.14}$$

where η_{HP} is an efficiency that takes into account real processes done by the working fluid inside the heat pump; it can be assumed to be 0.6; φ_t is a theoretical heat pump COP that can be determined by the following relation taking into account heat irreversibility in the evaporator and condenser:

$$\varphi_t = \left[1 - \frac{T_{ev}^{HP}}{T_c^{HP}}\right]^{-1} = \left[1 - \frac{273.15 + t_s^{out} - \Delta t_{ev}}{273.15 + t_c + \Delta t_c}\right]^{-1}, \tag{2.15}$$

where T_{ev}^{HP} is the working fluid evaporating temperature, K; T_c^{HP} is the working fluid condensation temperature, K; t_s^{out} is the surroundings temperature at the evaporator outlet, °C; Δt_{ev} is temperature difference between the heat pump surroundings and the working fluid at the evaporator outlet, °C; Δt_c is temperature difference between the heat pump working fluid and the water at the condenser outlet. According to standard practice, one can assume that $\Delta t_{ev} = 10$ °C, 5 °C, and 5 °C for air, water and brine solutions, respectively; $\Delta t_c = 5$ °C for low-temperature water-heating systems.

The water temperature t_c at the condenser outlet is specified by the needs of the heat consumer. In the case of a HP to supply a hot water heating system, the temperature of the heat carrier that is fed from the heat pump condenser into the low-temperature heating system can be found from the following equation [11]:

$$t_c = t_{hc}^c + \left(t_{hc}^c - t_a\right)\left[\frac{t_a - t_0}{t_a - t_0^c}\right]^{(1+n)^{-1}}, \tag{2.16}$$

where t_{hc}^c is the calculated temperature of the heat carrier being heated in the heating system with calculated ambient temperature t_0^c, °C; $n = 0$ for low-temperature heating systems.

Taking into account Eqs. (2.10)–(2.14), Eq. (2.9) for the specific external energy losses for heating takes the following form:

$$l_h = \frac{1}{\varphi}\left[1 + \frac{A(\varphi - 1)}{\left(t_s^{in} - t_s^{out}\right)\eta_s^{m.e}\eta_{dr}}\right], \tag{2.17}$$

where $A = \frac{\Delta p}{\rho_s c_p}$ is a defined factor, a function of given values, °C.

Thus, the total specific external energy loss for heating l_h depends on the defined function A, ambient temperature, the temperatures of a low-temperature energy source at the evaporator inlet and outlet, and the calculated temperature of the heat transfer fluid in the heating system, namely, $l_h = f\left(A, t_0, t_s^{in}, t_s^{out}, t_{hc}^c\right)$.

2.6.2 Optimum degree of cooling in heat pump evaporator

In the following sections, we consider this problem for systems that use air, water and soil as the low-temperature energy source for the heat pump.

2.6.2.1 Ambient air as heat source

Figure 2.17 shows the principal scheme of low-temperature water heating using an "air-water" heat pump. The principle of operation is the following: ambient air with the temperature t_0 and volume fan loss V_a is fed into the heat pump evaporator. In the evaporator the ambient air is cooled and at the outlet its temperature is t_a^{out}. The room being heated has heat losses into the environment Q_{RH}. The heat flow from the condenser of the heat pump Q_c with a working fluid temperature t_c at the heating system outlet is used to compensate for these heat losses.

It is necessary to determine the optimal air temperature at the heat pump evaporator outlet t_a^{out} for a given ambient temperature t_0.

A change in the air temperature at the heat pump evaporator output t_a^{out} with a given heat flow value Q_c results in changing the ambient air volume loss Va which in turn leads to a change in the fan drive energy losses. At the same time, the temperature t_a^{out} changes (at constant condenser outlet temperature t_c), thus changing the operating conditions and energy losses for the heat pump compressor drive. Since the changes in the energy losses for the heat pump compressor and fan drive are in opposite directions, there is an optimal air cooling level in the heat pump evaporator which minimizes the total energy loss for the overall heating system.

To determine optimal heat pump operating conditions, let us plot the specific external energy losses for the heating system l_h versus the evaporator outlet air temperature for an assumed working fluid temperature of $t_{hc}^c = 45°C$ in the low-temperature heating system. The defined factor A is taken as $A = 0.1$; 0.5; and 1 °C, taking into account real values of pressure loss in the evaporator (a convective heat exchanger) that are in the range of $\Delta p = 10-100$ Pa. The range of ambient temperature is from -20 °C to $+15$ °C. By substituting these values into Eq. (2.16) and taking into account of the values of a real heat pump COP that can be estimated using

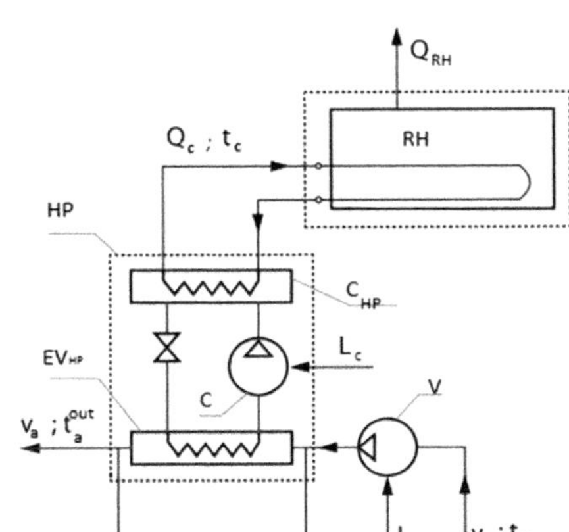

Fig. 2.17 Schematic flow diagram of low-temperature water heating using an "air-water" heat pump: *RH*, room heating; *HP*, heat pump; C_{HP}, heat pump condenser; EV_{HP}, heat pump evaporator; *C*, compressor; *V*, fan; L_c, heat pump compressor drive input; L_v, fan driver.

Equation (2.13), and the temperature of the working fluid being fed into the heating system from Eq. (2.15), we obtain the curves shown in Fig. 2.18.

It is clear from the graph that there are optimal air temperature values at the evaporator outlet t_a^{out} that lead to minimum specific external energy losses for heating l_h^{min}.

To determine the optimal air cooling level in the heat pump evaporator, let us represent the temperature T_{ev}^{HP} in Eq. (2.15) in the following way:

$$T_{ev}^{HP}=T_0-\Delta t_a-\Delta t_{ev}=273.15+t_0-\Delta t_a-\Delta t_{ev}\,, \tag{2.18}$$

where $\Delta t_a=t_0-\Delta t_a^{out}$ is the temperature difference between the air at the evaporator inlet and outlet, °C. Then, Eq. (2.17) in conjunction with Eqs. (2.15) and (2.18) can be written in a similar form as the Δt_a function as follows:

$$l_h=a+\frac{\Delta t_a}{T_c^{HP}\eta_{HP}}+\frac{Ab}{\eta_{ev}\eta_{dr}\Delta t_a}, \tag{2.19}$$

where

$$a=\frac{\Delta t_a}{T_c^{HP}\eta_{HP}}\left(T_c^{HP}-T_0+\Delta T_{ev}-\frac{A}{\eta_{ev}\eta_{dr}}\right)\quad \text{and}\quad b=1-\frac{1}{\eta_{HP}}+\frac{T_0-\Delta t_{ev}}{T_c^{HP}\eta_{HP}}.$$

Applying the calculus of variations to the function $l_h=f(\Delta t_a)$ allows one to find the optimal level of cooling of the ambient air in the heat pump evaporator, namely,

$$\Delta t_a^{opt}=\sqrt{\frac{A(273.15+t_c+\Delta t_c)}{\eta_{ev}\eta_{dr}}\left[\eta_{HP}-1+\frac{273.15+t_0-\Delta t_{ev}}{273.15+t_c+\Delta t_c}\right]}. \tag{2.20}$$

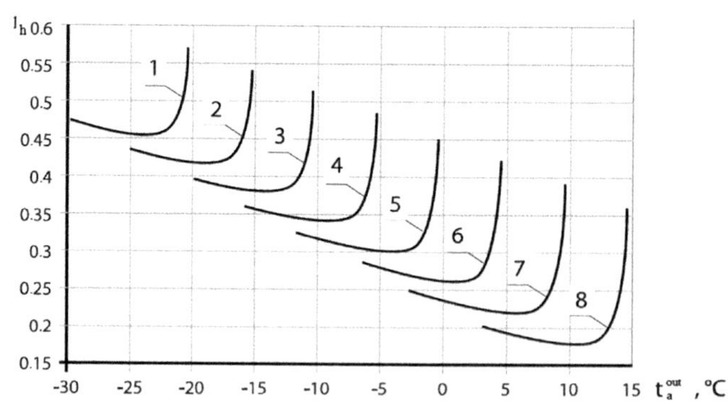

Fig. 2.18 Specific external energy losses for heating as a function of air temperature at the evaporator outlet at $t_{hc}^c=45°C$ (at $A=0.1°C$); 1, 2, 3, 4, 5, 6, 7, 8 correspond to ambient temperatures $t_0=-20, -15, -10, -5, 0, 5, 10, 15°C$, respectively.

Thus, the optimal air cooling level in the evaporator depends on the factor A, the ambient temperature t_0 and the calculated temperature of the working fluid in the heating system, t_{hc}^c , i.e., $t_a^{opt} = f(A, t_0, t_{hc}^c)$

The relationship between the optimal level of air cooling in the heat pump evaporator and the ambient temperature is shown in Fig. 2.19, for different values of A and t_{hc}^c. It follows from Fig. 2.19 that the optimal air cooling level in the heat pump evaporator increases with an increase in factor A (which basically depends on the pressure loss in the convective heat exchanger) and very weakly depends on the working fluid temperature for heating in the range of 30–50 °C.

Considering that the effect of the working fluid temperature is minimal, a cross-plot of Fig. 2.19 is presented in Fig. 2.20 for the average value of t_{hc}^c.

It is seen from Fig. 2.20 that the optimal air cooling level in the heat pump evaporator depends strongly on the factor A that characterizes the air flow pressure losses in the evaporator and depends weakly on the ambient temperature over the range of t_0 from −20 to +15 °C.

2.6.2.2 Natural or waste water as heat source

To find the best operating point for a heat source in a heating system, we will examine the variation of the specific external energy losses l_h as a function of the water temperature at the evaporator outlet for a working fluid temperature of $t_{w.f.} = 30$ °C. The value of the factor A is taken as 0.15, 0.05, and 0.01 °C, according to the range of values for Δp. The results are presented at an assumed ambient temperature of −20 °C. By substitution of the corresponding terms into Eq. (2.17), and taking into account Eqs. (2.14) and (2.16), we obtain results given in Fig. 2.21.

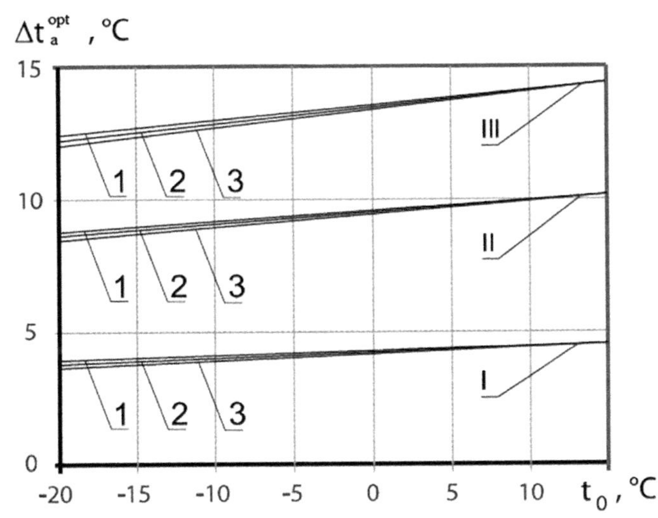

Fig. 2.19 Optimal air cooling level in the evaporator as a function of ambient temperature: I, II, III (at $A = 0.1, 0.5, 1$°C); 1, 2, 3 (at $t_{hc}^c = 30, 40, 50$°C).

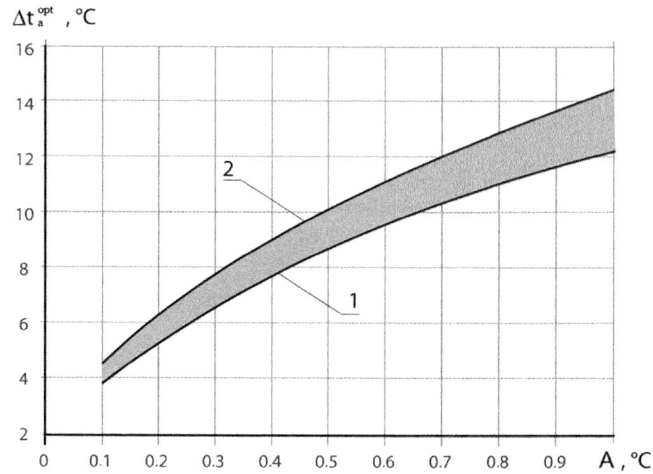

Fig. 2.20 Optimal degree of air cooling in the evaporator as a function of the factor A: 1, 2: Ambient temperature, $t_0 = -20$ to $+15°C$.

It is clear from the curves 1–5 that for each value of inlet water temperature there are optimal water temperature values at the evaporator outlet that yield minimum specific losses of external energy.

To determine the optimal water temperature difference between the evaporator inlet and outlet, Eq. (2.19) can be used by replacing the ambient air temperature t_o with the water temperature at the evaporator inlet t_{ev}^{in}.

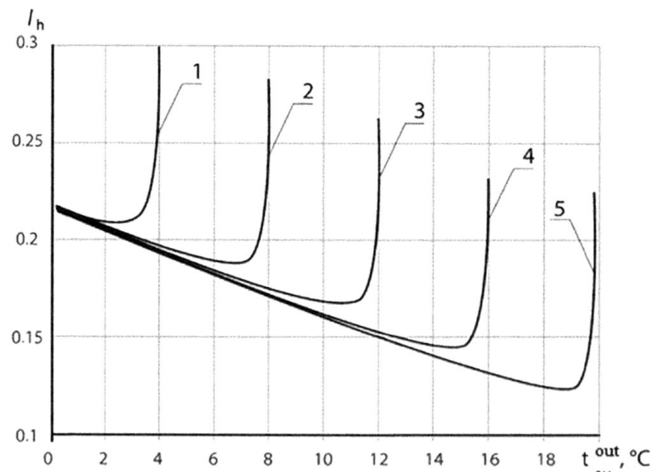

Fig. 2.21 Specific external energy losses for heating as a function of water temperature at the evaporator outlet at $t_{w.f.} = 30°C$ (at $A = 0.01°C$): 1–5 are water temperature at the evaporator inlet, $t_{ev}^{in} = 4, 8, 12, 16, 20°C$, respectively.

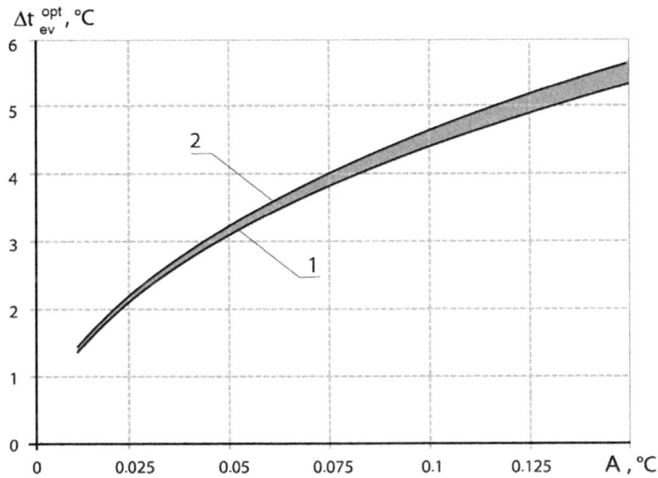

Fig. 2.22 Optimal amount of water cooling in the evaporator as a function of the factor A: *1,2*, water temperature at the evaporator inlet t_{ev}^{in} of 4, 20°C.

Thus, the optimal temperature of cooling water in the evaporator depends on the factor A, ambient temperature t_o, evaporator inlet water temperature t_{ev}^{in} and temperature of the working fluid $t_{w.f.}$, i.e., $t_{ev}^{opt} = f(A,\ t_0,\ t_{ev}^{in},\ t_{w.f.})$.

Figure 2.22 shows the optimal water temperature difference as a function of the factor *A* at an ambient temperature $t_e = 20$ °C. The optimal water cooling level is seen to depend greatly on the factor *A* and very weakly on the heat source temperature over the range from 4 to 20 °C.

2.6.2.3 Soil as heat source

Vertical and horizontal soil heat exchangers are used to extract heat from soil and put it to use as a low-temperature heat source for heat pumps. Horizontal soil heat exchangers are examined in the following analysis. A non-freezing working fluid receives heat from the soil, transfers it to the heat pump system and thence to a building. Figure 2.23 shows the results of analyses of varying heat collecting system parameters from the soil array during a heating period. It is seen that the temperature of the working fluid at the soil heat exchanger outlet (HEO) changes very little during a heating period and is about 3 °C [12]. However, in some geographical locations, the outlet temperature at the end of winter can be less than 0 °C.

The factor A was varied over the range 0.005–0.027 °C to account for real values of total pressure drop in the lower loop, i.e., in the heat pump and the soil heat exchanger. Accordingly, the evaporator pressure drop ranged from 10 to 40 kPa, and in the soil heat exchanger it ranged from 10 to 65 kPa.

To determine the optimal operating conditions of the heat source in the heating system, the specific external energy losses for heating l_h for the working fluid of the heat pump was plotted using the calculated temperature of a heating working fluid for

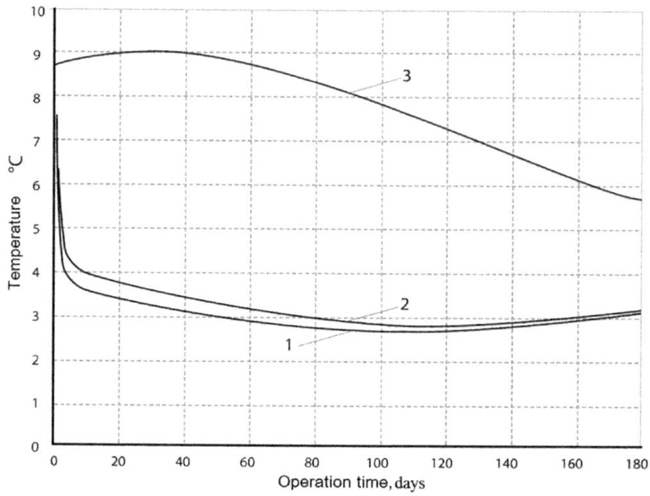

Fig. 2.23 Temperature variations of the heat collection system from a soil array during a heating period: *1*, working fluid temperature at the HEO; *2*, soil temperature close to HEO at 3 m depth; *3*, far-field soil temperature at 3 m depth.

$t_{w.f.} = 30$, 40 and 50 °C. The calculations were performed at an ambient temperature of −20 °C and a working fluid temperature at the evaporator inlet $t_{ev}^{in} = 3$ °C. Substituting the corresponding values into Eq. (2.17) and taking into account Eqs. (2.14) and (2.15), we obtain the results graphically presented in Fig. 2.24.

It is seen that for each value of the factor A there is an optimal value of working fluid temperature at the evaporator outlet at which the specific external energy losses are minimized.

The qualitative results for the soil heat source are very similar to the previous cases of air and water heat sources. Figure 2.25, for a working fluid temperature $t_{hc}^{c} =$ 40 °C, displays the same trends as Figs. 2.20 and 2.22 for air and water heat sources, respectively. Approximation of the curves shown in Fig. 2.25 can be obtained using logarithmic coordinates, and yields the following correlation:

$$\Delta t_{cp}^{opt} = 13.5A^{0.5} \tag{2.21}$$

2.7 Summary

This chapter reviews the main systems using heat pumps in heat supply systems for buildings using various heat sources. Analytical and practical methods together with numerical examples of typical calculations are given.

It is possible to use heat pumps effectively in heating and air conditioning systems in residential, administrative and industrial buildings, in shopping centers, sporting

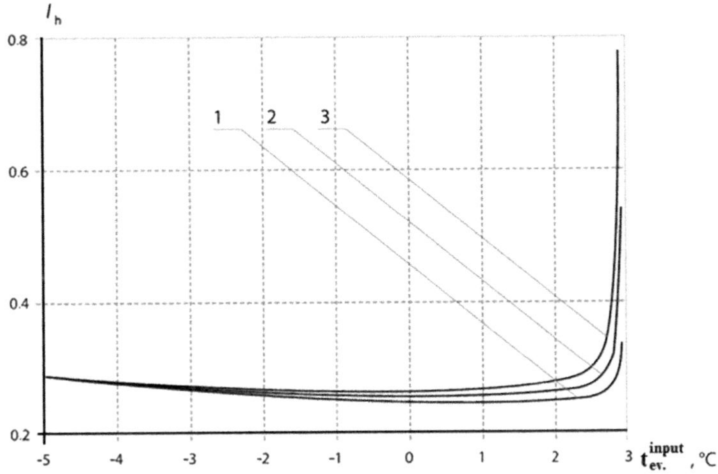

Fig. 2.24 Specific external energy losses for heating as a function of the temperature of the working fluid (brine solution) $t_{ev.}^{input}$ at the evaporator inlet (temperature of the working fluid $t_{w.f.} = 40°C$): curves 1, 2, 3 correspond to $A = 0.005, 0.015, 0.027°C$.

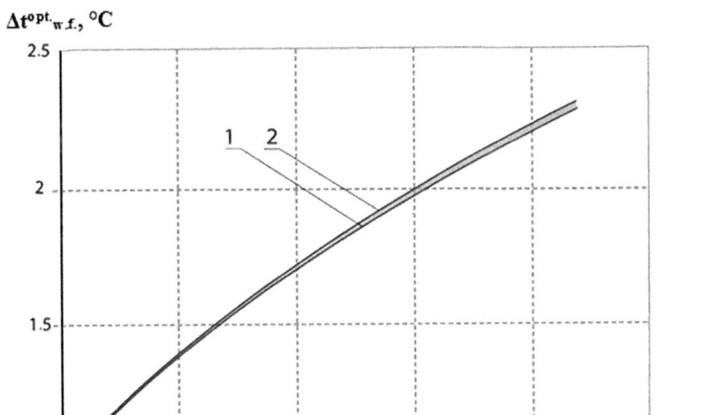

Fig. 2.25 Optimal working fluid temperature difference in the evaporator as a function of the factor A: 1, 2, temperature of working fluid (brine solution) at evaporator input $t^{in}_{w.f.} = 2, 5°C$.

arenas, concert halls, and medical complexes. The sources of heat can be the following: ambient air (with temperatures as low as −25 °C); air of ventilation systems (+15 to 25 °C); well water and urban water effluents and ones from industrial enterprises (from 5 to 20 °C); water of open reservoirs and upper layers of oceans (up to 20 °C); return water of thermal power plants and boiler heating plants (from 30 to 50 °C); and soil heat at a depth of 100 m (from 4 to 10 °C). Each of these sources has its advantages, disadvantages, and limitations, i.e., temperature level and volume, design features of the plant, cost, and reliability. Nevertheless, it has been demonstrated that the use of heat pumps can significantly reduce the overall energy consumption of buildings in an economically favorable manner.

Atmospheric air is the most available source of low-temperature heat. In the US, the most widely used type of heat pump is the air-to-air design, with a share of more than 80%. About 77% of heat pumps installed in Europe use atmospheric air. In Southern Europe, air-to-air split systems are used in the heating/air-conditioning mode. However, air-to-air heat pumps are characterized by a seasonal load factor or seasonal performance factor (SPF), which reduces their efficiency on average by about 10–30% compared to water-to-water heat pumps. This is caused by:

- Significant reduction in capacity and productivity with decreased temperature of atmospheric outside air
- Relatively large difference in evaporation and condensation temperatures during the lowest winter temperatures
- Energy costs for defrosting the evaporator surface (at temperatures from 0 to −6 °C) and fan power consumption.

In Northern Europe, heat pumps are used for heating and hot water supply. In Norway, the share of air heat pumps is 67%, while 19% of them use water and ground heat. In Finland, water-to-water systems predominate; the heat source is soil and lakes. In Sweden, heat pumps are the main technology in heating. Ground-source models prevail with a large number of air heat pumps [13]. In Switzerland and Austria, heat pumps using the heat of soil (40% and 82%, respectively) predominate.

According to the European Heat Pump Association (EHPA), the number of heat pumps installed in Europe is about 10 million units [14]. In 2015 the sales level was 880,179 units; a 25% growth is observed in the air-to-air segment as well, but air-to-water heat pumps show the most significant growth, namely, the number of sales has increased by 6–7 times compared to year 2014. The use of heat pumps is mainly concentrated in the following countries: France (209,000), Italy (121,000), Sweden (103,000), Spain (83,000) and Germany (70,000), using data for 2015. Asia remains the driving force of the heat pump market, accounting for 78% of the world sales: about 1.9 million units, mostly air-to-water units.

It is clear that heat pump technology is being widely developed, and new construction and reconstruction of buildings offer great opportunities for the use of heat pumps in the future.

Nomenclature

Symbols and acronyms

A	defined factor, a function of given values
APS	atomic power stations
C	compressor

CCS	conditioner cold supply
C_{HP}	heat pump condenser
CHPS	combined heat and power generating stations
COP	coefficient of performance
c_p	specific heat
CPSES	central power station extraction steam
EV_{HP}	heat pump evaporator
HN	heat network
HP	heat pump
HPHS	heat pump heat supply
HPS	heat pump station
HPU	heat pump unit
HS	heating system
HWS	hot water supply system
HX	heat exchanger
L	length of polyethylene pipeline
L_c	heat pump compressor drive input
L_c, $L_s^{m.e}$	energy losses for a heat pump compressor and ambient coolant driver, respectively
L_v	fan driver
l_h^{min}	minimum specific external energy losses for heating
OTEC	Ocean Thermal Energy Conversion
Δp	aerodynamic and hydraulic pressure losses at the heat pump evaporator
Δp_{evap}	evaporator pressure drop
$\Delta p_{s.e.}$	pressure drop in soil heat exchanger
Q_{ev}	heat flow in heat pump evaporator
Q_{HP}	heat pump thermal power
Q_c	heat flow from the heat pump condenser
Q_{RH}	heat losses to the environment
q_s	specific heat flow
q_w	specific heat flow relating to a 1 m well
RH	room heating
S	sewer
SHE	soil heat exchanger
T_c^{HP}	working fluid condensation temperature
T_{ev}^{HP}	working fluid evaporation temperature
$t_{h.c}^c$	calculated temperature of heat carrier
t_s^{in}, t_s^{out}	surroundings temperatures at evaporator inlet and outlet, respectively
t_c	mean fluid temperature that cools the soil
$t_{w.f.}^{out}$	working fluid temperature at the evaporator outlet
$t_{w.f.}$	working fluid temperature
TES	thermal electric stations
t_{ev}^{in}	water temperature at evaporator inlet
t_s	soil temperature in natural state
t_s^{out}	surroundings temperature at evaporator outlet
t_0	calculated ambient air temperature
Δt	difference between water temperature in pond and average glycol temperature

Δt_{ev}	temperature difference between heat pump surroundings and working fluid at evaporator outlet
Δt_a	temperature difference between air at the evaporator inlet and outlet
Δt_c	temperature difference between heat pump working fluid and water at condenser outlet
Δt_{io}	fluid temperature difference at soil collector input and output
Δt_w	water temperature difference at evaporator inlet and outlet
V	fan
V_a	volume fan loss
V_s	volume loss
W_c and W_w	amount of cooling and heating used per year
WP	water pipe

Greek symbols

$\eta_s^{m.e}$ and η_{dr}	efficiency of ambient coolant driver and its motor, respectively
η_{HP}	efficiency accounting for real processes of the heat pump working fluid
λ	heat conductivity
ρ_f	fluid density
ρ_s	surrounding density
φ	calculated COP
φ_t	theoretical heat pump COP

Review questions

1. Discuss the problems of using ambient air in typical winter conditions in the northern hemisphere.
2. Discuss the characteristics of ventilation wastes and their use in domestic and municipal buildings.
3. What air velocities should be assumed in the "air-air" type heat pumps?
4. Which ventilation wastes from a building can be used in heat pumps? What ventilation frequencies are appropriate for different building purposes? What are the advantages and disadvantages of the heat pumps of "ventilation air-water" type?
5. What system of heating devices is used with heat pump heating supply?
6. What are the peculiarities of using ventilation air from mines and metropolitan infrastructure in heat pumps?
7. What are the requirements for well siting when using artesian waters?
8. Describe the construction of the heat collector using open pool water.
9. What are the advantages and problems of using the heat of sewer waters in heat pumps?
10. How may the heat of turbine condensers be used? Include a discussion of efficiency values.
11. In what cases is it desirable to use the recycled water from district heating networks in heat pumps?
12. What are the advantages of using a heat pump instead of alternative systems?
13. Explain the designs of vertical and horizontal soil heat exchangers.
14. Explain the principle of using the heat of solidification of water (i.e., phase transition from liquid to solid) for air heating. Give the value of water crystallization heat.

Example: sizing a European residential ground-source heat pump

Case A. Determine the length of the ground collector of a heat pump for heating a new residential building in Europe with dimensions: 20 m × 20 m = 400 m^2. You may assume that a radiant floor heating system is employed. The soil at a depth of 5 m is wet clay ($\rho = 600$ kg/m^3, moisture content is 30%); coefficient of performance, $\varphi = 3.5$; $W_c/W_w = 0.4$; $W_{ec} = 4.9$ kW; soil temperature is 8 °C; and brine temperature is 0 °C, soil coefficient of thermal conductivity $\lambda = 3.56$ W/mK (Table 2.3 or standards VDI 4640).

Solution. Let us determine the required heat output of the heat pump. According to EU-Norm EN 12831, the heat consumption for new buildings is 30–50 W/m^2. So the required capacity of the heat pump is 400 × 50 = 20,000 W = 20 kW.

Horizontal collector

The specific heat flux from the ground to the horizontal pipe, from Eq. (2.4), is

$$q_0 = [(1.4 \times 3.56) - 0.5)](0.4 + 1)(8 - 0) = 50.2\ \mathrm{W/m}$$

The thermal capacity of the reservoir is $Q_0 = Q_{hp} - W_{ec}$, where Q_{hp} is heat pump power, and W_{ec} is electric power of the compressor.

The required collector power is $Q_0 = 20–4.9 = 15.1$ kW.

The length of the collector pipes, from Eq. (2.3), is

$$L = \frac{15100}{50.2}\left(\frac{3.5 - 1}{3.5}\right) = 214.9\ \mathrm{m}.$$

Thus, for the construction of a ground collector, two loops of 108 m length are required.

The necessary area of the site to accommodate the collector is

$$A = L \times a\frac{n - 1}{n}$$

where a is the distance between the pipes (assuming $a = 0.75–1$ m), and n is the number of loops.

$$A = 215 \times 0.75\frac{2 - 1}{2} = 81\ \mathrm{m}^2$$

If three contours of length of 40 m are used, the area of the plot will be

$$A = 215 \times 0.75\frac{3 - 1}{3} = 108\ \mathrm{m}^2$$

Next, let us consider the inverse problem. The area of the site should not exceed 100 m^2; it is necessary to determine the number of loops.

$$A = 100 = 215 \times 0.75 \frac{n-1}{n}$$

Hence,

$$\frac{1}{1 - \frac{100}{215 \cdot 0.75}} = n = 2.63$$

Since the area of the site should not exceed 100 m^2, we round off n on smaller side, i.e., $n = 2$. Then the real area of the section will be

$$A = 215 \times 0.75 \frac{2-1}{2} = 81 \text{ m}^2 \leq 100 \text{ m}^2$$

Vertical collector

The specific heat flux from the soil to the pipes of the vertical collector, using Eq. (2.7) and $k = 1.2$ is

$$q_0 = 1.2[(1.4 \times 3.56) - 0.5)][(0.4 + 1)](8 - 0) = 60.26 \text{ W/m}$$

The length of vertical manifold pipes is

$$L = \frac{15100}{60.26}\left(\frac{3.5-1}{3.5}\right) = 179.0 \text{ m.}$$

We accept the design of two well collectors, each with a depth of 90 m.

Case B. This example follows the specifications for Case A except that the soil is sand: $\rho = 1400$ kg/m^3, humidity is 5%. Determine the length of the ground collector for the heat pump.

Solution.

Horizontal collector

The specific heat flux from the ground to the horizontal pipe is

$$q_0 = [(1.4 \times 1.57) - 0.5](0.4 + 1)(8 - 0) = 19.0 \text{ W/m}$$

The thermal collector capacity is $Q_0 = Q_{hp} - W_{ec}$, where Q_{hp} is HP power and, W_{ec} is electric power of the compressor.

The required compressor power is

$$Q_0 = 20.0 - 4.9 = 15.1 \text{ kW}$$

The length of the collector pipes is

$$L = \frac{15100}{19.0}\left(\frac{3.5 - 1}{3.5}\right) = 568 \text{ m}$$

For the construction of a ground collector, three 190 m long loops are required.

The required area of the site to accommodate the collector is

$$A = L \times a\frac{n - 1}{n}$$

$$A = 568 \times 0.75\frac{3 - 1}{3} = 284 \text{ m}^2$$

Vertical collector

The specific heat flow from the soil to the pipes of the vertical collector is

$$q_0 = 1.2[(1.4 \times 1.61) - 0.5](0.4 + 1)(8 - 0) = 23.5 \text{ W/m}$$

The length of vertical manifold pipes is

$$L = \frac{15100}{23.5}\left(\frac{3.5 - 1}{3.5}\right) = 459 \text{ m}$$

We accept the collector design of five wells, each with a depth of 92 m.

Exercises

1. Determine the length of a horizontal ground heat exchanger for a heat pump with a capacity of: A: 25 kW; B: 10 kW. The heat exchanger is laid in soil consisting of clay with a moisture content of 10%; the soil temperature at a depth of 2 m is 5 °C.
2. Determine the length of the vertical ground heat exchanger for a heat pump: A: 250 kW; B: 100 kW. Geological data: at a depth of up to 100 m, the soils are represented by sand with a moisture content of 5%; soil temperature is 10 °C.
3. Determine the coefficient of performance of a heat pump system with a ground heat exchanger. The heating system is radiant floor heating. R410a (Case A) and R407a (Case B) are used as heat pump working fluids.

System data are:
- Thermal load is 1 kW.
- Temperature of the glycol at the heat pump inlet: $t_{p1} = 3$ °C.
- Glycol temperature at the outlet of the heat pump: $t_{p2} = -1$ °C.
- Water temperature at the heat pump inlet: $t_{w1} = 27$ °C.
- Water temperature at the outlet of the heat pump: $t_{w2} = 32$ °C.
- Temperature difference at the evaporator and condenser outlet: $\Delta t_{min} = 5$ °C.
- Ambient temperature is -5 °C.

4. Determine the thermodynamic efficiency of an air-air heat pump with a thermal capacity of 3 kW. Ambient air temperature is: -1 °C (Case A); $+2$ °C (Case B). Air temperature in the room is 18 °C; outside air temperature is -20 °C; air temperature difference in the evaporator and condenser is $\Delta t = 10$ °C. Assume that $\Delta t_{max} = 10$ °C and heat pump working fluid is R152a.

References

[1] Pisarev VE. Heat pumps and refrigerators. Manual – K. KNUCA. 2002.
[2] 2.04.05-91 "construction standards and heating, ventilation and condition rules".
[3] 2.01-99 "energy saving in buildings, norms of heat protection and heat, water and energy supply".
[4] Bezrodny MK. Power efficiency of heat pump schemes of heat supply. M.K. Bezrodny, H.O. Prytula. K. NTU «KPI»; 2012.
[5] Gershkovich VF. The peculiarities of designing building heat supply systems with heat pumps. K. Ukrainian Architecture Academy PE "Energominimum"; 2009.
[6] Bezrodny MK. Energy efficiency of the heat pump system of ventilation with heat recuperator and recirculation of exhaust air. M.K. Bezrodny, A. Galan Energy and Power Generation Technologies 2011;(2):16–9.
[7] Fialko NM. The estimation of heat pump using efficiency under metropolitans and coal mine conditions. N.M. Fialko, Z.B. Zimin Industrial Heat Technique 2006;(2):111–9.
[8] Sorokin OA. The use of heat pump units for combined low potential heat utilization at thermal power stations. O.A. Sorokin Industrial Power Engineering 2005;(6):36–41.
[9] Nicolaev YE. The determination of heat pump efficiency using the. Heat of Recycling Power Engineering 2007;(9):14–7.
[10] Zimin LB. The analyses of heat pump system of sewer drain heat utilization for heat supply of social objects. L.B. Zimin, N.M. Fialko Industrial Heat Technique 2008;(1):77–85.
[11] Nekrasova OA. The investigation of heating heat-pump systems (a model approach). O.A. Nekrasova, Ju.V. Sinyak Heat Power Engineering 1986;(11):30–4.
[12] Kostikov AO. The effect of soil heat state on the efficiency of a heat pump unit with the soil heat-exchanger. A.O. KostiKov, D.H. Harlampidi Power Engineering: Economy, Technology, Ecology 2009;(1):32–40.
[13] The Heat Pump Centre. http://heatpumpingtechnologies.org/.
[14] The European Heat Pump Association. www.ehpa.org.

Effective use of heat pumps for various heating applications

3

3.1 Heat pumps in individual and multi-family residences

Heat pumps are now recognized as one of the most efficient means of providing year-round comfort in homes, buildings of various purposes, and industrial applications. In this chapter we will focus on their strengths, while showing how certain drawbacks can be mitigated.

3.1.1 *Energy efficient houses*

An energy efficient house is a house with low energy consumption per living space area, the optimal microclimate usually achieved using ground heat and heat pump units.

In Europe residential buildings are categorized according to their level of energy consumption:

1) Old houses (built before 1970) with a heating system consuming ~ 300 kWh/(m$^2\cdot$y);
2) Newer houses (built from 1970 to 2002) with consumption of ~ 150 kWh/(m$^2\cdot$y);
3) Houses with low energy consumption of 60 kWh/(m$^2\cdot$y);
4) Passive houses with very low energy consumption of 15 kWh/(m$^2\cdot$y);
5) Zero-energy buildings with no net energy consumption;
6) Energy-plus houses whose engineering equipment (solar panels, heat pumps, energy recovery units, etc.) generate more energy than is consumed.

Under actual conditions in northern countries, e.g., United States, Canada, Scandinavia, and Russia, energy consumption can typically be 30–40 kWh/(m$^2\cdot$y).

The concept of the "passive house project" appeared in Germany in the 1990s. In the passive house, the heat losses through the enclosing structures, which are oriented to the south, are compensated by solar energy in winter due to architectural and planning decisions. Also, energy-saving measures are provided by such means as:

- dividing the house into residential and buffer areas using volumetric planning
- locating auxiliary rooms in the buffer zone on the north side
- locating the active living areas in the southeastern part of the house
- using external protection against solar radiation in summer (for southern countries)
- using external protection from winds on the northern side.

A reduction in energy costs is achieved by decreasing ventilated air, but the microclimate inside the house may not match comfortable living conditions. As examples,

Low-Temperature Energy Systems with Applications of Renewable Energy
https://doi.org/10.1016/B978-0-12-816249-1.00003-0

the quality of indoor air in schools increases the productivity of school work with an increase in the flow rate of ventilated air from 5 to 10 L/s; an increase in air flow rate to 25–30 L/s increases labor productivity in an office; and an increase in air flow rate in living quarters reduces the number of asthmatic and allergic diseases by 2–4 times. Therefore, to ensure sanitary and hygienic living conditions, good ventilation ($\sim 30\ m^3/h$ per person) with fresh, cold, outside air is required.

Worldwide practice of specifying requirements for energy efficiency in buildings varies widely. The following are features of implemented building projects: complex designs, including modern architectural solutions and technologies of construction, construction materials and effective climate technology, and intelligent control and management systems. For example, in Scandinavian countries the requirements for energy-efficient heating and ventilating systems in buildings are as follows:

- Use of low-temperature heat of return water from the heat supply system for floor and wall panel radiant heating.
- Use of heat of exhaust air.
- Individual mechanical ventilation with separate heat recovery in each apartment of a multi-unit, multi-story building.
- Increase in efficiency of natural ventilation.
- Ventilation of premises with preheated outdoor air coming through specially designed windows.
- Use of low-temperature heating systems.
- Use of solar collectors to heat water.
- Individual control of air temperature in each room.

In individual houses and multi-unit apartment buildings, heat pumps are mainly used in water heating and hot water supply systems. In this case, natural sources of energy such as ambient air, water, ground, and solar radiation can be mainly used as low-temperature heat sources for heat pumps systems along with captured heat from ventilation air.

The realization of the "energy efficient house" concept is possible by applying modern technologies of architecture and construction, various renewable energy sources (sun, wind, ground heat, etc.), and high-tech engineering equipment (Fig. 3.1). The main feature underlying the concept of a passive house is the use of heat pump technology.

Actual energy efficient buildings meet these requirements. Figures 3.2 and 3.3 show the architectural solutions of energy-efficient buildings resulting from high-technology construction.

However, there arises a dilemma between the quality of the architectural environment and the creation of healthy and productive workspaces, and the creation of low-temperature energy consumption systems. This requires compromise and optimization in the building design.

World experience in selling heat pumps shows that about 70% of them are installed in individual residences. When heating individual houses, various types of heat pumps can be used, such as air-to-air, air-to-water, ground-to-water, and water-to-water systems.

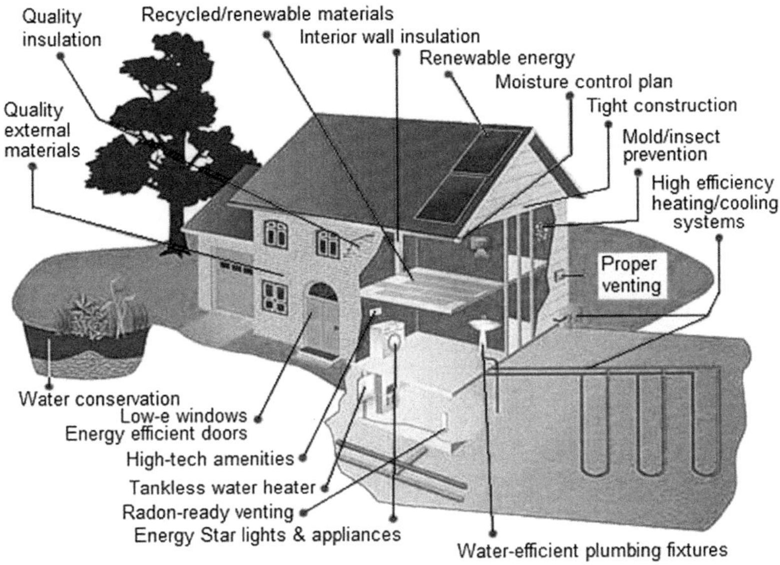

Fig. 3.1 An energy efficient house [Hibbs homes: green home builders, https://www.hibbshomes.com/green-home-builders-st-louis/].

Fig. 3.2 The European Parliament building in Strasbourg [From Wikimedia Commons, the free media repository, Photo by Octavio Espinosa Campodonico].

In Germany, according to the BWP (Budesverband Warmepumpe e.V.) in 2017, 78,000 heat pumps were sold and installed in the building heating sector. The greatest number of installed heat pumps are air-to-water type (55,000 units), i.e., 71% of the market. As for heat pumps of the "ground-to-water" type, 23,000 units were installed, i.e., 29% of the market. There were 91,500 heat pump units were in operation in Germany in 2017.

The "20-20-20 Program" adopted by the European Union stimulates the use of heat pumps in building heating systems. The targets to be achieved by the year 2020 for this

Fig. 3.3 London City Hall [From Wikimedia Commons, the free media repository, Photo by Garry Knight].

ambitious program are: 20% reduction in greenhouse gas (GHG) emissions; 20% contribution from renewable sources as a proportion of energy consumption; and 20% reduction in energy consumption, all relative to baseline projections [1].

3.1.2 Addressing the low ambient temperature problem

Heat pumps that use outside air as the heat source are generally used in places where moderate climate conditions are found, namely, in southern and temperate latitudes. The difficulty of using them in more harsh climates arises whenever the outside temperature is lower than the evaporation temperature of the heat pump working fluid, thus preventing heat transfer from the outside source to the heat pump. Under such conditions, supplementary heat must be obtained to maintain comfortable indoor temperatures, usually from electric resistance heating – a very expensive alternative method.

A recent development by a Japanese manufacturer addresses this problem. The patented Zubadan heat pump employs multiple liquid expansions and internal heat recuperation to reduce the evaporation temperature such that the evaporator can maintain full capacity down to ambient temperatures of −10 °C. Below that temperature the system loses thermal capacity linearly as the temperature drops, but is still able to operate at 75% capacity even when the ambient temperature is −25 °C. A simplified schematic flow diagram of the Zubadan heat pump is shown in Fig. 3.4; a pressure-enthalpy process diagram is shown in Fig. 3.5 [2].

With reference to Fig. 3.4, indoors is to the left and outdoors is to the right. Instead of the usual single liquid expansion (throttle valve), the Zubadan Inverter has three such expansions, shown as LEV's and processes $2 \rightarrow 3$, $4 \rightarrow 9$ and $5 \rightarrow 6$ in Fig. 3.5. Normally, only LEV B is found in a simple heat pump, and the working fluid would pass directly from there to the evaporator via a vertical process from state $2 \rightarrow 3 \rightarrow 6'$. Moreover, the simple plant would have the outlet from the evaporator (state 7) go straight to the compressor.

With reference to Fig. 3.5, the initial expansion ($2 \rightarrow 3$) lowers to pressure only slightly and controls the pressure in the "power receiver." The fluid is cooled to state 4 while heating the fluid from $7 \rightarrow 8$; this allows any liquid present to evaporate

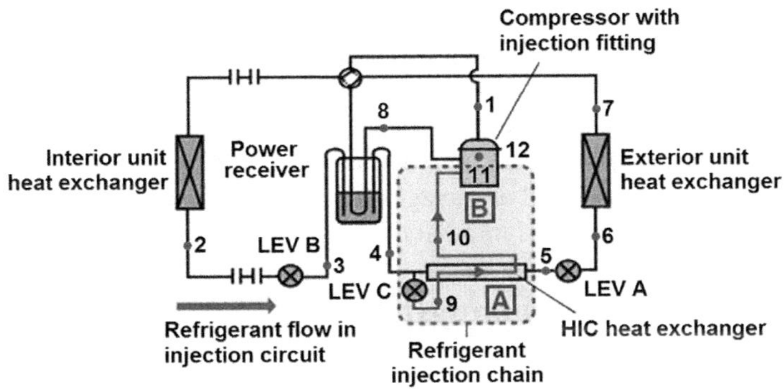

Fig. 3.4 Industrial heat pump - Zubadan Inverter (patented): A, heat recuperator; B, integrated mixer-compressor; LEV, liquid expansion valve.
Redrawn from [2].

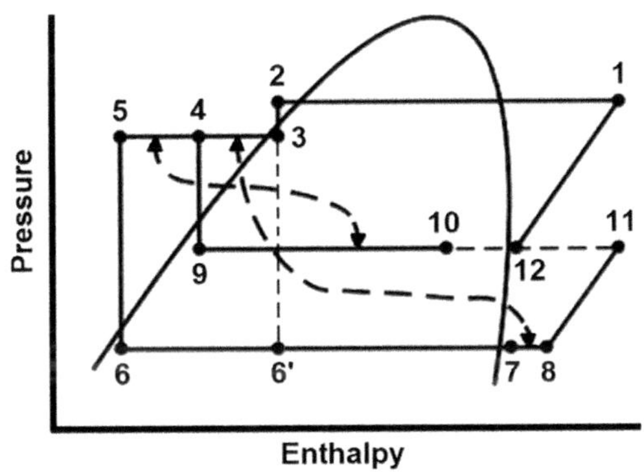

Fig. 3.5 Zubadan Inverter heat pump processes (heating mode) with state-points from Fig. 3.4.
Redrawn from [2].

before it enters the first stage of compression (8 → 11). This is of practical concern because any liquid present in the working fluid when it enters the compressor can result in damage to the impeller.

From state 4 the fluid separates into two streams: one passes through LEV C and emerges at state 9 at an intermediate pressure, and one continues through a heat recuperator (A) where it cools further to state 5 before it undergoes the final expansion in LEV A, where it is at a very low temperature. The cooling from 4 → 5 provides some of the latent heat needed to vaporize the fluid at state 9, moving it to state 10 closer to the vapor side of the 2-phase region. By controlling the pressure drops at the LEV's the temperature at state 6 can be made lower than the ambient temperature, for a wide range of low ambient temperatures. The direct mixing of fluid at states 10 and 11 occurs in the compressor housing (B), resulting in a slightly superheated state 12 which is injected into the second stage of compression (12 → 1).

Thus, triple-expansion of the liquid refrigerant results in a very low fluid temperature as it enters the evaporator, allowing heat transfer from the surroundings even at low ambient temperatures. Furthermore, using a 2-stage compressor with effective interstage cooling results in less power needed for the compressor. Both of these effects extend the range of the heat pump and increase its COP compared to a simple system. There are other practical advantages such as guaranteeing that only vapor enters the compressor impeller and that only liquid enters the expansion valves [2].

3.1.3 Combined heat pump systems

In temperate and northern climatic zones, heat pumps of the "earth-water" (or "ground-to-water") and "water-to-water" type are used to avoid the low ambient air temperature problem. Combined heat pump units include various energy sources such as solar energy, heat of waste water and sewages, heat of water crystallization, ventilated air, etc. When using heat pumps, various low-temperature renewable energy sources are used to reduce capital and operating costs. However, the use of various energy sources reduces the energy capacity of heat pump units by 25–30% compared to conventional energy sources. Figure 3.6 shows a diagram of a heat pump using the heat of comparatively clean sewage (waste water) with a temperature of about 28–32 °C to preheat ambient fresh air which is then mixed (MC) with the vented room exhaust air. The warmed mixture passes through the HP evaporator where it cools as it heats and boils the refrigerant in the HP.

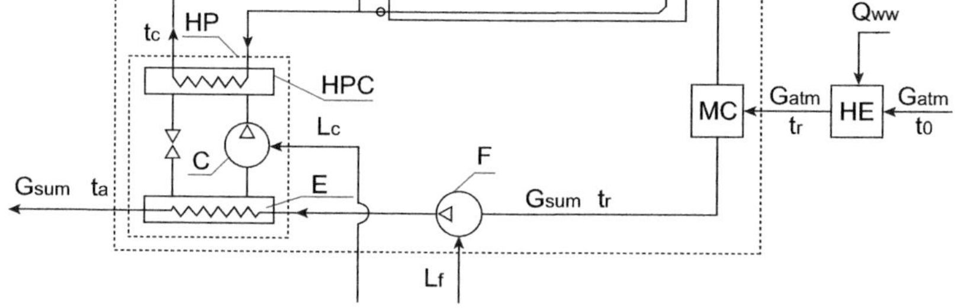

Fig. 3.6 Schematic diagram of a combined heat pump system for low-temperature water heating and ventilation using the waste heat of the ambient ventilation air heated by wastewater. *C*, compressor; *E*, evaporator; *F*, fan; *H*, air heater; *HE*, waste water heat exchanger; *HP*, heat pump; *HPC*, condenser; *MC*, mixing chamber. Other terms: G_{atm}, ambient air flow rate; G_{sum}, total air flow for heating and ventilation; G_v, air flow rate for ventilation; L_c, compressor drive work; L_f, fan drive work; Q_h, heat flow for heating; Q_{h+v}, total heat flow for heating and ventilation; Q_v, heat flow to ventilation; Q_{ww}, heat flow from wastewater; t_a, air temperature after evaporator; t_c, indoor condenser temperature; t_o, outside temperature; t_r, room temperature.

In Fig. 3.6, an air-water heat exchanger is used for heating the ambient air before entering the mixing chamber (MC). Thus, after the MC, a mixture of ventilating and ambient air with a temperature being approximately equal to the temperature in the room (20 °C) is supplied to the evaporator of the heat pump. Figures 3.7 and 3.8 show ways of using the heat of waste water.

3.1.4 Heating with ice: An efficient and inexpensive source of energy for heat pumps

An ice storage system consists of a tank for water and ice storage, a heat exchanger, and a solar collector or a solar-air absorber (Fig. 3.9). The cylindrical tank is made of concrete, and when installed it does not require a large land area as would horizontal ground heat exchangers or the drilling of deep wells. The ice storage system uses the heat of water crystallization during the phase transition from a liquid to a solid (~334 kJ/kg). Thus, a system with a thermal power of 20 kW consists of one or two bunkers with a volume of 10 m^3 of water. The system uses the heat of various sources, namely, solar energy, the heat of the ground and groundwater. Heat is supplied by means of a regenerative heat exchanger located on the outer wall of the bunker. In summer, the ice storage is used as a natural source for cooling the premises. Ice may be produced at night when electricity rates are low and used during the day for air conditioning.

The efficiency of several heat pump units under a variety of conditions is shown in Table 3.1.

3.1.5 Heat pump performance calculations

The relative energy efficiency using primary energy of fuel for the indicated schemes in Table 3.1 was analyzed. The basic equation comparing the performance of a heat pump to the performance of a conventional fuel-fired power plant is given as Eq. (3.1):

$$\eta_{rel} = \frac{\eta_{hps}}{\eta_{hb}} = \frac{\eta_{pp} \cdot \eta_{hp} \cdot COP_{HP}}{\eta_{bi} \cdot \eta_{hg}}, \tag{3.1}$$

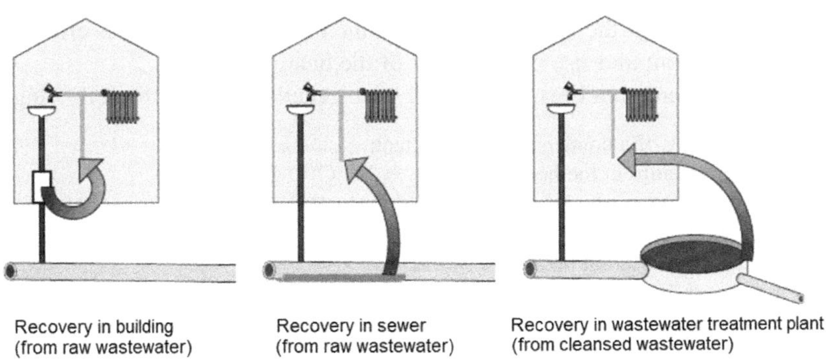

Fig. 3.7 Possibilities for energy recovery from wastewater [3].

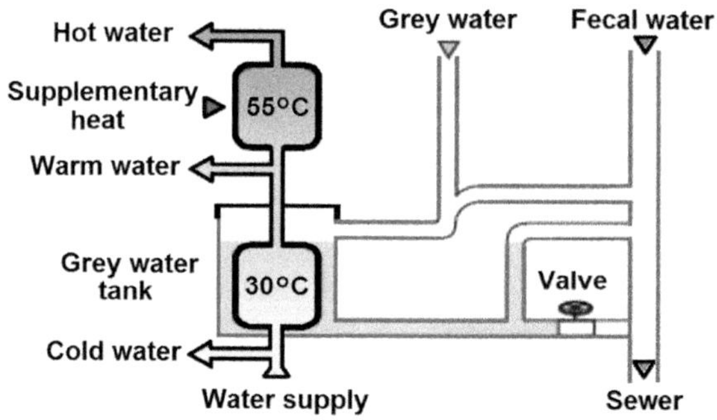

Fig. 3.8 System for heat recovery from wastewater in single-family homes [3].

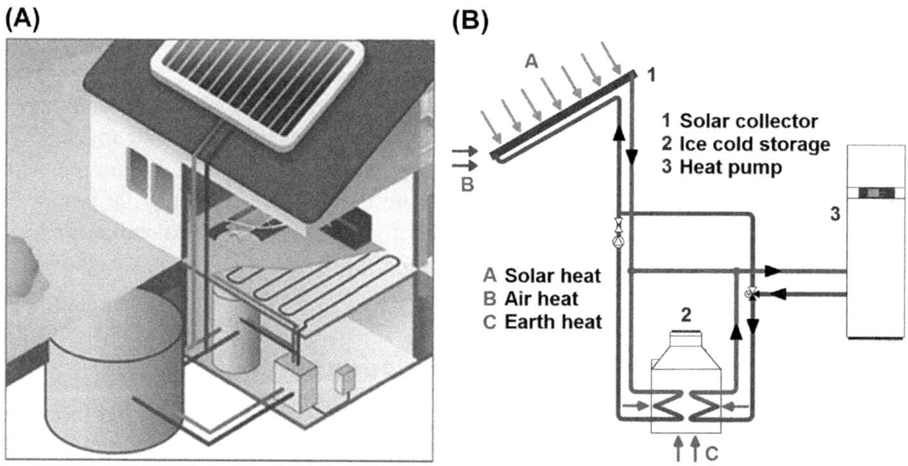

Fig. 3.9 Schematic of a residential heat pump system with ice maker and solar and earth heat sources: (A) Ref. [4]; (B) Ref: [5].

where η_{hps} and η_{hb} are efficiencies of using primary energy of fuel for a heat pump system and a heating boiler, respectively; η_{pp} is the thermal efficiency of the electric generating power plant; COP_{HP} is the theoretical coefficient of performance of the heat pump; η_{hp} is efficiency (the loss coefficient) of the real heat pump; η_{bi} is efficiency of the boiler installation; and η_{hg} is efficiency of the heating grid.

The results presented in Figs. 3.10–3.12 were calculated using the following data:

- Thermal efficiency of a condensing power plant: $\eta_{pp} = 0.38$
- Estimated temperature in the heated room: $t_r = 20$ °C
- Calculated temperature of a heat transfer fluid in the heating system: $t_{h.t.} = 40$ °C
- Assumed average ambient temperature for heating: $t_h = -20$°C
- Efficiency of water-heating boiler: $\eta_{bi} = 0.85$
- Using the lower heating value (LHV) of combustion of the fuel, the efficiency of the condensing boiler is $\eta_{c.b.} = 1.05$. Using the higher heating value (HHV), $\eta_{c.b.} < 1$.
- Efficiency of the thermal network: $\eta_{hg} = 0.95$. For a condensing boiler, $\eta_{hg} = 1$.
- Efficiency of a real heat pump: $\eta_{hp} = 0.6$.

Table 3.1 COP of heat pumps for various heating purposes and heat sources.

Heat source & temperature	Radiant heat "warm floor" 25–30 °C	Hot air system 25–30 °C	Hot water "radiators" 70–95 °C	Hot water supply 40–55 °C
Air −5 °C to −10 °C	4.0	3.9	–	3.15
Ground 5 °C–10 °C	4.0	3.9	2.0	3.15
Groundwater 8 °C–15 °C	4.4	4.0	2.2	3.6
River, lake 4 °C–17 °C	4.4	4.0	2.2	3.6
Wastewater 10 °C–17 °C	4.7	4.2	2.2	3.8

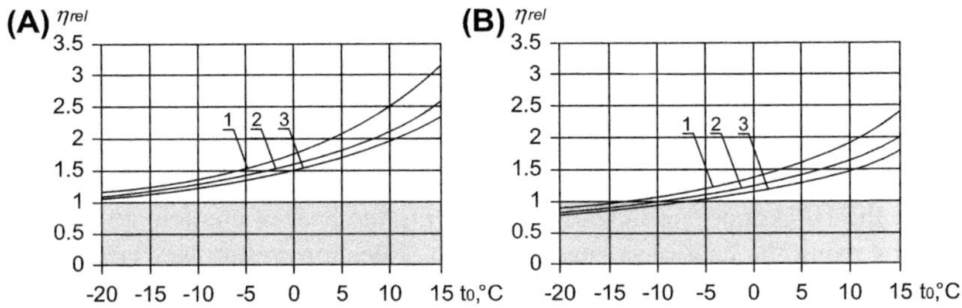

Fig. 3.10 Relative energy efficiency as a function of ambient temperature when comparing the HPU using ambient air: (A) with traditional water heating boilers; (B) with condensing boilers; 1, 2, 3 – correspond to $A = 0.1, 0.5, 1$ °C, where A is a defined factor,$A \equiv \Delta P/\rho_s c_p$, a function of given values and which depends mainly on the aerodynamic resistance of the low-temperature tubing circuit.

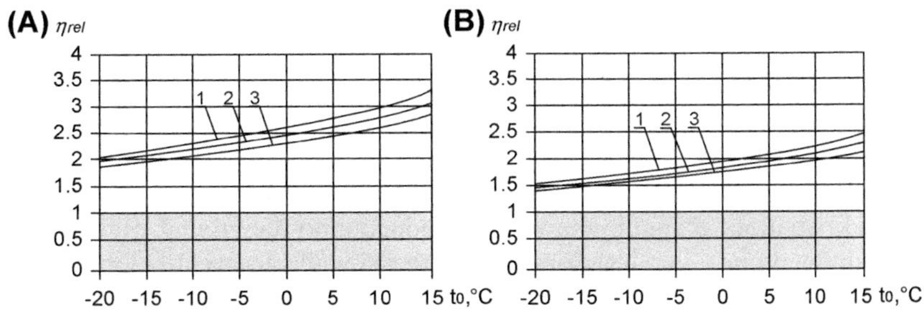

Fig. 3.11 Relative energy efficiency as a function of ambient temperature when comparing HPU using water: (A) with traditional water heating boilers; (B) with condensing boilers; 1, 2, 3 - correspond to $A = 0.01, 0.05, 0.15$ °C.

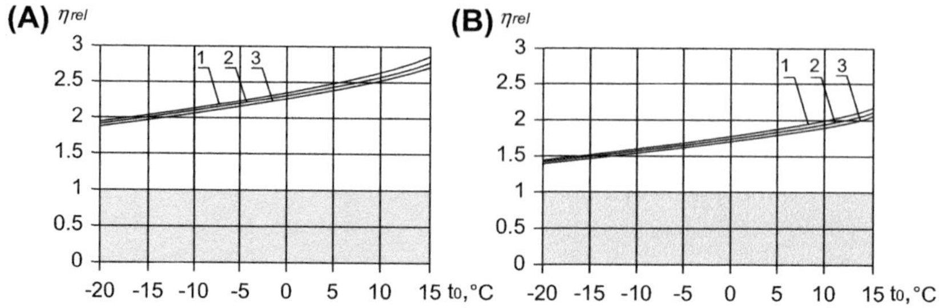

Fig. 3.12 Relative energy efficiency as a function of ambient temperature when comparing HPU using ground heat: (A) with traditional water heating boilers; (B) with condensing boilers; 1, 2, 3 – correspond to $A = 0.005$; 0.015; 0.027 °C.

The results of the calculations using Eq. (3.1), taking into account the pre-calculated COP of the heat pump under the given conditions, are presented in Figs. 3.10–3.12. From an energy point of view, the heat pump heating system will be rationally superior in those cases where the relative energy efficiency is greater than one, $\eta_{rel} > 1$.

These plots show that, with optimal cooling of the medium in the evaporator of the heat pump, the relative energy efficiency for all the above cases is more than one, $\eta_{rel} > 1$, except for the case of an ambient air heat source compared with a condensing boiler (Fig. 3.10B). In that case, the results indicate that the HPU holds an advantage over the conventional condensing boiler as long as the ambient air temperature is greater than −13, −8, or −5 °C, when $A = 0.1$, 0.5, 1 °C, respectively. It is possible to increase the HPU energy efficiency while using the heat of ambient air at low temperatures by using the heat of crystallization of water for preliminary heating of air.

Nevertheless, hot water supply systems with HPU's using natural sources of energy always require the use of a supplementary peak heat source (a gas-fired or electric heater) to raise water temperature to the required value in very cold winter periods.

3.2 Heat pumps for indoor and outdoor pools

3.2.1 Indoor pools

An important feature of indoor pools is significant evaporation from the water surface. It is believed that about 20% of the heat for evaporation is taken from the air and 80% from the water. Heat losses consist of pool capacity losses and losses from the building that houses the pool. Heat capacity losses of the pool are caused by: evaporation, convection and radiation; changing the water; conduction to the ground. Building losses are the traditional ones, i.e., heat losses through the walls and ventilation.

The amount of evaporated moisture depends on the surface area of the pool and the difference in the partial pressure of the saturated vapor near the surface of water and in the pool area. This is also affected by the number of patrons. In calculations it is assumed that the relative humidity of air in the room should be within the range of 60–80%, the room temperature is 27 °C, and the water temperature is 25 °C. In the absence of patrons, at a relative humidity of 70%, about 64 g/h of water will be evaporated from each square meter at the temperatures mentioned above, i.e., 64 g/(m^2 h), whereas at a relative humidity of 60%, 91 g/(m^2 h) will evaporate. A higher humidity in the room will result in vapor condensation on the building walls.

When there are bathers in the pool, the evaporation will be increased by increasing the evaporation surface because of people's bodies, wet walls and floors, water agitation, people's breathing, etc. At 70% relative humidity and a small number of people (15–20) in the pool, evaporation increases to 162 g/(m^2 h), and up to 475 g/(m^2 h) assuming each person in the filled pool occupies 1.5 m^2. Thus, the average evaporation rate ranges from 100 to 300 g/(m^2 h).

The amount of heat (thermal power) consumed for evaporation will be found as the product of the moisture amount for the whole area of the pool times the heat of evaporation, namely, 2444 kJ/kg at 25 °C. Heat losses of the pool area into the environment are determined by traditional methods. It should be noted that air warms the pool water, because it has a higher temperature. The heat transfer coefficient may be taken as 6 W/(m^2 K). When the water in the pool is colder, which occurs mainly at night when the system power is lowered, the heat transfer coefficient may be taken as 20 W/(m^2 K). Therefore, it is important to cover the pool when not in use. When changing water in the pool (e.g., according to the norms of France, 5% of water should be changed every day, and all of it should be changed out once every three months), the power for water heating is determined by the water heat capacity and the difference in water temperature in the pool and at the entrance to the heating plant. Heat losses to the soil at constant pool filling are small and can be found according to empirical studies using a theoretical coefficient of heat transfer in the "wall-ground" system of about 5 W/(m^2 K).

To reduce the humidity of air in the pool area, conventional ventilation of the room is often used. The volume flow rate of fresh air (m^3/s) is estimated by the following relation:

$$\dot{V} = \dot{m} / [(d_{room} - d_{input}) \cdot \rho_{air}] \tag{3.2}$$

where $\dot{m}$ = evaporated water rate, kg/s; d_{room} = absolute air humidity in the pool area; d_{input} = absolute humidity of atmospheric air entering the room, kg_{ha}/kg_{da} [ha = humid air, da = dry air]; ρ_{air} = density of fresh air from outside, kg/m^3.

The air flow rate reaches 20,000 m^3/h for a pool with an area of 250 m^2. This air requires heating which consumes a significant amount of thermal energy. To reduce its consumption, heat recuperators are installed using thermosiphons or a heat pump (HP) may be used when the pool is occupied. Fresh air is warmed by heat from a HP condenser. In order to maintain comfortable conditions, it is recommended installing one more HP for heating fresh air both in combination with thermosiphons and without them. The HP, which dries the air, sucks it out of the pool area and ejects it into the room. The air passing through the evaporator cools down and the moisture condenses at the evaporator. Dried air from the evaporator, equipped with a condensate collector, returns to the condenser of the HP. When installing a thermosiphon unit, the evaporator is in the air direction in front of the HP evaporator, and the condenser is in front of the HP. The amount of heat generated in a HP condenser installed to reduce the humidity of the pool air is greater than the amount required to heat the dried air. Therefore, excess heat can be used to heat the water in the pool.

In addition, one should note that a rule requiring 5% daily water changing at 25 °C makes it logical to use the heat of this water in a heat pump (before the water is drained) to heat the replacement water. Retrofitting heat pumps in existing pools can achieve a 2–3 year payback period for the capital investment; for new pools, the payback is rapid, only 3–6 months. To maintain air parameters in pools regulated by sanitary standards, it is necessary to perform the following actions: remove a significant amount of water vapor formed as a result of evaporation from the surface of the pool and as a result of respiration of swimmers and spectators; compensate for heat effects through enclosing structures, i.e., heat losses in winter and heat gain in summer; maintain the composition of air as regulated by sanitary standards, especially in the presence of a large number of people in the swimming pool area.

All these measures require high power costs considering that in winter with ventilated air being emitted into the atmosphere, a huge amount of low-temperature heat is lost and a large amount of energy is consumed for heating inflow air. Figure 3.13 illustrates the concept of a system of heat recovery from discharged ventilation air by means of heat exchangers and an "air-to-air" heat pump.

In this basic arrangement during winter, it is necessary to supply only 8.5% of the required amount of energy from an external source. About 38% of the total amount of heat is returned by the heat exchanger, 13% by means of the heat pump, and 32% due to recirculation. In summer, the built-in heat pump allows air in the pool to be dried under any weather conditions. The system has regulation, manually adjusted for a given mode of operation (swimmer training, sports competitions, classes with fitness groups, etc.) and automatically adjusts depending on external weather conditions. The long-term operation of the system has shown its energy efficiency is characterized by an 8-fold reduction in peak thermal load. The introduction of the system has removed a cause of corrosion of metal and concrete building structures. The payback period of the system was 18 months. Such experience can be used in systems of heat utilization of discharged ventilation flows of movie and concert halls, sports complexes, industrial enterprises, and buildings in farming and industrial complexes.

3.2.2 Outdoor pools

The heating of water in outdoor pools from an energy point of view is feasible only at an ambient temperature of 10–12°C. In other colder periods heat consumption is too large. Heat is used to compensate for heat losses and daily heating of the 5% of water that is changed. The water temperature should be higher than 20 °C and lower than 23 °C. The water temperature in the shower used by swimmers leaving the pool should

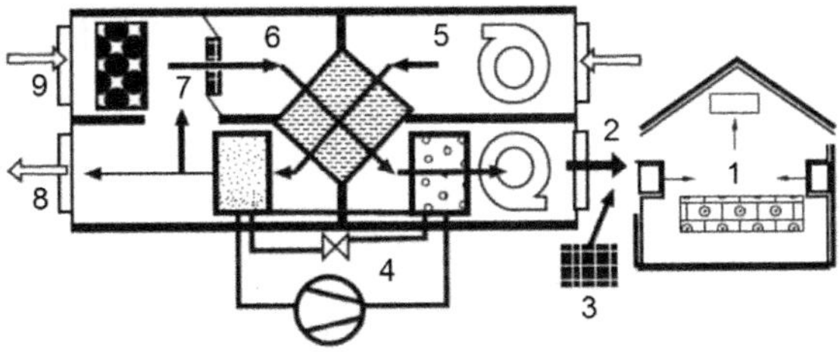

Fig. 3.13 Schematic of a system of inflow and exhaust ventilation of a pool with air recirculation and a heat pump. *1*, pool area; *2*, air supplied to the pool; *3*, source of external heating of air; *4*, heat pump; *5*, exhaust air from the pool; *6*, heat exchanger; *7*, recirculation flow; *8*, exhaust air; *9*, ambient air. [6].

not be lower than 30 °C. Heat losses from the surface of an outdoor pool are about 170 W/m^2 which causes a daily consumption of thermal energy of about 2 kWh/m^2. Therefore, a small investment for the protection of the pool surface from excessive heat losses is quickly compensated by less required power of the HP and reduced energy consumption. Heat pumps for outdoor pools are used in summer and therefore the "ambient air-water" heat pump does not require defrosting of the evaporator. The evaporator works with free air convection.

For individual residential consumers who have a pool, there is a good opportunity to operate the heat pump year-round, using it in winter for heating their home, and in summer for heating water in the pool. Since in summer one can heat water in flat solar collectors, including simple low-cost ones which are quite efficient up to 30°C, the combination of solar collectors with a HP is an attractive and efficient solution. Such a system can even be used in cloudy weather and at dusk. For large outdoor pools, it is expedient to use water from rivers, lakes and wells as a cold source for the HP. This allows the conversion ratio of the HP to be increased by 4–5 times.

An outdoor swimming pool heating system using a heat pump is shown in Fig. 3.14A [7]. It consists of a heat pump, filter, water pump, and chlorinator. Figure 3.14B shows an outdoor pool where the heat losses are due mainly to convective heat removal and evaporation from the water surface [8].

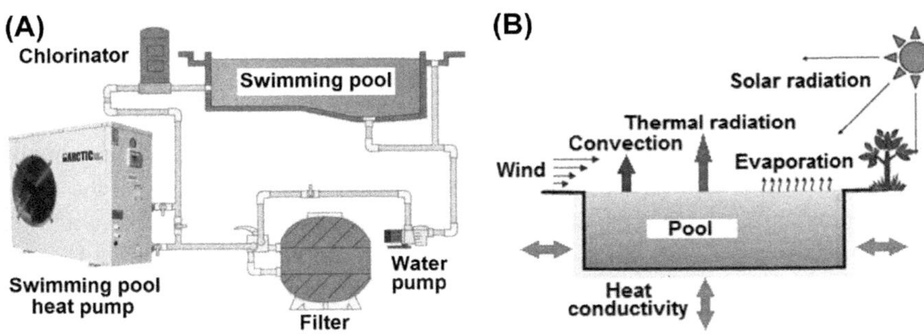

Fig. 3.14 Outdoor pool: (A) schematic system design [7]; (B) heat losses [8].

3.3 Heat pumps for heating buildings and public premises

The specific nature of facilities such as theaters, movies, restaurants, exhibition centers, supermarkets, train stations, hotels, hospitals, etc. requires a significant amount of ventilation air when rooms are filled with people. Given that warm air concentrates at the top of large rooms, mixing this air with the lower cold air that surrounds people is an important opportunity for energy savings, especially in auditoriums and supermarkets. Also, it is important to adjust the frequency of ventilation depending on the number of people filling the premises, given that an average person needs about 0.005 m^3 of fresh air per second.

Heat sources for heat pumps in the above-mentioned premises are ventilation emissions that can be used for heating fresh air, and in some cases, for hot water supply, as for example, in heating water for kitchens in restaurants. Of course, in summer there is a need for air conditioning, so the heat pump can be used both as a refrigerating machine and as a way to obtain heating. The latter version is suitable for a small heat demand (hot water in bathrooms, kitchens, etc.). In grocery supermarkets, heat from refrigerator condensers can cover 50–70% of space heating demands, but in many cases this heat is simply discharged to the environment, thereby necessitating additional consumption of heat energy from boilers and heating networks.

In hospitals, various areas should be categorized according to their functions and then the corresponding temperature requirements should be specified. In order to facilitate operation, heat pumps should be designed and built with a power being sufficient for centralized heat supply with well-isolated heat networks. The heat pump can simultaneously cool water needed for air conditioning, and heat water needed for sanitary purposes in summer. A traditional boiler should be used as a peak source for winter heating. In some cases, due to the varying functionality of rooms and their occupancy with patients or visitors, local individual heat pumps should be used. With the introduction of lower night rates for electricity, it makes sense to use cold and heat storage units to cover peak loads and save money.

3.3.1 *Ventilation systems*

Among the most effective facilities for the application of heat pump technologies are the ventilation systems of industrial, public and residential premises. Due to the small difference in temperatures of the inflow and exhaust air, it is possible to attain high energy efficiency using heat pumps in ventilation systems at moderate ambient temperatures. However, when the temperature of ambient air is lowered, the efficiency of simple heat pump ventilation circuits decreases significantly, which results in the use of more complex circuits. In this regard, excellent results are achievable with heat pump and recuperative ventilation systems, in which the final energy effect is determined by a combination of an exhaust air heat recovery unit and a heat pump.

3.3.1.1 Ventilation with an exhaust air heat recovery unit

In order to have air at the required temperature supplied to a room, radiators (heaters) of different models are used to heat the inflow air in the simplest ventilation systems. They can be water (steam), gas, or electric depending on the source of heat. The decision concerning the use of a particular type of heater for a particular system is made based on specific conditions, possibilities, and economic expediency. Hot water or electric heating is preferred because they offer the opportunity to control the temperature of air more precisely. When exhaust air at the temperature in a room is simply discharged to the environment, the efficiency of such systems is very low. So, exhaust air heat recovery units are used in ventilation systems (Fig. 3.15).

Heat recovery from air prior to being discharged into the atmosphere by ventilation systems is the most practical way of using low-potential thermal secondary energy resources (SER). Ventilation air should be considered a main source of reducing operating costs for heat treatment of ambient air inflow in heating, ventilating, and air-conditioning systems.

The heat utilization of ventilation emissions can be carried out in the following ways: by recirculation of a part of exhaust air; by using recuperative heat exchangers; by using regenerative heat exchangers; by using two recuperative heat exchangers with an intermediate heat carrier; and by using heat pipes.

Let us consider the application of heat recovery units of ventilation emissions in detail. The schematic diagram of the use of recuperative heat exchangers of exhaust air in a ventilation system is shown in Fig. 3.16.

The intake air passing through the heat exchanger 3 is heated (or cooled) by the outflow air. The heat recovery coefficient η_p in such devices reaches 75%.

It has been shown in Ref. [7] that even at high values of η_p, the temperature of exhaust air at the outlet from the recovery unit $t_{cooling}$ significantly exceeds t_0, which indicates the possibility of using the heat of ventilation emissions to finish heating fresh air with the temperature t_{env} to the required level at the entrance to the room. This testifies to the *a priori* efficiency of heat pump and recuperative ventilation systems (Fig. 3.17), which are becoming increasingly popular.

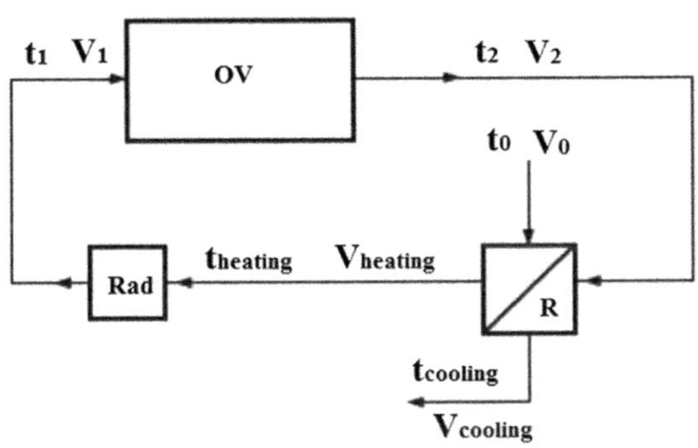

Fig. 3.15 Ventilation system using a radiator and heat recovery units of exhaust air. *OV*, ventilator; *R*, heat recovery unit; *Rad*, radiator.

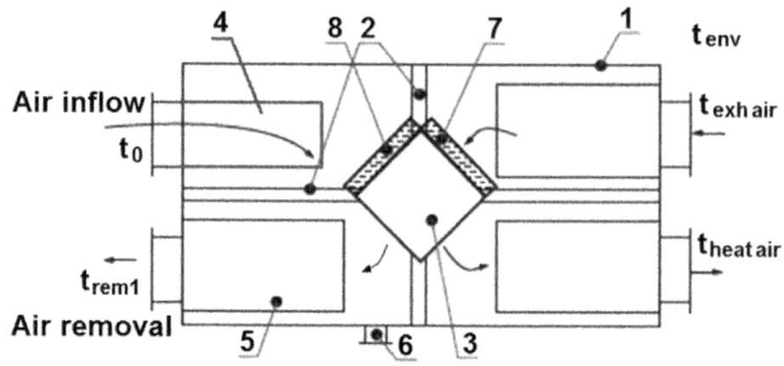

Fig. 3.16 Block diagram of inflow and exhaust ventilation with a plate-type heat exchanger. *1*, frame; *2*, partition; *3*, heat exchanger; *4*, intake fan; *5*, exhaust fan; *6*, condensate drain; *7, 8*, filters.

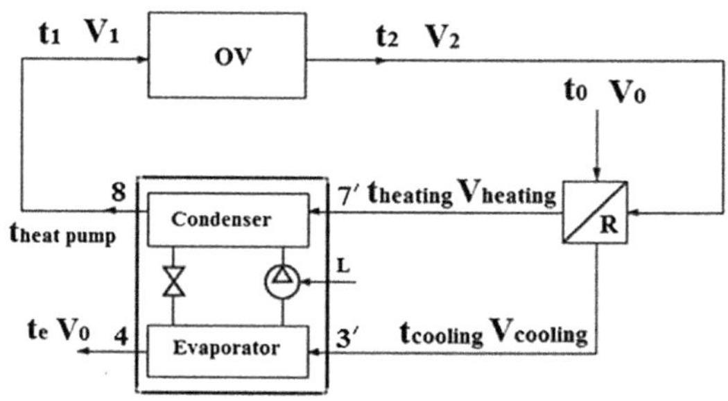

Fig. 3.17 Ventilation system using an exhaust air heat recovery unit with a heat pump; cf. Fig. 3.15.

With this arrangement, a heat recovery unit is installed in front of the heat pump, in which ambient air is heated by the heat of exhaust air, and only then it enters the heat pump, thereby reducing the energy consumption of a heat pump drive. Characteristic temperatures of air flows and a HP working fluid at the cycle points are shown in Fig. 3.18.

It was shown in Ref. [9] that the efficiency of HP operation depends on the temperature of ambient air t_0, but also somewhat decreases with an increase in the recuperation factor η_p, provided that it is used with a heat recovery unit. However, the influence of the value of η_p is insignificant and, therefore, the specific energy consumption of a heat pump and recuperator ventilation system in general is significantly reduced by using both the recovery unit and the heat pump.

3.3.1.2 Ventilation with heat recovery and exhaust air recirculation

In cases where air removed from the premises has a rather high temperature and does not contain harmful substances, part of it is not discharged outside in winter, but, after filtering, is mixed with inflow air for its final heating, and the resulting mixture is fed into the room. This results in savings related to the cost of installation and operation

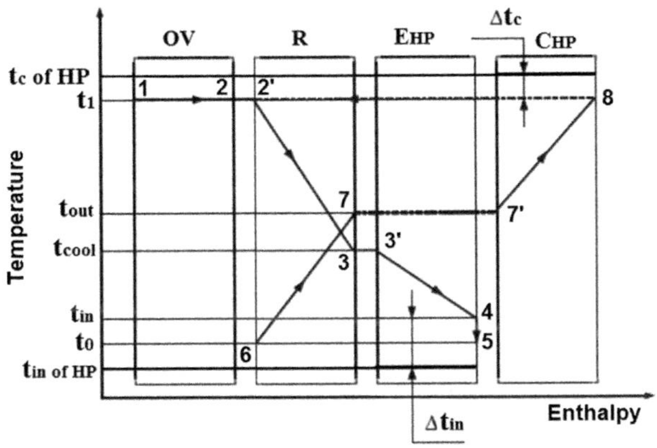

Fig. 3.18 Temperature changes of air in the elements of the system in Fig. 3.17. C_{HP}, heat pump condenser; E_{HP}, heat pump evaporator.

since the costs for heating the ambient air are reduced. The energy spent on heating the inflow air during the cold period can be reduced to 60–85%, compared to the usual inflow plant [7].

Recirculation is also widely used when air is cooled in summer. It can be seen from Fig. 3.19 that the flow of exhaust air is divided into two streams: one part, after passing through a filter, is combined in a mixing chamber with the inflow air; the other is directed to the heat recovery unit and HP. Due to the recovery unit and the HP, air leaving the room is transferred to the fresh air coming from the environment.

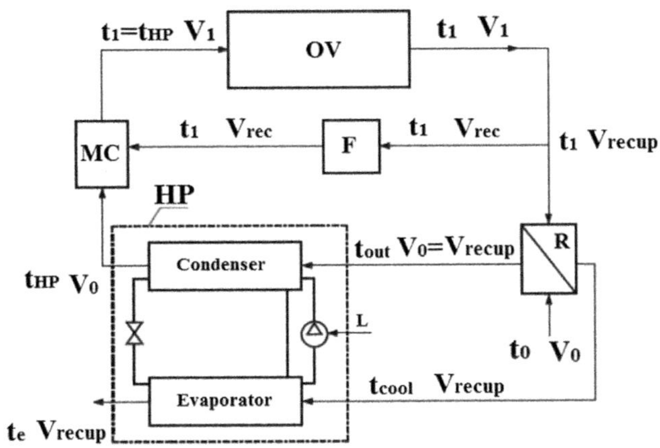

Fig. 3.19 Ventilation system using a heat recovery unit and recirculation of exhaust air. *F*, filter; *L*, heat pump compressor driver; *MC*, mixing chamber.

3.3.2 Air heating systems

One of the main features of air heating systems is the lack of an intermediate heat-transfer fluid or heat carrier. This has both positive and negative aspects. First, the problem of "defrosting" the system disappears with its long-term shutdown in winter. In addition, due to the high rate of hot air circulation, the inertia of an air system is much lower than that of the water system, where the boiler first heats the heat carrier, then the metal pipes and radiators, and - only later - the air indoors. The air heating system can operate efficiently in the ventilation mode. It has a simpler design than the traditional water system owing to the lack of a special boiler room, piping systems, radiators, pumps and complex control automation. The main disadvantages of air heating are due to the movement of large volumes of air in the room which is heated. This reduces comfort (drafts), leads to the movement of dust, and promotes the spread of bacteria throughout the room being heated or ventilated. It is also necessary to take into account that air ducts have a larger cross-sectional area than the internal pipe lines of water heating systems, and therefore large-scale openings in the walls are required for their installation [7]. But despite these disadvantages, air heating systems based on a heat pump offer lower energy consumption, and they are environmentally friendly since they operate without fuel burning on the premises, and therefore do not produce harmful emissions into the on-site atmosphere.

3.3.2.1 Air heating using the heat of ambient air

The two simplest heat pump circuits for air heating using ambient air heat are considered below [6]:

1) Air heating without ventilation; a so-called "split system" with a remote (relative to heated space) compressor, evaporator or condenser, operating in a heating or air-conditioning mode; and
2) Air heating with room ventilation and a heat pump completely separate from the room being heated.

The effectiveness of the first circuit, which is presented in Fig. 3.20, is due to the insignificant heating of internal air in the HP condenser and the lack of heating external ventilation air to the temperature in the room. However, taking heat out of ambient air requires air cooling at the HP evaporator outlet to sufficiently low temperatures, which worsens the working conditions of the heat pump, especially under conditions of low ambient temperatures.

The operational efficiency of the second heating system, shown in Fig. 3.21, depends, to a large extent, on the heating of external ventilation air from the ambient temperature to the temperature at the entrance to the heated room, which can reach significant values, especially at low ambient temperatures.

Moreover, in this system it is possible to use the heat of exhaust air that increases the air temperature at the outlet of the HP evaporator and which improves the working conditions of the HP. In this connection, in Ref. [7], the thermodynamic efficiency of the above-mentioned heating circuits was determined and a comparison was made, showing that these systems differ in their effectiveness from each other but only by a little.

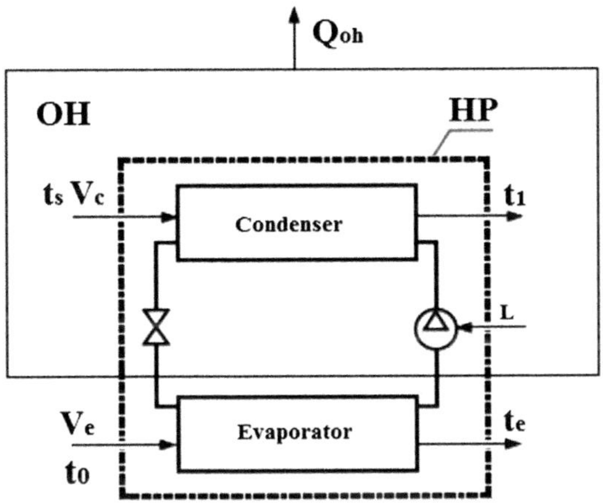

Fig. 3.20 Heat pump air heating system. *OH,* heated object; Q_{oh}, heat losses from OH.

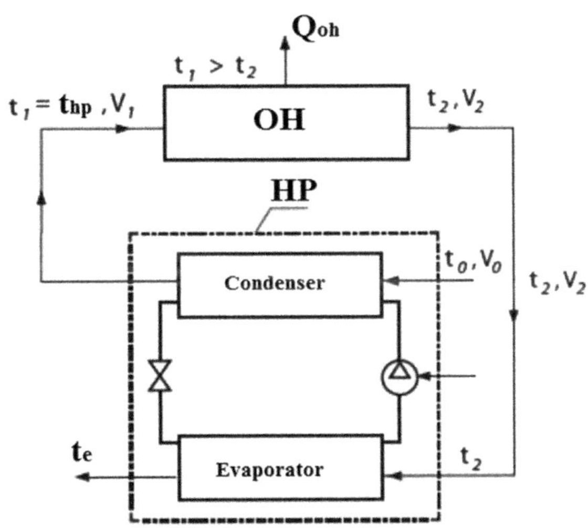

Fig. 3.21 Heat pump air heating system with ventilation.

3.3.2.2 Air heating and ventilation with heat recovery unit and recirculation of exhaust air

In a previous subsection we discussed "ventilation with heat recovery and exhaust air recirculation" and displayed Fig. 3.19. That same figure is appropriate to illustrate the content of this subsection; the reader may wish to refer back to it. There are some differences in notation but the flow directions are the same. The simultaneous use of a heat pump with a recovery unit and recirculation represents the most efficient level of heat utilization of exhaust air.

The exhaust air at the exit (ventilator) of the room has a high enough temperature to be efficiently used for heating. The exhaust air flow is divided into two streams. One part is sent to a filter, and then to the mixing chamber (recirculation process). The second part is fed to the heat exchanger (heat recovery heat exchanger) in which ambient air is heated by the exhaust air. This reduces the energy costs of heating the ambient air in the heat pump.

The analysis of such a system showed that the use of air recirculation in the heating and ventilation system led to a reduction in the air flow through the heat pump, caused the necessity to reduce the air temperature at the outlet of the evaporator and increase the temperature at the outlet from the HP condenser to obtain air with a given temperature at the entrance of a room being heated. As a result, the thermal conditions of the heat pump become worse, which manifests itself in a significant reduction in the HP COP with an increase in the coefficient of recirculation C_{rec}. Nevertheless, the specific external energy consumption in the air recirculation circuit is somewhat reduced due to the dominant effect of utilizing the heat of exhaust air. The use of a recovery unit in a heat pump heating and ventilation circuit is a more effective means of improving the thermodynamic efficiency of the energy supply system than that of using exhaust air recirculation. At the same time, it is interesting to use both means of heat usage, at the expense of heat recovery and at the expense of recirculation of exhaust air.

On the basis of an analysis of the thermodynamic efficiency of the heat pump air heating and ventilation systems with a heat recovery unit and recirculation of exhaust air, the following conclusions were obtained: (1) the recirculation of air in a heat pump air heating and ventilation cycle has its limitations and can only be applied with relatively small coefficients of recirculation ($C_{rec} \leq 0.5-0.6$) due to the sharp increase in air temperature at the outlet of the HP condenser; (2) the application of a heat recovery unit of exhaust air leads to a more significant increase in the COP of the heat pump circuit of heating and ventilation compared with that of exhaust air recirculation alone; (3) the additional use of air recirculation in the heat pump and recuperator circuit of air heating and ventilation gives a positive effect only at relatively low values of the recuperation factor and at high values ($\eta_r = 0.8$), and (4) the circuit with recirculation (Fig. 3.19) has a lower efficiency than a simpler one having only the heat recovery unit.

3.3.3 Hot water heating systems, hot water, and air-conditioning facilities

Heat pump systems for water heating and hot water supply for various objectives can be carried out using various low-temperature sources of energy. The advantages of such systems are illustrated by the example of a heat pump system using earth (or ground) heat for the heat supply of a railway station. The system utilizes the natural heat of ground and provides water heating for a passenger waiting room and offices in the cold season, air conditioning in office rooms in summer, and hot water supply throughout the year. The schematic diagram of the heat supply system is shown in Fig. 3.22.

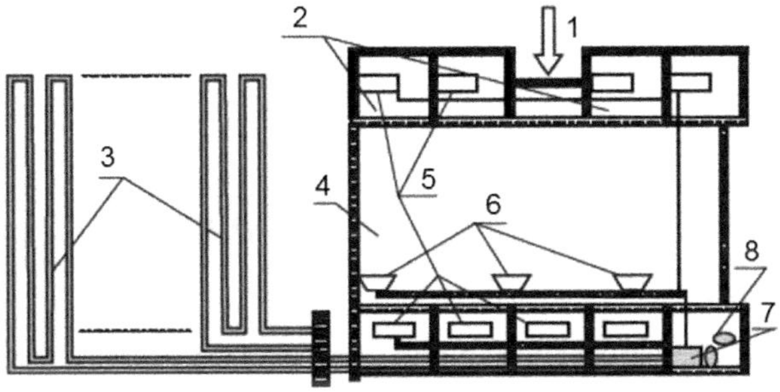

Fig. 3.22 Schematic plan view of a heat supply system for a railway station. *1*, entrance to the premises; *2*, office space; *3*, ground heat exchanger; *4*, customer hall; *5*, fancoil; *6*, air heaters; *7*, heat pump with a hot water tank; *8*, storage tank.

An important condition for the efficient use of heat pump heat supply of buildings is to reduce the thermal losses of the building structures to a rational minimum. Therefore, repair of the building wall was carried out beforehand in order to reduce heat losses and eliminate unacceptably excessive infiltrations in the passenger hall and office premises. The heat pump system was designed according to a bivalent scheme, i.e., using two sources of energy, with a nominal heat pump capacity of 38.6 kW and the power of an electric boiler-finishing device of 12 kW. Hot water with a temperature of 45 °C is prepared and stored in accumulator tank.

The thermal capacity of the heat pump in such cases is designed for 70−85% of the maximum load required in winter. This allows meeting the heating objective down to an ambient air temperature of −15 °C by providing up to 90−95% of the required amount of heat. The additionally required amount of heat (5−10%) in the coldest days of the year is covered by an electric boiler-finishing device. Thermal conditioning can be supported in two ways, namely, active and passive. The transition from one method to another occurs automatically with a room temperature sensor. To capture the ground heat, a horizontal ground heat exchanger was deployed, located in the area adjacent to the railway station building in a trench with a depth of 1.5−1.8 m. The total length of a plastic pipe, having a diameter of 42 mm, laid in five loops was 1200 m. The low-temperature system of water heating using fancoils automatically kept the temperature at given levels: 18 ± 2°C in the offices and 16 ± 2°C in a passenger waiting room.

In summer at an ambient air temperature of 33°C, the heat pump unit operated in the air conditioning mode keeping the temperature in office premises at 23−25°C. Positive properties of heat pump systems are the possibility of using passive air conditioning in summer when the compressor of the heat pump is off and electricity is consumed only for pumping water cooled in the ground heat exchanger by circulating pumps. At the same time 30 kW of cooling can be obtained for an expenditure of 1 kW of electricity; this is 10 times more efficient than a conventional air conditioner. If the power of passive air conditioning is not enough, then the compressor is switched on and the heat pump operates as a conventional air conditioner. In this case, the cooling of the premises becomes active. The results of the operation of the heat pump system showed that the total operating costs during the heating period were reduced to 20% of heating from a boiler house.

3.3.4 Examples of effective use of heat pumps

3.3.4.1 Airports

In 2008 construction began on a heat pump to provide Orly Paris Airport with heat for all its buildings and a future business district; see Fig. 3.23. In 2011 the geothermal heat pump with biomass augmentation went into operation. Hot geothermal water is obtained at 74 °C from the 1.8 km-deep Dogger formation via a doublet – one production well and one injection well. The heat pump is an absorption unit that gets its motive energy from the burning of straw in a boiler to allow a thermal capacity of 135 MWt for the heat pump. The maximum temperature in the district heating system is 105 °C; the geofluid is reinjected at 40 °C after dropping by 34 °C in passing through the heat exchangers. The volumetric flow rate of the geofluid is about 300 m^3/h. The total length of the piping in the district heating system is 35 km that encompasses 108 substations. From an environmental perspective, the new system reduces carbon dioxide emissions by 9000 t/y [10].

Zurich (Switzerland) and Heathrow (London) airports use heat pumps with ammonia (R717). Systems with heat pumps provide thermal energy to buildings to cool aircraft while they are parked at the gate. The systems have been in operation for more than 10 years. Geothermal heat pumps (with well depths of about 300 m) provide thermal energy to the airport (60%) and cold production for air conditioning systems (40%). At night, ice generators produce liquid gel-like ice, which is stored in the accumulator (220 m^3). The accumulated cold is enough for cold supply for one working day. Geothermal heat pumps are planned to be used to prevent the (freezing) icing of the runway at the airport.

At St. Petersburg airport (Russia) there are three absorption refrigerators with a thermal capacity of 4 MWt/unit. Heat sources are hot water from the exhaust gas heat

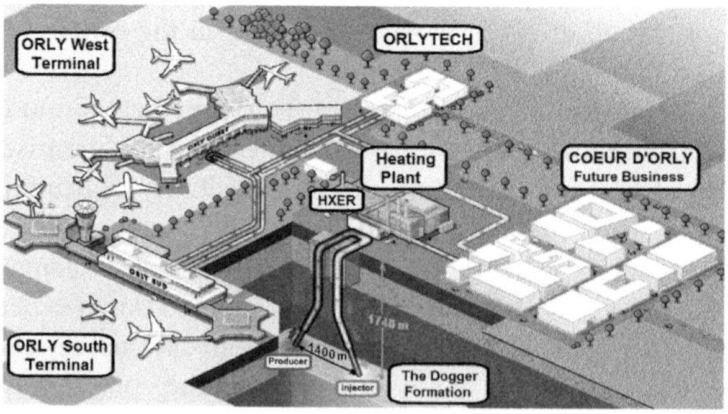

Fig. 3.23 Geothermal district heating system at Orly - Paris Airport. Redrawn from [10].

recovery system of two gas turbines with a capacity of 5.25 MWt/unit. The purpose: the absorption refrigeration machine is part of a trigeneration complex which provides the airport terminals with electricity, heat energy (heating, ventilation, hot water), and cold for air conditioning.

Other airports with heat pumps include the following:

- Airport Bangkok - Refrigeration capacity = 30,000 kW.
- John Wayne Airport, Orange County, CA, US - Refrigeration capacity = 3700 kW.
- Airport Madrid - Refrigeration capacity = 19,800 kW.
- Changsha Airport, China - Refrigerating capacity: 12,200 kW.
- Tianjin Binhai Airport, China - Chiller = 10 MW.
- Canberra Airport, Australia - Refrigeration capacity = 2100 kW.
- Sabiha Gokcen Airport, Istanbul - Refrigerating capacity = 3000 kW.

3.3.4.2 Office buildings

In 2008, a multi-story office building was put into operation in the center of the Austrian city of Linz; see Fig. 3.24. This was the first passive building in the world. There are no city heating systems or city gas mains connected to this building. One of the largest photovoltaic systems in Austria is installed on the building. Heat pumps provide 100% heating and cooling. The heat source is groundwater from a depth of 150 m at a temperature 6–10 °C. The water discharges from the heat pump at a temperature of 30–35 °C for use in the building. The thermal capacity is 337.4 kWt. The heat pump cycle uses R134a as the working fluid and consists of a compressor with one screw-type and one turbo-type machine [11].

Fig. 3.24 Linz, Austria, geothermal heat pump heated and cooled passive office building [11].

3.3.4.3 Lotte World Tower

In Seoul, capital of South Korea, the tallest building in the country and the fifth tallest in the world opened in April 2017: the Lotte World Tower (Fig. 3.25). At 123 stories and 554.5 m in height, it has 12 high-power heat pumps to provide clean energy for heating and hot water supply of the facility. Six earth-coupled, brine-water heat pumps each with a thermal output of 1.7 MWt and a cooling capacity of 1.9 MWt are fed from 720 wells drilled to 200 m depth. The other six heat pumps use river water as the heat source. The total installed heating capacity is 22.2 MWt, and the cooling capacity is 20.4 MWt [12].

Other examples of heat pump installations for heating and cooling in buildings include:

- Grand Hotel Dolger, Zurich - Refrigeration capacity = 315 kW
- Hotel Ibis, Sydney - Refrigeration capacity = 1163 kW
- Marriot Kauai, Hawaii - Cooling capacity = 442 kW
- Alex Hotel, New York - Refrigeration capacity = 1745 kW
- Hotel Bushan Hyatt, Dubai - Refrigeration capacity = 3512 kW
- Hotel Canto do Sul, Brasilia - Refrigeration capacity = 582 kW
- Hotel Vitoria Jardines Uleta, Barcelona - Refrigeration capacity = 280 kW

Fig. 3.25 Lotte World Tower, Seoul; © Kohn Pedersen Fox Associates, used with permission [12].

- Basurto Hospital Madrid - Refrigeration capacity = 1000 kW
- Hospital Echuka, Sydney - Refrigeration capacity = 500 kW
- Semmelweiss Hospital, Budapest - Refrigeration capacity = 1128 kW
- Radcliffe Hospital, Oxford, UK - Refrigeration capacity = 1000 kW
- Medical Resort Chungnam, Yessan, South Korea - Refrigeration capacity = 3489 kW
- St. Vincent's Hospital, Jacksonville, Florida - Refrigeration capacity = 3489 kW
- DLF Cyber City, New Delhi - Refrigeration capacity = 70,769 kW
- Beijing West Railway Station - Chiller capacity = 12 MW
- University Hospital, Bern - Dual-effect chiller capacity = 1.5 MW; uses hot water from geothermal sources.

3.4 Water-loop heat pump systems

Multi-zone buildings are characterized by different thermal loads that change seasonally. The design and selection of equipment for centralized air conditioning systems in multi-zone buildings is often based on the maximum potential load, leading to low efficiency when conditions of reduced load exist. An effective measure in energy saving is the creation of a closed water circuit with heat pumps. A closed water circuit provides energy transfer over long distances in a more economical way than air ducts.

The advantages of water-loop heat pump systems (WLHPS) are the following (see Fig. 3.26): (1) constant power consumption and high efficiency usage of the equipment; (2) high capability to provide the required conditions in all areas of the building, with heating equipment being automatically switched on in each room at any time; (3) reliability; (4) durability; (5) security; (6) ecological cleanliness; (7) hot water supply, heating, and air conditioning are provided simultaneously; (8) thermal surpluses put to use by redistribution to other sectors; (9) the low-temperature heating and climate system complies with Directive 2002/91/EC, which grants a higher energy class to the building; (10) in comparison with a 4-pipe system using fancoils (or with variable

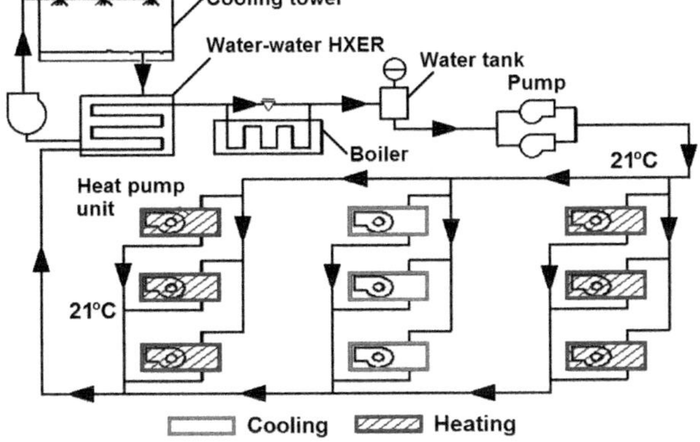

Fig. 3.26 The structure of a water-loop heat pump system (WLHPS), after [13].

air flow-rate), the WLHPS is the most inexpensive investment, 25% lower than the cost of fancoils; (11) high energy efficiency, so heating costs are 50% lower than in centralized heat supply systems; and (12) the possibility of combining separate systems for utilizing the heat of ventilated air and that of waste water into one system. Therefore, using WLHPS as local units in comparison with fancoils is preferable.

A water-loop circuit receives thermal energy from condensation and yields it to evaporation in reversible heat pumps that address the thermal loads of different zones of a building. Heat pumps using a water loop as a heat source have very good efficiency. One important advantage of these systems is the transfer of energy between zones of the building. The net energy necessary to keep the water loop temperature in a proper range can be obtained from gas-fired boilers or other energy production systems.

The heat output of a WLHPS can supply up to 50% of the thermal energy required for building heating while compensating for heat losses through the enclosing structure. Such a design approach is realized due to the possibility of using the power of heat pumps, which is transferred to the water circuit. An important requirement is the need to ensure the flow rate of water. Therefore, automatic flow regulators operating on the principle of dynamic control are used, i.e., a constant flow rate of water in the system is ensured. WLHPS were first presented in the 1960s. WLHPS have been paid more and more attention as the energy consumption of air conditioning has increased since the 1990s and have been widely applied in the United States, Japan, and other countries [14–18].

A detailed analysis of WLHPS is given in Refs. [14–18]. The results of a feasibility study for applying WLHPS in European climatic conditions are presented in Ref. [17]. The criterion for partitioning Europe into zones is Heating Degree Days (HDD), given in kelvin·days (K·d). Climatic zones of Europe and HDD index are shown in Table 3.2 and Fig. 3.27.

The annual specific heat consumption for 32 cities in Europe (HDD 2500–3500 K.d) vary from 20 to 60 kWh/(m^2 y) and the energy saving of WLHPS depends on the value HDD index at different internal loads. As a case study, a typical office building was considered in cities on the Iberian Peninsula: Spain - Madrid, Barcelona and Zaragoza; and Portugal - Porto. The gross total inhabited area of the building was 918 m^2, and it had internal zones with four orientations as well as an inner zone; see Fig. 3.28. Calculations show that the thermal capacity varies from zone to zone, ranging from 93% to 107%.

Two different systems were compared: a WLHPS and a conventional 4-tube fancoil water heating, ventilation, and air conditioning (HVAC) system. Figure 3.29 shows the simplified schematic configuration for these systems, while the results of comparing the efficiency of the two systems are given in Tables 3.3 and 3.4.

The climate zones of the USA are shown in Fig. 3.30.

In Ref. [20], a detailed study was reported of WLHPS in comparison with coal-fired and electric boilers in 12 cities in China (Harbin, Urumqi, Beijing, Jinan, Lanzhou, Xi'an, Shanghai, Chengdu, Wuhan, Fuzhou, Kunming, Guangzhou). The study looked at four common building types, a multi-zone floor plan, and covered a wide variation in climate conditions. The results showed that WLHPS were not competitive

Table 3.2 Climatic zones, HDD index and HDD zones for some European countries.

Country	Climatic area	Latitude, N	HDD, K.d	HDD zone
UK	Lerwick	60°08′	4024	F
	Eskdalemuir	55°19′	3970	F
	Aberporth	52°08′	3178	F
	Kew (London)	51°28′	2900	E
Denmark	Copenhagen	55°46′	3696	F
Ireland	Dublin	53°26′	3133	F
	Valentia	51°56′	2741	E
Netherlands	Eelde	53°08′	3427	F
	De Bilt	52°06′	3194	F
	Vlissingen	51°27′	2877	E
Belgium	Oostende	51°12′	3147	F
	Uccle (Bruxelles)	50°48′	3020	F
	Saint Hubert	50°02′	4188	F
France	Trappes	48°46′	3069	F
	Nancy	48°41′	3245	F
	Macon	46°18′	2980	E
	Limoges	45°49′	2899	E
	Carpentras	44°05′	2266	E
	Nice	43°39′	1650	D
Italy	Bolzano	46°28′	3087	F
	Venezia	45°30′	2317	E
	Milano	45°26′	2551	E
	Genova	44°25′	1560	D
	Roma	41°48′	1663	D
	Foggia	41°31′	1759	D
	Cagliari	39°15′	1349	C
	Crotone	39°04′	1409	C
	Trapani	37°55′	976	C

with coal-fired boilers anywhere in the country, but that WLHPS achieved significant energy savings over electric boilers in all places, except warm areas especially near the southeast coast. In Fig. 3.31, all locations north of the red line were areas where

Fig. 3.27 Climatic zones, HDD index and HDD zones for central Europe [18].

WLHPS held at least about a 10% advantage in reduced energy consumption over electric boilers. For all cases studied, the WLHPS had a maximum energy savings over electric boilers of about 19%. As can be seen, WLHPS are particularly advantageous along the Yangtze River and areas north of it.

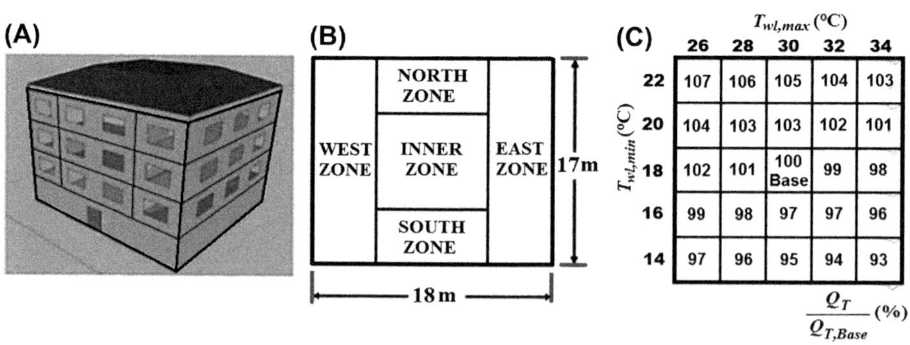

Fig. 3.28 (A) External view of the building; (B) Distribution of thermal zones in the occupied floors; (C) Effect of variations in minimum and maximum water-loop set-point temperatures relative to annual total thermal consumption [18].

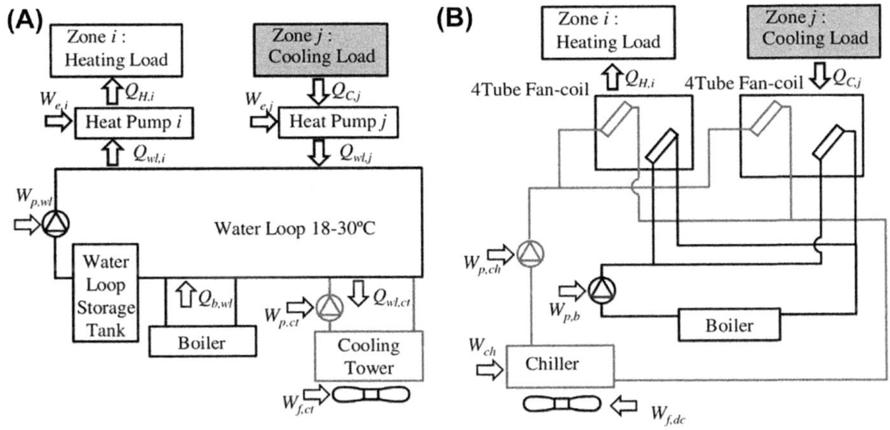

Fig. 3.29 (A) Water-loop heat pump (WLHP) system; (B) 4-Tube fan-coil water heating, ventilation and air conditioning (HVAC) system. $Q_{wl,ct,}$ energy discharged, cooling tower; $W_{p,wl}$, water loop pump consumption; $W_{p,ct}$, cooling tower pump consumption; $W_{f,ct,}$ cooling tower fan consumption [18].

Table 3.3 WLHP system: Annual energy consumption and mean efficiency parameters.

WLHP system	Madrid	Barcelona	Zaragoza	Porto
Total thermal consumption (Q_T), kWh	7279.7	2443.3	6786.6	1751.6
Total electrical consumption ($W_{e,T}$), kWh	12,974	13,550	14,729	10,140
Heat pumps seasonal COP	4.05	4.07	4.05	4.08
Heat pumps seasonal energy efficiency ratio (EER)	3.82	3.82	3.82	3.82
Energy evacuated, cooling tower ($Q_{wl,ct}$), kWh	46,337	55,287	54,400	41,553
Water loop pump consumption ($W_{p,wl}$), kWh	335.2	362.9	383.7	272.5
Cooling tower pump consumption ($W_{p,ct}$), kWh	362.5	432.5	425.6	325.1
Cooling tower fan consumption ($W_{f,ct}$), kWh	318.8	434.0	405.2	300.6
Max. boiler power ($P_{b,max}$), kW	23.3	16.3	25.8	12.5
Max. cooling tower dissipation rate ($P_{ct,max}$), kW	80.6	98.7	104.9	79.5

Table 3.4 4-Tube fan-coil system: Annual energy consumption and mean efficiency parameters.

4-Tube fan-coil water system	Madrid	Barcelona	Zaragoza	Porto
Total thermal consumption (Q_T), kWh	9876.3	3501.1	9332.3	2543.0
Total electrical consumption ($W_{e,T}$), kWh	12,629	14,307	15,002	9856
Chiller seasonal (EER)	3.23	3.38	3.20	3.67
Heating circuit pump consumption ($W_{p,b}$), kWh	19.6	6.9	18.5	5.0
Cooling circuit pump consumption ($W_{p,ch}$), kWh	230.6	275.2	271.0	206.9
Dry condenser fan consumption ($W_{f,dc}$), kWh	969.1	1004.0	1163.8	641.1
Max. boiler power ($P_{b,max}$), kW	32.6	22.8	36.1	17.4
Max. chiller cooling rate ($P_{ch,max}$), kW	22.8	24.8	29.3	21.8

The choice of WLHPS is determined by climatic conditions. In southern latitudes, air cooling is effectively provided by a chiller-fancoil system. In the northern latitudes a bivalent system is applied more effectively. In a temperate climate of middle latitudes, it is advisable to use a WLHPS. However, whether a WLHPS is appropriate or not for any project will be determined by a feasibility study that accounts for technical and economic considerations [20].

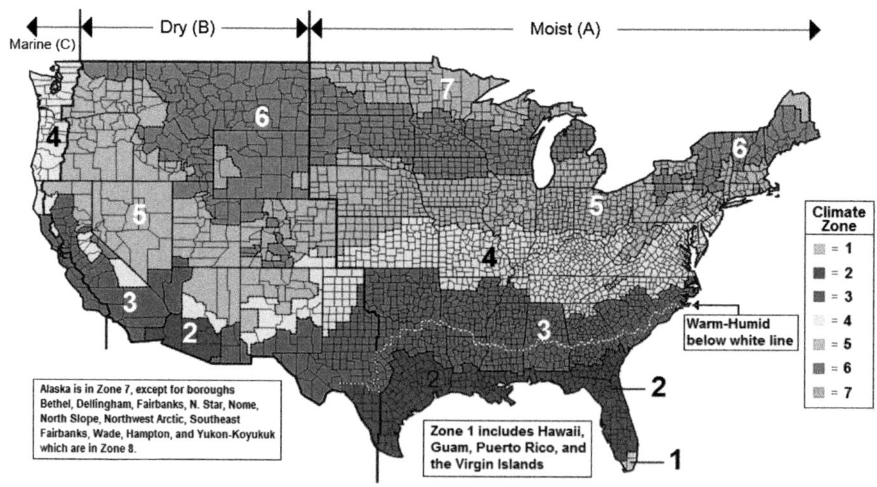

Fig. 3.30 International Energy Conservation Code (IECC) climate regions [19].

Fig. 3.31 Favorable areas, north of the red line, where WLHPS offer energy savings compared to electric boilers used in China, after [20].

The effect of variable frequency-drive (VFD) loop pumps was evaluated in Ref. [15] together with geothermal heat pump facilities. It is common for WLHPS to be combined with geothermal energy in ground-source water loop heat pump applications [16,21] and also with solar collectors.

3.5 Heat pumps in district heating systems

Modern district heating systems are developing in the direction of building energy-efficient residential and administrative buildings, justifying moving toward construction of so-called fourth Generation District Heating (4GDH) systems; see Fig. 3.32.

The 4GDH plan provides the ability to supply a low-temperature energy source in conjunction with a smart thermal grid that would result in (1) a reduced temperature range across the system e.g., 50/20 °C for heating and hot water supply; (2) the possibility of using renewable energy (solar and geothermal), and (3) the use of heat pump technology.

Stockholm's district heating system uses heat pumps with a total capacity of 320 MWt (12% of the total heating load). The heat of the Baltic Sea water with a temperature of 2–8 °C is used as a heat source. Household sewage is also used in Stockholm with a capacity of 220 MWt. Two other district heating systems in Sweden are: Gothenburg (142 MWt) and Orebro (40 MWt). Industrial wastewater is used as a heat source in the cities of Karskar, Sweden (28 MWt) and Yöngöping, China (25 MWt).

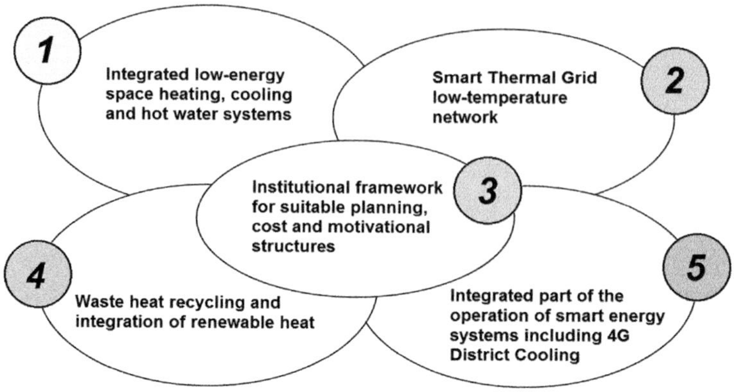

Fig. 3.32 Elements of the 4GDH plan. After [22].

Large heat pump systems are installed in several other European countries: Denmark, Netherlands, Germany, Switzerland, Finland, Norway and Italy. In Japan, more than 50 heat pumps operate at various enterprises. In the heat supply system of the Koraku 1-Chome Office Building in Tokyo, a heat pump station with a capacity of 30.6 MWt is in operation which uses waste water as a source of heat. In the US and China, absorption heat pumps are used in many industrial plants.

The world's largest heat pump using a natural working fluid with a thermal capacity of 13.2 MW is installed in a district heating system in the city of Drammen, Norway. The Drammen municipality has more than 63,000 inhabitants. Three heat pumps use the heat of sea water with a temperature of 8 °C. Heat transfer fluid temperature is ensured up to 90 °C, with a COP_{HP} value of more than 3.05 with ammonia (R717) as the refrigerant. R717 is environmentally friendly, having a very low GWP. The heat pumps reduce CO_2 emissions by 4500 t/y while meeting about 85% of the heat load on the system, the rest being covered by boilers during peak demand [23].

3.5.1 Heat pumps with electric-powered compressor

In this and the next two sections, we consider three different means of supplying the motive power to drive the compressor in a vapor-compression heat pump. This exercise will yield the optimal method among these three alternatives.

The schematic diagram in Fig. 3.33 shows a heat pump system (HPS) with a vapor compression unit for centralized heat supply. In this case, energy conversion occurs in two plants: in the heat pump unit (HPU) and in hot water boilers (WB). The heating of water occurs in series in the HP condenser and in the water boiler. The heat pump station heats a given flow rate of circulating water from the temperature T_{cw1} to T_{cw2}.

The overall effectiveness of the HPS will depend not only on the efficiency of HPU and WB operation, but also on the distribution of water heating in them. It is obvious that when the HPU and the WB have different efficiencies, there should be a certain optimal distribution of water heating between the HPU ($T_{HP} - T_{dcw}$) and the WB

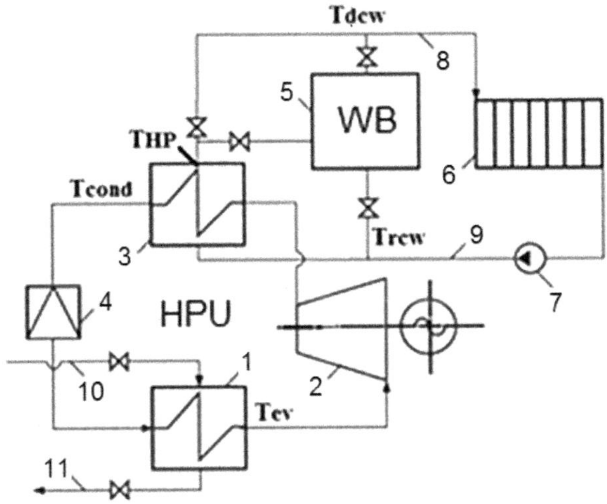

Fig. 3.33 Flow diagram of a heat pump station HPS with integrated boiler. *1,* HPU evaporator; *2,* compressor with an electric drive; *3,* HPU condenser; *4,* throttle valve; *5,* hot water boiler; *6,* heat consumers; *7,* circulating pump; *8, 9,* pipelines of direct and return water, respectively; *10, 11,* pipelines of a low-temperature energy source; T_{ev}, T_{cond}, temperatures of evaporation and condensation of refrigerant in *HPU;* T_{HP}, temperature of heated water in HPU; T_{dcw}, T_{rcw}, temperature of direct and return water in heat supply system, respectively.

(T_{rcw} $-T_{HP}$). By changing the intermediate temperature value, there will be a redistribution of the amount of heat supplied in the separate plants, and the efficiency of the operation of each plant will be changed according to different governing equations.

Thus, the solution of the problem of effective implementation of the HPS is to determine the optimal load distribution between the HPU and WB for different temperature modes of the heating network. In numerical studies of the problem, the water temperature in the supply pipeline varied in the range from 90°C to 150°C, and the temperature of the return circulating water from 45°C to 70°C. Waste water from an industrial plant with a temperature of 25 °C was used as a low-temperature source of heat. The drop in water temperature in the HPU evaporator was assumed to be 8 °C. Parametric calculations were carried out under the conditions of given specific equipment of the HPS and the temperature mode of the heating network. The independent variable was T_{HP} at the outlet of the HPU condenser and at the input to the WB.

The specific consumption of equivalent fuel "*b*" in the boiler and the equivalent fuel savings $\Delta B = b_{boiler} - b_{HP}$ by using a heat pump system instead of a conventional boiler are shown in Fig. 3.34 as functions of the temperature in the condenser.

The results confirm the notion of an optimum intermediate temperature of the HPS, in which there is a minimum value of the total specific consumption of conventional fuel, i.e., the corresponding maximum fuel saving due to combining a HPU with a hot water boiler in the HPS. For this example, the optimum temperature is about 72 °C.

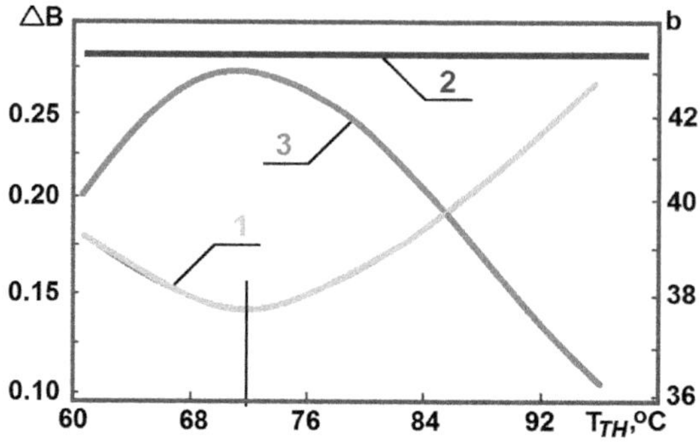

Fig. 3.34 Energy performance of the HPS for T_{dcw}, 100 °C; T_{rcw}, 60 °C: 1, b_{HPS}; 2, b_{boiler}; 3, ΔB.

The studies carried out in Ref. [24] allow us to draw the following conclusions:

1) The application of the heat pump station with a hot water boiler provides up to 15% fuel saving, reduces the consumption of electricity for hot water boilers, and the amount of harmful emissions to the atmosphere;
2) Optimal circulating water heating temperatures in heat pump units depend on the temperature schedule of the heating system operation and are in the range 70–80 °C;
3) Operating modes with almost the same heating loads for the heat pump units and the hot water boilers should be considered as optimal conditions of operation.

3.5.2 Heat pumps with diesel driven compressor

HPUs with a gas or diesel engine drive can lead to energy savings relative to the previous electric-motor driven compressor because it allows the use of waste exhaust gases passing through the combustion engine as well as use of jacket cooling water. General possibilities of using HPU with a gas engine are considered in Ref. [24]. The specific advantages of HPS that use HPU with a diesel engine are shown in Ref. [24]. In this case, diesel power plants with an electric generator power of 2.5 MW and 37% engine efficiency were assumed to power the HPU compressor.

The schematic diagram of a HPS with a diesel generator drive for the HPU compressor and the heat usage of the engine exhaust gases and jacket cooling water is shown in Fig. 3.35. HPU compressor is driven by a diesel engine, an electric generator, an electric motor, and a gear box. The HPU scheme involves heating one portion of return water in the HPU condenser, raising the circulating water temperature to T_{HP}, and the other portion in the jacket cooling system of the engine, followed by a heat exchanger where exhaust gases provide heating to the temperature T_{HP}. Furthermore, depending on the temperature requirement of the heating network, hot water may additionally be heated in a hot water boiler or supplied directly to heat consumers. Thus the system in Fig. 3.35 has an abundance of flexibility to handle a wide variety of heating demands.

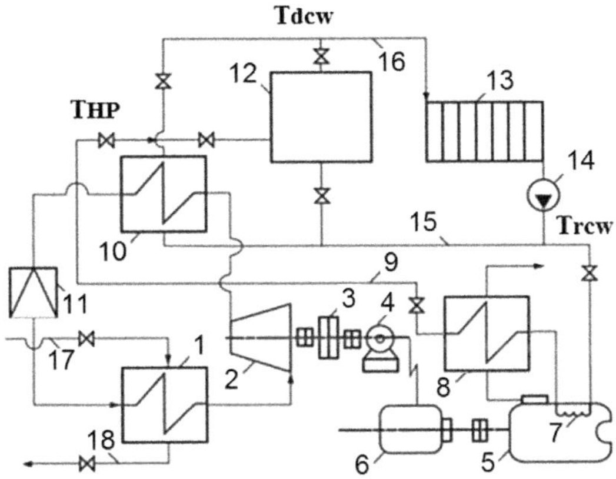

Fig. 3.35 Schematic diagram of a HPS with a diesel generator-drive of the HPU compressor and with waste heat recovery from the diesel engine exhaust and jacket cooling water. *1,* HPU evaporator; *2,* compressor; *3,* gear box; *4,* electric motor; *5,* diesel engine; *6,* electric generator; *7,* heat-exchanging surface of the diesel water cooling system; *8,* waste heat recuperator of diesel exhaust gases; *9,* hot water circuit from the recuperator; *10,* HPU condenser; *11,* throttle valve; *12,* hot water boiler; *13,* heat users; *14,* heating system circulating pump; *15, 16,* direct and return water; *17, 18,* supply and removal of "cold" heat-carrier at HPU evaporator; T_{dcw} and T_{rcw}, temperatures of direct and return water in the heat supply system; T_{HP} is the temperature of circuit water after heating in the HPU.

Research was carried out by changing the temperature of the circulating water in the range of 90–150 °C and return water from 40 to 70 °C. The results of the research were presented in the form of relative indicators of the HPS operation (conversion factor, exergy efficiency, specific conventional fuel consumption) compared to the corresponding indicators for the HPS with an electric-driven HPU compressor, on the condition that the circulating water is heated to the optimum T_{HP}, which was determined earlier for the system with the electric drive. The change in these relative parameters was investigated depending on the temperature of the circulating water T_{cw2} while holding constant the temperature of the return water T_{cw1}. It was shown that the HPS performance was better with a diesel drive than with an electric drive for the HPU compressor. Also, the performance decreases with an increase in the circulating water temperature, which is related to a decrease in the share of load of the HPU and, therefore, an increase in the share of hot water boiler load.

Figure 3.36 shows the saving of conventional fuel due to the introduction of the HPU with a diesel-driven compressor and waste heat recovery from the exhaust of the engine.

In Fig. 3.36, curve 1 describes the fuel saving in comparison with the HPU having an electric-driven compressor; curve 2 describes the fuel saving compared to the use of a hot water boiler. It is seen that additional fuel savings due to the use of heat pump unit with a diesel drive can be up to 21% (at $T = 90$ °C). These indicators decrease with an increase in the temperature of direct circuit water, which occurs owing to a decrease in the share of heat pump unit load, β, shown by curve 3, and read from the right-hand axis.

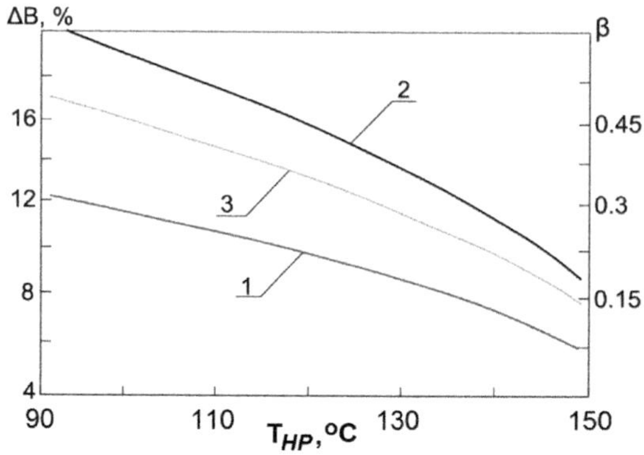

Fig. 3.36 The value of the savings in conventional fuel ΔB in relation to the operation of the HPS with an electric drive and the load fraction of the heat pump unit β. *1*, fuel saving for HPU; *2*, fuel saving for HPS; *3*, HPU share of heat load β.

Based on the research carried out in Ref. [24], the following conclusions may be drawn:

1) The efficiency of HPU with a diesel compressor drive is higher than that of the HPU with an electric drive due to the utilization of exhaust gases leaving the engine;
2) The application of HPS with diesel engines determines fuel saving in comparison with hot water boiler houses of up to 21%;
3) Fuel saving leads to a reduction of oxygen consumption for fuel combustion, a decrease in harmful emissions into the atmosphere, and lowering the energy consumption for hot water boilers.

3.5.3 Heat pumps with gas turbine driven compressor

One of the options for cogeneration units involves the use of gas turbines (GT) for electric power generation and waste heat recovery from the GT exhaust for producing steam or hot water. When combining such plants with a heat pump unit, a possible variant of the HPS is having the HPU compressor driven from a GT unit and using the GT exhaust gas to make hot water. The analysis of the efficiency of such a HPS was conducted under conditions of using gas turbines with electric generator power generating capacity of 1.5 MW and an efficiency of 31.5% at a nominal mode [24].

As in the previous cases, the HPS operation was investigated by varying the temperature of direct circuit water from 90 to 150 °C, and that of the return water from 40 to 70 °C. The HPS operation was analyzed under the condition of heating the circuit water in the HPU to the previously defined optimal temperature. The results were compared to using a hot water boiler. The effectiveness of the HPS operation with GT-driven compressor was also compared with the operation efficiency of the hot water boiler with the GT and using GT exhaust heat in the furnaces of the boilers.

Figure 3.37 shows how the load share factor β of the HPU varies with the circulating water temperature of the HPS for the following systems:

1) HPS with an electric-drive HPU compressor (Curve 1);

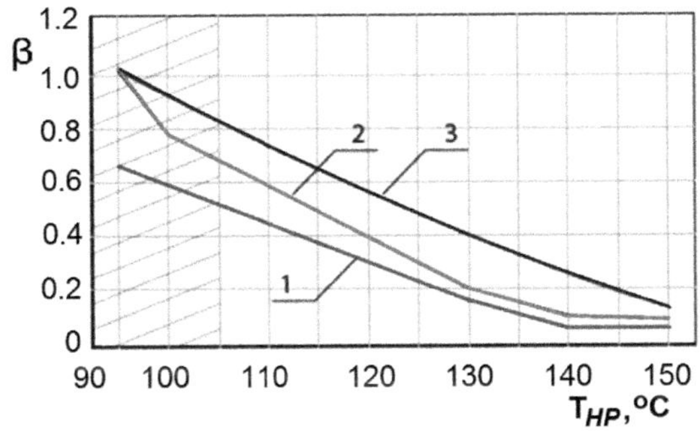

Fig. 3.37 Variation in the load fraction of the HPU part of the HPS as a function of the circulating water temperature.

2) HPS with a diesel engine-driven HPU compressor (Curve 2);
3) HPS with a gas turbine-driven HPU compressor (Curve 3).

It can be seen that under the optimal conditions of the HPU operation there is an increase in the share of the HPU load for the HPS using a GT-driven compressor compared with other HPS options. The decrease of β as the circulating water temperature increases is explained by an increase in the load share of hot water boilers in the range where the HPU is less efficient.

Figure 3.38 shows the variation in the specific consumption of conventional fuel as a function of the circulating water temperature. It is evident that the lowest specific consumption of conventional fuel is provided by the HPS with the GT and use of exhaust waste. Somewhat higher fuel consumption takes place for the HPS with an electric-drive HPU compressor and a diesel-driven compressor, and the highest fuel consumptions occurs at the cogeneration unit with GT using the exhaust gases in a water boiler. However, at high values of circuit water temperature, 140–150 °C, the specific consumptions of conventional fuel for all schemes are almost identical. This is explained by an increase in the load share of hot water boilers (WB), which operate less economically than the HPUs.

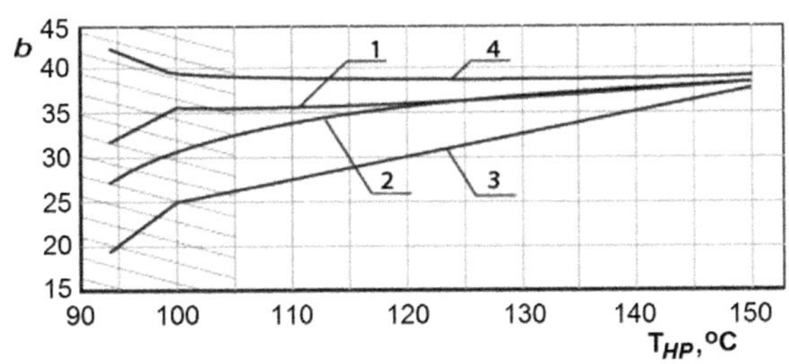

Fig. 3.38 Change of specific consumption of conventional fuel as a function of the temperature of direct circuit water: *1*, HPS with an electric drive; *2*, HPS with an ICE drive; *3*, HPS with a GT drive; *4*, WB with a GT.

On the basis of the research carried out in Ref. [24], the following conclusions were drawn:

1) HPS with a GT compressor drive and waste heat recovery for use in the furnaces of water boilers provides the HPU a greater share of load β of the HPS than one with an electric or compressor drive from diesel engine.
2) The lowest specific consumption of conventional fuel is provided by the HPS with a GT driving the compressor. The HPS with an electric drive and the one with a diesel approach it in terms of efficiency. The water boiler with a GT is characterized by the largest specific consumption of conventional fuel.
3) The highest saving of conventional fuel is observed for the HPS with a GT driving the compressor (up to 45.3%). The savings of conventional fuel for the HPS driven by a diesel is up to 36.7%. The savings of conventional fuel at the HPS with an electric drive does not exceed 26.2%, but is higher than that of the conventional fuel of the WB with a GT.
4) Thus, the HPS scheme with the compressor driven by a GT has the highest overall effectiveness.

3.5.4 Impact of condensers and evaporators on heat pump efficiency

The arrangement of condensers and evaporators in heat pump units can have a significant effect on the thermal performance the HPU. The main alternatives are series and parallel connections. Although a series connection of condensers of several HPU involves higher energy consumption for pumping the working fluid, this higher power consumption is offset by significant savings on power to drive the compressors. Using a series connection of HPU condensers with a parallel evaporator connection ensures working fluid heating to a higher temperature with lower energy consumption (a higher COP) or allows the working fluid to be heated to a higher temperature with the same energy consumption as with a parallel condenser connection. The analysis of various arrangements of HPU connections was performed in Ref. [24].

Figure 3.39 shows the circuit of a HPU with series connections for the condensers and boilers, and parallel connections for the evaporators.

The temperatures of direct and return circuit water in the system of heat supply are determined by the temperature schedule that depends on the temperature of ambient air. The value of the HPU load fraction in the HPS (β) depends on the temperatures of the direct and return circuit water, which, in turn, depend on the temperature of ambient air. The optimal values of β for HPS circuits with different types of drives and arrangements of evaporators and condensers at different modes of heat network operation were determined in Ref. [25]. A method was proposed for determining the proportion of thermal power load on the HPU, β, that depends on the temperature of ambient air, by taking into account the complex influence of HPU condenser connection circuits, the type of the HPU compressor drive, and the modes of HPS operation.

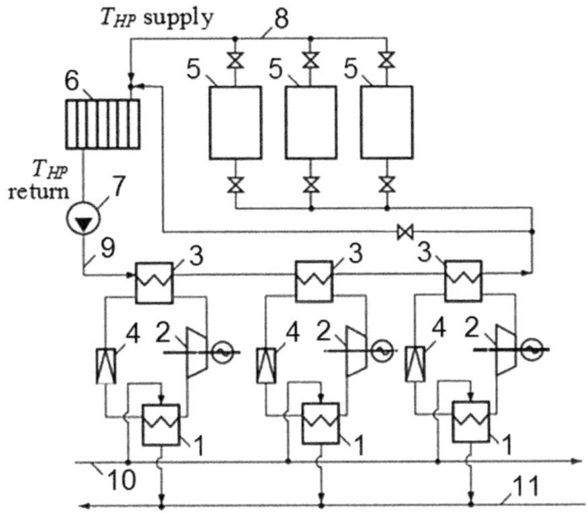

Fig. 3.39 Flow diagram for a heat pump system with parallel evaporators, series condensers and boilers, with electric-driven compressors. *1,* HPU evaporators; *2,* electric-driven compressors; *3,* HPU condensers; *4,* throttle valves; *5,* hot water boilers; *6,* heat users; *7,* circulating pump; *8, 9,* direct and return water; *10, 11*, low-temperature energy source, sink.

3.5.5 Absorption refrigeration and absorption heat pumps in heat supply systems

The principal difference between vapor-compression heat pumps (VCHP), presented in the previous section, and absorption heat pumps (ABHP) is the supply of mechanical work for VCHP, and the supply of heat for ABHP. Relatively little electromechanical power input is required for an ABHP while a significant amount is needed for VCHP. Since the cost of electricity differs from the cost of heat (and is usually higher), the economic performance of the two types of heat pumps is different as well.

ABHP may be used in combined heat-power (CHP) systems to improve energy usage. In this context, the ABHP is driven by two heat input streams - one a hot source and one a low-temperature source (Fig. 3.40). Extraction steam from a steam turbine or engine jacket cooling water may serve the first purpose, while cooling water from the power plant condenser system may serve the latter purpose (Fig. 3.41). For a typical ABHP system, a COP of about 1.7 can be achieved from 100 energy units of input high temperature heat and 70 energy units of low-temperature heat, plus a small amount of electric power to run the pumps in the ABHP.

Steam with pressure of 0.4−0.6 MPa, combustion exhaust gases, water with temperature above 140 °C, and organic fuel can be used as the hot source for the ABHP. Liquid flows with temperature of 15−30 °C may be used as a low-to-moderate heat source. The district heating fluid leaving the ABHP can have a temperature of 80−90 °C. For bivalent absorption systems, an ABHP can operate in summer in the mode of a refrigeration machine and in air conditioning systems (ABRM), and in winter in the mode of a heat pump with costs being 40% lower than with other heat generating systems [26,27].

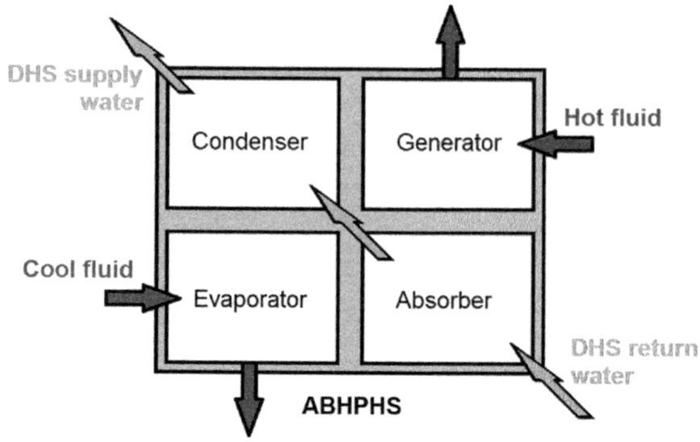

Fig. 3.40 A simplified block diagram for an absorption heat pump heating system ABHPHS, used in conjunction with a district heating system and an electric generating station.

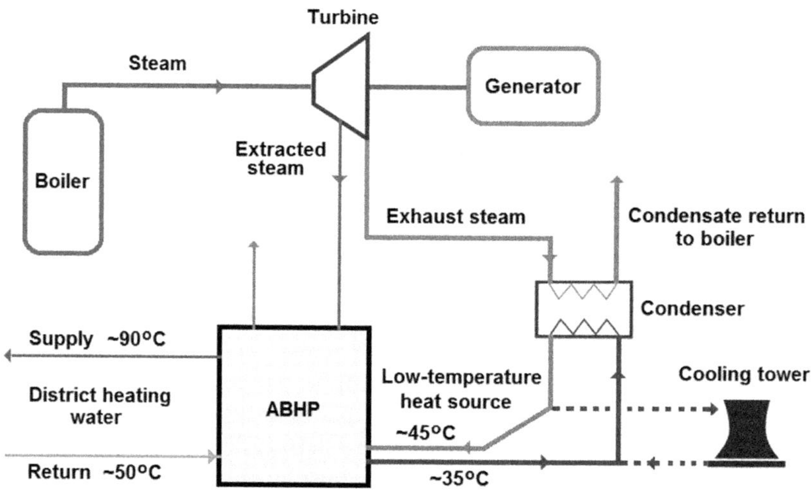

Fig. 3.41 Simplified flow diagram for a CHP system augmented by ABHP.

In a single-stage ABHP, the COP $= Q_s/Q_g$ is 1.65−1.75, where Q_s is the amount of heat produced from the absorber and condenser and Q_g is the amount of heat required in the generator, and the relative energy consumption for heat production is 0.67−0.76. In a 2-stage ABHP, the COP is 2.05−2.15, and the relative energy consumption for heat production is 0.55−0.67. Specific fuel consumption for heat production in an ABHP is 40−45% lower than in boilers. ABHPs are widely used in heating systems in Japan, South Korea (share of ABHPs is 70%), China, India, Canada (up to 50%), and Sweden. At the same time, secondary thermal energy resources in industrial plants are effectively put to use.

3.5.6 Economic comparison of the efficiency of various heat sources

A comparison of different types of heat pumps is carried out according to an economic criterion, i.e., by the equivalent fuel consumption B_b in kg/s for the generation of the specified thermal power Q_0 in kJ/s or kWt. Fuel consumption (in coal equivalent) for a boiler is:

$$B_b = \frac{Q_0}{29.3 \cdot \eta_b}, \tag{3.3}$$

where 29.3 MJ/kg is the heat of combustion of coal fuel equivalent; η_b is the boiler efficiency.

For a vapor-compression heat pump (VCHP) with an electric drive:

$$B_{HP} = \frac{Q_0}{29.3 \cdot \eta_{el} \cdot \eta_e \cdot COP_{HP}} \tag{3.4}$$

where η_{el} is efficiency of an electric generating plant; η_e is efficiency of power transfer in transmission lines and networks.

For an absorption heat pump (ABHP):

$$B_{ABHP} = \frac{Q_0}{29.3 \cdot M \cdot \eta_g}, \tag{3.5}$$

where $M = Q_0/Q_g$; Q_g is the heat added in the generator; η_g is generator efficiency.

For the purpose of comparison, noting that the quantity $Q_0/29.3$ is a constant, we obtain:

Boiler:

$$B_b = \frac{1}{\eta_b} = \frac{1}{0.90} = 1.11 \text{ kg/s}. \tag{3.6}$$

Vapor compression heat pump:

$$B_{HP} = \frac{1}{\eta_{el} \cdot \eta_e \cdot COP_{HP}} = \frac{1}{0.33 \cdot 0.95 \cdot 3} = 1.06 \text{ kg/s}. \tag{3.7}$$

Absorption heat pump:

$$B_{AHP} = \frac{1}{M \cdot \eta_g} = \frac{1}{1.7 \cdot 0.9} = 0.65 \text{ kg/s}. \tag{3.8}$$

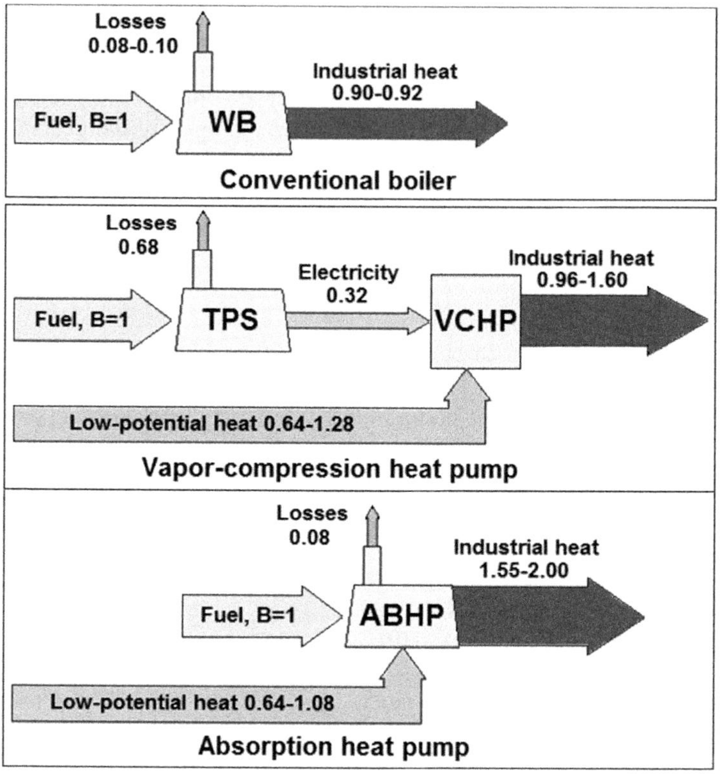

Fig. 3.42 Energy balances of two types of heat pumps in comparison with a conventional boiler for the production of heat based on one unit of input fuel energy.

As can be seen, the equivalent fuel consumption for the absorption heat pump is 40–50% less than for the boiler and the vapor compression heat pump. Therefore, the use of large ABHP with thermal power of tens of MW is economically more feasible than the use of boilers or vapor-compression heat pumps. The results of the comparison are shown graphically in Fig. 3.42.

3.6 Summary

This chapter covers a wide range of heat pump applications from individual private residences to modern skyscrapers to airports and other public facilities. Although mainly technical topics are presented, the economics of various systems is covered in the last section where equivalent coal consumption is used as the basis for comparison. Heat pumps are shown to be very effective in controlling the climate in pools, particularly indoor units where humidity can pose challenges for designers. Water-loop systems are becoming popular in many places owing to their inherent flexibility to allow heating and cooling simultaneously in building with variable thermal loads. In the case of a hot-water heating system, detailed assessments are carried out considering a conventional water boiler and heat pumps with three different means of powering the heat pump compressor, namely, electric, diesel or gas turbine. The heat pump with a gas turbine compressor drive is the most efficient on several thermodynamic and practical measures. An absorption heat pump is shown to be the most effective kind of heat pump.

Nomenclature

B	consumption
COP	coefficient of performance
d	humidity
Q	energy discharged
T	temperature
V	velocity
W	consumption

Greek letters

β	load
η	efficiency
ρ	density

Subscripts

4GDH	fourth generation district heating
ABHP	absorption heat pump
GHG	greenhouse gas
HP	heat pump
MC	mixed chamber
F	filter
HDD	heating degree days
HPU	heat pump unit
LHV	lower heating value
L	heat pump driver
rec	recirculation
recup	recuperation
WLHPS	water loop heat pump system
WB	water boiler
GT	gas turbine
VCHP	vapor compression heat pump

Review questions

1. Why is the efficiency of using the HP higher in buildings with a "warm floor" type heating?
2. In which cases is the HP more efficient when using a heating boiler (consider the efficiency value)?

3. What heat sources are used as reserve and peak ones in systems with the HP?
4. Name the components of heat losses of indoor pools.
5. What factors influence the amount of moisture evaporating into the indoor pool premises?
6. Name the value of heat transfer coefficients from air to the surface of water in the pool in the day time and at night.
7. Explain the scheme of the series use of heat recovery units and HPs.
8. How does the humidity in the pool decrease when using the HP?
9. Name the loss of heat per square meter of the outdoor pool.
10. Why are HPs more cost effective when used as air conditioners in summer?
11. What are the advantages and disadvantages of centralized heat supply using the HPs in hospitals.
12. What are specific factors of using HPs in public buildings.
13. What is the essence of heat-pump-recovery systems of ventilation?
14. Explain the value of the coefficient of recovery and multiplicity of ventilation.
15. In which cases is recirculation of air in ventilation systems used?
16. Explain the operation of the air heating circuit with the HP.
17. What is more effective - heat recovery or recirculation of air in ventilation systems with the HP?
18. What is the advantage of using heat pump stations with centralized heat supply?
19. Explain the advantages of the series connection of HP condensers in the heat pump station.
20. Show the processes of the absorption refrigerating machine in the ξh-diagram (concentration-enthalpy).
21. How does the absorption refrigerating machine operate?
22. Compose the heat balance of the absorption refrigerating machine and determine the value of the thermal coefficient?
23. How does the absorption refrigerating machine of periodic action operate?
24. How does the absorption refrigerating machine of continuous action operate?
25. Is the efficiency of the HPU a function of the heat source?
26. How do the temperatures of condensation and evaporation of the refrigerant affect the heat pump COP?
27. Compare and discuss the energy efficiency of vapor compression and absorption heat pumps.
28. Which component in a vapor-compression heat pump is responsible for heating a desired space: (a) absorber; (b) condenser; (c) evaporator; (d) regenerator?
29. Which component in an absorption heat pump is responsible for heating a desired space: (a) condenser and generator; (b) throttle and pump; (c) generator and regenerator; (d) evaporator and absorber; (e) condenser and absorber?

Exercises

1. Using as examples of refrigerants CO_2 and R152a, determine the effect of these different fluids on a heat pump COP?
2. The *P-h* diagram in the figure below indicates the cycle for a heat pump operating between 100 kPa and 1000 kPa R152a is the refrigerant (Fig. 3.E1).

   Part A. The compression work (kJ/kg) is nearest: (a) 117.7; (b) 110; (c) 100; (d) 130.

   Part B. The cooling load (kJ/kg) is nearest: (a) 215; (b) 220; (c) 230; (d) 213.2.

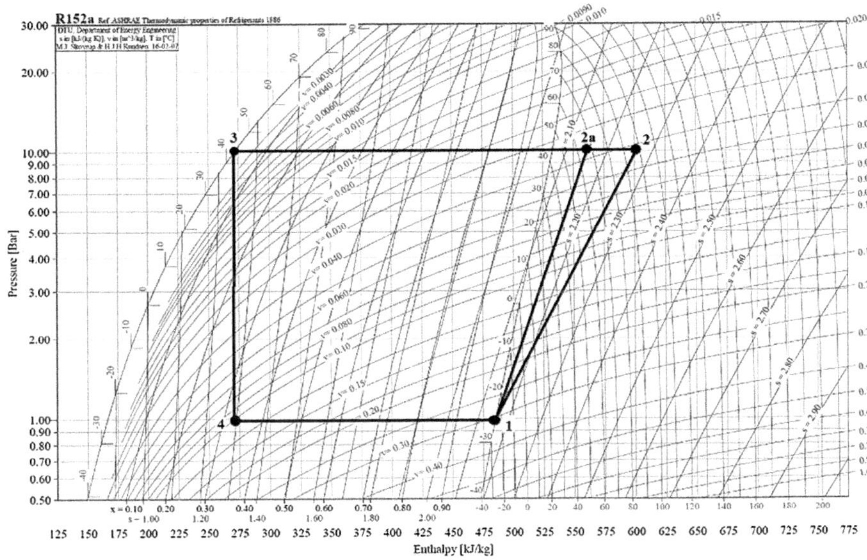

Fig. 3.E1 Heat pump cycle in the psychrometric chart.

Part C. The heating load (kJ/kg) would be nearest: (a) 355; (b) 305.5; (c) 308.7; (d) 330.9.
Part D. If the compressor is 80% adiabatic efficient, the enthalpy (kJ/kg) of the R152a at the compressor outlet is nearest: (a) 601.5; (b) 590.2; (c) 595.5; (d) 588.7.
Part E. If the average logarithmic temperature of the coolant is $T_e = 273$ K, the ambient temperature is $t_0 = -10$ °C, and the heat load of the evaporator is $q_e = 210$ kJ/kg, the exergy (kJ/kg) given to the low-temperature R152a in the evaporator is: (a) 6.75; (b) 8.41; (c) 7.69; (d) 7.32.
Part F. If the average logarithmic temperature of the high-temperature refrigerant is $T_k = 313$ K, the ambient temperature is $t_0 = -10$ °C, and the thermal load in the condenser is $q_c = 335$ kJ/kg, then the exergy (kJ/kg) obtained by the coolant in the condenser is: (a) 53.6; (b) 48.8; (c) 48.0; (d) 49.8.
Part G. If the efficiency of the electrical network is $\eta_{el} = 0.95$ and the efficiency of the compressor motor $\eta_{el.c.} = 0.80$, then the specific energy (kJ/kg) consumed by the compressor motor is: (a) 119.5; (b) 164.5; (c) 120.1; (d) 128.8.
Part H. For the conditions of Parts E and F, if the efficiency of the electrical network is $\eta_{el} = 0.95$ and the efficiency of the compressor motor $\eta_{el.c.} = 0.80$, then the exergic efficiency of the heat pump is: (a) 0.311; (b) 0.350; (c) 0.382; (d) 0.348.

References

[1] European Environmental Agency. Part 1. Overall progress towards the European Union's '20-20-20' climate and energy targets. 2017. https://www.eea.europa.eu/themes/climate/trends-and-projections-in-europe/trends-and-projections-in-europe-2016/1-overall-progress-towards-the.

[2] Heat Pump Mitsubishi Electric ZUBADAN. http://www.mitsubishielectric.com.ua/zubadan.html.
[3] Schmid F. Sewagewater: interesting heat source heat pumps and chiller. In: Procs. 9th IEA heat pump conference, Zurich, May 19–23, 2008. pp. 1–12.
[4] Mit SOLAREIS heizen. http://www.peter-solar.de/web-startseite/mit-solareis-heizen.html.
[5] Viessmann, Heating with airborne and geothermal energy, VITOCAL, Viessmann Technology Brochure, 9449 593 GB, 03/2019, p. 30, Allendorf (Eder), Germany.
[6] Matsevity YM, Chirkin NB, Bogdanovich LS, Klepanda AS. Introduction of heat pump technologies. Ecotechnology and Resource Saving 2008;(3):4–10.
[7] Arctic Heat Pump https://www.arcticheatpumps.com/pool-spa-heat-pumps.html.
[8] Sadovnikov AA. The use of heat pumps for heating water in swimming pools. Plumbing 2013;(2). https://www.abok.ru/for_spec/articles.php?nid=5498 [in Russian].
[9] Redko AO, Bezrodny MK, Zagoruchenko MV, Ratushnyak GS, Redko OF, Hmelnûk MG. Low potential energy, textbook LLC. Kharkov: Typography Madrid; 2016. 412 pp.
[10] Developing geothermal district heating in Europe. GEODH; 2014. http://geodh.eu/wp-content/uploads/2012/07/GeoDH-Report-2014_web.pdf.
[11] Ochsner heat pumps. http://www.geotherm.com.ua/overview.html [in Russian].
[12] Lotte World Tower, Seoul, South Korea: http://www.worldofarchi.com/2012/09/skyscrapers-lotte-world-tower-seoul.html.
[13] Zhiwei Lian, Seong-ryong Park, Henian Qi, Analysis on energy consumption of water-loop heat pump system in China, Applied Thermal Engineering, 25 (2005) 73–85.
[14] Pietsch JA. Water-loop heat pump systems assessment. ASHRAE Transactions 1990;96: 1029–38.
[15] Howell RH, Zaidi JH. Analysis of heat recovery in water-loop heat pump systems. ASHRAE Transactions 1990;96:1039–47.
[16] Kush EA. Detailed field study of a water-loop heat pump system. ASHRAE Transactions 1990;96:1048–63.
[17] Cooper WS. Operative experience with water-loop heat pump system. ASHRAE Transactions 1994;100:1569–76.
[18] Henderson HI, Carlson SW, Khattar MK, et al. The implications of the measured performance of variable flow pumping systems in geothermal and water loop heat pump applications. ASHRAE Transactions 2000;106:533–42.
[19] Buonomano A, Calise F, Palombo A. Dynamic Building Simulation: Applied Energy 2012;91:222–34.
[20] Fernández FJ, Folgueras MB, Suárez I. Energy study in water loop heat pump systems for office buildings in the Iberian Peninsula. Energy Procedia 2017;136:91–6.
[21] USDOE, Energy Efficiency & Renewable Energy, Int'l. Energy Conservation Code, Ch. 3, General Requirements, Sect. R301, Climate Zones, Commercial Scope and Envelope Requirements, PNNL-SA-132967, 2018.
[22] Tsinghua University. Manual for designer's simulation toolkit (DeST 2.0). Tsinghua Tongfang, Ltd.; 2000.
[23] Lund H, Werner S, Wiltshire R, Svendsen S, Thorsene JE, Hvelplunda F, Mathiesen BV. 4th Generation District Heating (4GDH): integrating smart thermal grids into future sustainable energy systems. Energy 2014;66:1–11.
[24] Tkachenko SY, Ostapenko OP. Steam-compression heat-pump installations in heating systems. Vinnitsa, Vntu.; 2009. p. 175.

[25] Gorshkov VG. LLC "OKB Teplosibmash," Institute of Thermal Physics, Siberian Department Russian Academy of Science, "Russian lithium-bromide absorption heat converters of new generation – practice and prospects of application". In: International conference: heat pumps in CIS countries, Alushta, May 14–17, 2013.

[26] Romanyuk VN, et al. ABHP in the thermal scheme of the HEGS for increasing its exergic efficiency. Energie und Management 2013;(1):14–9.

[27] Dyachek PI. Perspectives of the ABRM use. Energie und Management 2007;(2).

Heat pumps in the drying industry[(1)]

4

4.1 Introduction and overview of drying using heat pumps

One of the main thrusts of this book is the use of heat pumps for a broad range of applications. Currently the range of heat pump capacities are typically from 1 to 5 MWth, with the highest thermal power of a single heat pump being 35 MWth. Heat pumps are capable of providing working fluids at temperatures up to 100 °C for many applications including heating, hot water supply, and various technological needs in industry. Using heat pumps for applications above 100 °C is problematic, and the industry has obstacles to overcome in this area. Even with these limitations, more than 10 million heat pumps units, from very small to very large, have been installed so far in Europe.

The main obstacles limiting the use of heat pumps in industry are high initial capital cost which might lead to uneconomic returns on investment and perceived risk on the part of industrialists who may view heat pump technology as new and not sufficiently proven in commerce and having limited examples of successful applications. Thus, to expand the adoption of heat pumps for a wide application of technologies, it is necessary to remove structural barriers in industry; apply intensive energy saving methodology; integrate capabilities and responsibilities in their implementation from an overall system perspective that will optimize industrial processes and commercial applications.

Figure 4.1 shows the wide spectrum of thermal demand (heat energy) in various industries. The demand is broken down into heat over temperature bands. Of this huge energy demand, heat pumps provide 174 TWh-th or 8.7% of all industrial needs [1,2].

According to Ref. [3], among the first applications of heat pumps for drying and dehydration was the work of Sulzer on the dehydration and drying of underground cavities in Germany in 1943 and a grain dryer developed in the USA in 1950. A significant part of the work involving the use of heat pumps is connected not only with convection drying, but also with freeze-drying [4,5], in which refrigerating equipment is used to freeze the product. Particularly important drying materials with heat pumps are described in Refs. [6–11]. The most famous and widespread application of heat pumps in convection drying is the drying of wood (sawn timber) in chambers with closed air circulation and condensation moisture removal.

There is no doubt that drying is one of the most energy intensive industrial processes. In developed countries, about 10% of fuel consumption is used for drying various products of industry [12]. One of the common processes of drying plants is

(1) Written by Andriy and Oleksandr Redko; edited by Ronald DiPippo.

Low-Temperature Energy Systems with Applications of Renewable Energy
https://doi.org/10.1016/B978-0-12-816249-1.00004-2

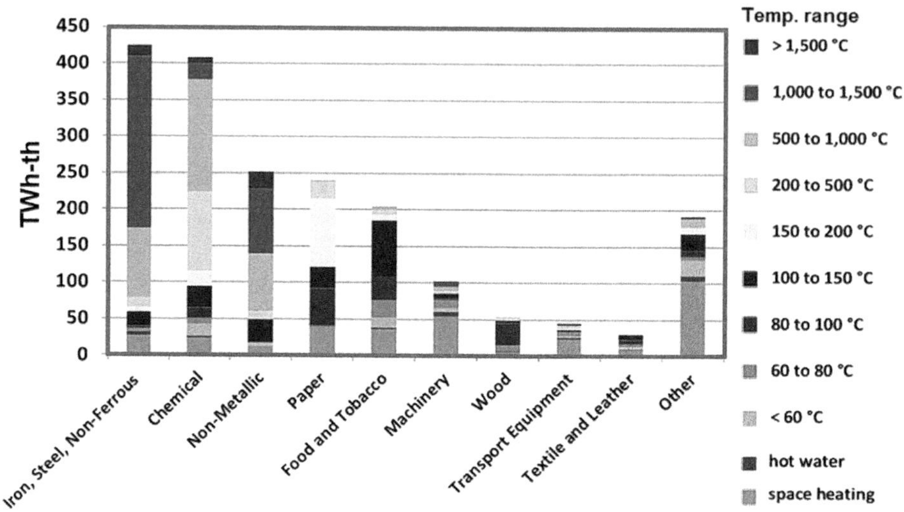

Fig. 4.1 Spectrum of heat demand in industry by sector and temperature range, after [1].

the use of convection in which preheated air passes over the surface of a product. In an open system, a mixture of air with moisture taken from the product is discharged to the atmosphere, carrying away the heat expended on evaporation of the moisture. The outlet air temperature may range from 30 to 70 °C.

Dryers can be either (1) continuous, i.e., the product is fed into the chamber and is removed continuously, or (2) periodic, i.e., the product is loaded into the chamber, dried, and after reaching a certain moisture level, the material is removed, new material loaded, and the process is repeated. Method (2) is often called "batch" processing. Theoretically, it is possible to reduce energy consumption by increasing air recirculation, which will be discussed later. There are trade-offs involved that form the basis of an optimization study.

In addition to heating the air for the drying process, the supplied energy is spent to compensate for heat losses through the enclosing structures, the losses of hot air due to poor seals, and losses when heating is done in radiators. Energy is also spent on the motor drive of the fan. According to Ref. [13], the effectiveness of such dryers is very low. Energy consumption reaches 23 MJ/kg of evaporated moisture, which is almost 10 times more than the heat of evaporation.

We now present eight systems for carrying out the drying process: Cases A–H.

4.1.1 *Case A: Basic open system using ambient air*

The simplest unit is an open, flow-through arrangement where ambient air is introduced to the drying chamber, passes over the moist product, evaporates the moisture, and is discharged to the surroundings. A heater is built in to boost the ambient air

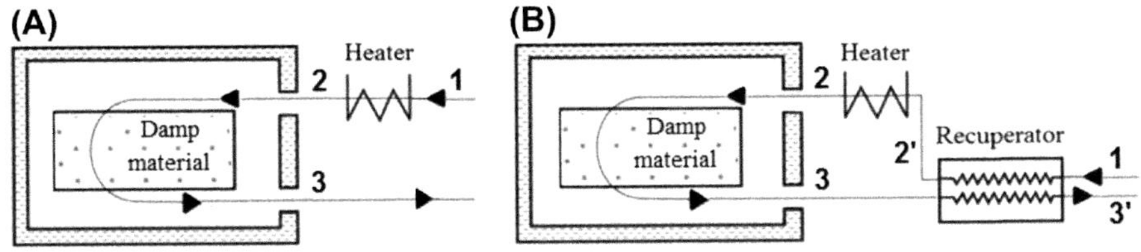

Fig. 4.2 Schematic diagram of a (A) basic drying plant and (B) one with a heat recuperator.

temperature as needed. See Fig. 4.2A. The figure also shows a recuperated system (B) that will be described shortly.

Before moving on to more complex systems, we present a graphical method to analyze the basic case. Consider the adiabatic drying process which takes place in a dryer operating on the basic scheme (Fig. 4.2A), i.e., the air is heated in an air heater and then makes a single pass through the drying chamber. The processes may be visualized in a psychrometric chart (PC) (Fig. 4.3) that presents the properties of moist air in coordinates of moisture content (ordinate) versus the dry-bulb temperature (abscissa) [14,15]. The curved lines that fan out and upward from the lower left represent lines of constant relative humidity, with the highest one being 100%. In this figure, point 1 is determined from the properties of the ambient air. Assume that the ambient air temperature, $t_1 = 20$ °C and moisture content is 0.10 kg/kg dry air; from the PC one can read that the relative humidity is ~70%. The process of heating the air prior to the dryer is represented by line 1 → 2′, where point 2′ is determined by the assumed

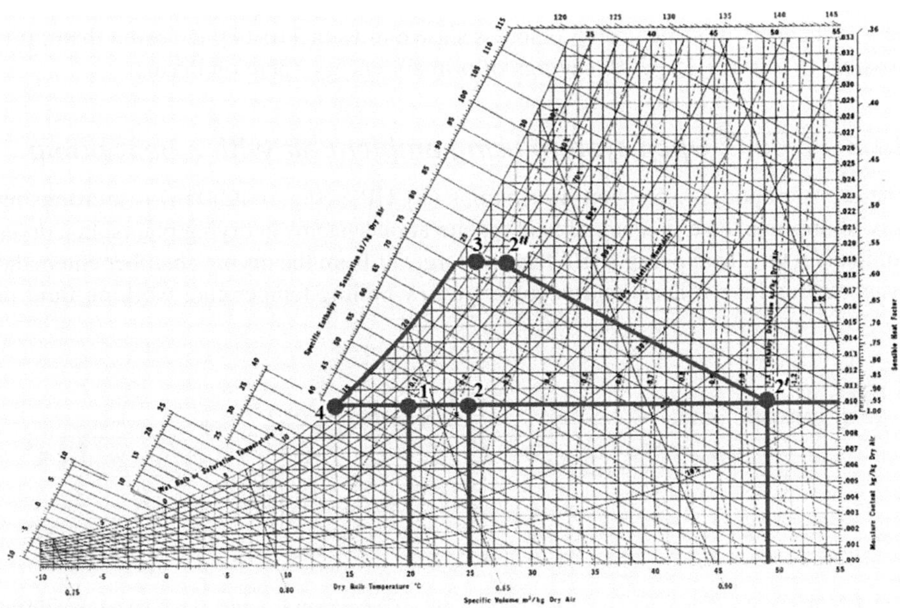

Fig. 4.3 Psychrometric chart for typical values of a basic dryer: processes 1 → 2′→2″.

temperature $t_2 = 49$ °C and a constant moisture content of 0.10 kg/kg dry air. Process $2' \rightarrow 2''$ is an adiabatic process of cooling and humidifying the air in the dryer that follows a line of constant enthalpy to the outlet temperature of the air, in this example, ~28.3 °C. It can be seen that the outlet relative humidity is ~78% and the absolute moisture is 0.19 kg/kg dry air. In an ideal adiabatic saturation process, point $2''$ would move to the 100% relative humidity line. In practice, the dry- and wet-bulb temperatures of the incoming air can be measured using a "sling psychrometer," and the PC can then be used to determine the relative humidity, absolute humidity and specific enthalpy.

4.1.2 Case B: Open system using ambient air with a heat recuperator

The next simplest system involves the use of a heater and a heat exchanger (recuperator) in which the inlet air is preheated by the outlet flow of drying air (Fig. 4.2B). The psychrometric chart (Fig. 4.3) shows the process of preheating the air in the recuperator $1 \rightarrow 2$ and continued heating in the heat exchanger $2 \rightarrow 2'$. For the example chosen, the recuperator covers the temperature increase from 20 to 25 °C while the heater (using external heat) takes care of the heating from 25 to 49 °C. Without the recuperator, heat energy of about 30.5 kJ/kg of dry air is needed to reach the desired dryer inlet condition; with it, the energy requirement is reduced to about 17.5 kJ/kg, a significant energy reduction of about 13 kJ/kg, or a 43% reduction. Process $3 \rightarrow 4$ is explained below.

Rotary regenerative heat exchangers – typical air preheaters used in many industrial facilities – can also be used. The efficiency of using recuperative and regenerative heat exchangers in some drying plants is shown in Refs. [16,17]. By using them, it is possible to achieve 15–20% savings of primary energy spent on drying.

4.1.3 Case C: Open system using ambient air with a heat pump

The use of vapor-compression heat pumps (VCHPs) (Fig. 4.4C) [18] is a promising way to improve the efficiency of heat regeneration systems in drying plants. In a dryer equipped with a heat pump (HP), the discharge air from the drying chamber enters the evaporator, where it heats and vaporizes a low-boiling-temperature working fluid (a

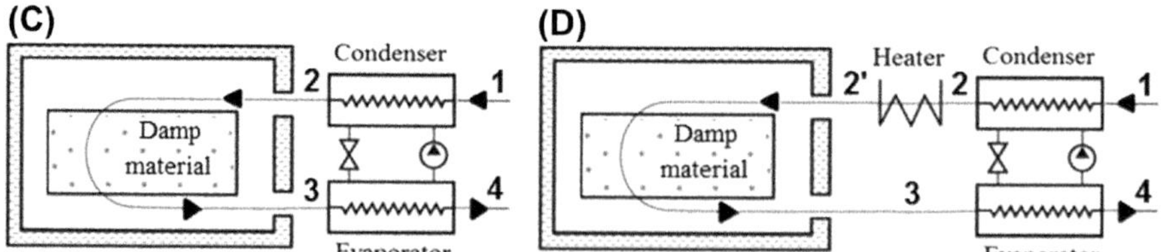

Fig. 4.4 Schematic diagram of a drying plant with a (C) heat pump and a (D) heat pump and an auxiliary heater.

refrigerant). The refrigerant vapor is compressed in the compressor and enters the condenser. As the vapor condenses, it heats up the ambient air entering the dryer.

4.1.4 Case D: Open system using ambient air with a heat pump and auxiliary heater

If it is impossible to heat the air to the required temperature in the condenser using the VCHP system in Case C, a supplementary electric heater is installed (Fig. 4.4D).

4.1.5 Case E: Closed-air system with a HP and dehumidification-recirculation

Dryers with VCHP may be configured so that the drying air is not discharged but recirculated. Such closed circuits are the most widely developed and implemented. The system also incorporates dehumidification (Fig. 4.5E) which removes the moisture acquired by the air as it passes over the product in the drying chamber. In the evaporator, the humid air cools down below the dew point, condensing the moisture which is drained away. The dried air is then blown across the condenser of the heat pump where it heats up to the required drying temperature and repeats the cycle. The process involved with condensation of moisture for this case is shown on the PC in Fig. 4.3 as process 3 → 4.

4.1.6 Case F: Closed-air system with a HP, dehumidification-recirculation and bypass

An alternative to Case E includes a bypass line that allows a portion of the air to bypass the HP evaporator, Case F (Fig. 4.5F). Since the rate of moisture release from the product, i.e., the drying rate, is less than the drying speed of the circulating air, Case F may be used with partial bypass of air in the heat pump. The main part of the air (3′) enters the HP evaporator, where it cools, its moisture condenses and is removed. The two air

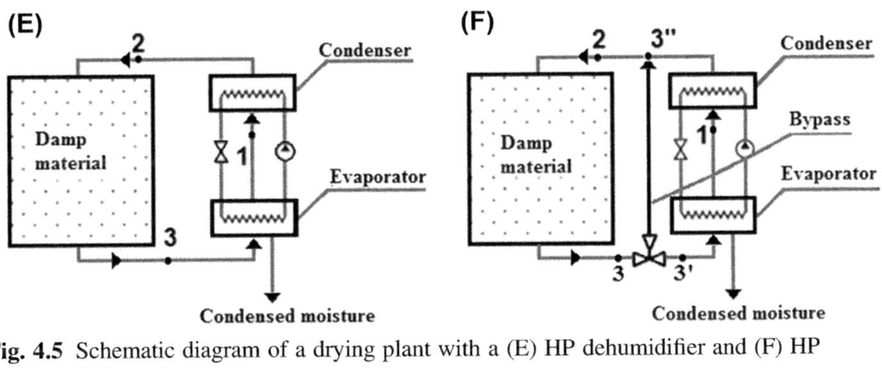

Fig. 4.5 Schematic diagram of a drying plant with a (E) HP dehumidifier and (F) HP dehumidifier with a bypass controlled by the 3-way valve.

streams (3′ and 3″) remix before passing over the condenser. As a result of the bypass, the air temperature entering the condenser is higher since the bypass air is not cooled in the evaporator; this reduces the heating load on the condenser. However, only a portion of the moisture acquired by the drying air is removed in this case which means that the air returning to the dryer carries more moisture than in Case E.

The efficiency of Case F operation is determined by the value of the bypass ratio α, defined as the volume ratio of the bypass air to air passing through the evaporator, i.e.,

$$\alpha = v_{3''}/v_{3'}, \tag{4.1}$$

with reference to Fig. 4.5F.

The heat pump dryer with recirculation of the drying agent provides a significant reduction in the specific energy consumption (SEC) compared to the traditional dryer Case A.

The SEC of a heat pump dryer with recirculation, i.e., Case C, Fig. 4.4C with a flow line from state 3 to state 2, is lower than the SEC of a traditional dryer with recirculation (Case G) by about 50%. From Figs. 4.8 and 4.9, bypassing part of the air around the heat pump evaporator (Case F) increases the energy efficiency of the drying plant by 30–40% for different types of wood (Fig. 4.7).

4.1.7 Case G and Case H: Basic open system using ambient air with partial recirculation (Case G) and partial recirculation and HP bypassing (Case H)

A peculiarity of this scheme (Case G, Fig. 4.6(G)) is that, with the fact that under constant temperature operation of the heat pump, the operation of the dryer is possible only with a certain value of the recirculation coefficient that depends on the maximum air temperature that can be provided by the heat pump at the outlet of the condenser.

Bypassing the heat pump (Case H, Fig. 4.6(H)) is an effective means of increasing the energy efficiency of the heat pump with recirculation and reducing the specific energy consumption for evaporation of moisture from 40% for pine to 60% for oak (Fig. 4.7).

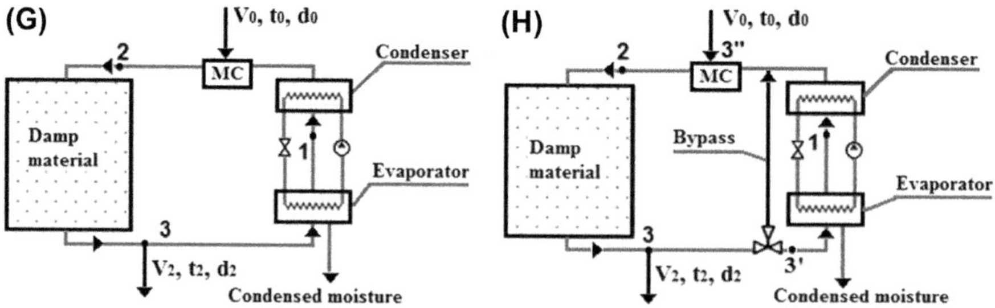

Fig. 4.6 Schematic diagram of a basic open system drying plant with partial recirculation (G), and partial recirculation and bypassing (Case H). V_O, t_O, d_O – entrance of atmospheric air; V_2, t_2, d_2 – removal of exhaust air; MC – mixing chamber.

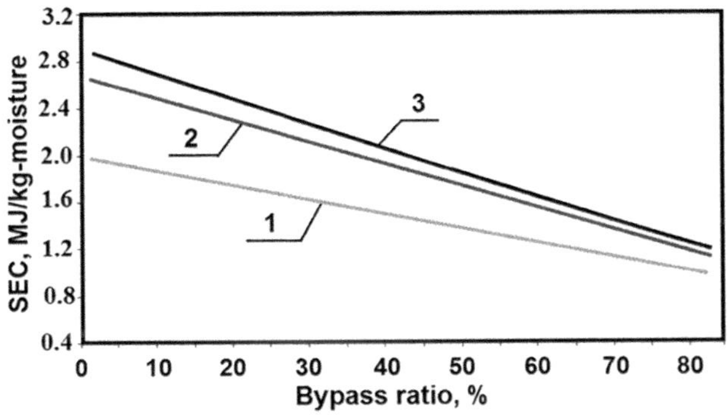

Fig. 4.7 Dependence of the specific energy consumption on the bypass ratio α for wood drying using Case F: 1 = pine at $\phi = 75\%$; 2 = oak with $\phi = 70\%$; 3 = larch at $\phi = 85\%$.

Cases E and F are used when the amount of moisture to be removed is small and the air is completely dried when cooled in the heat pump evaporator.

If the removed amount of moisture is large, then Cases G and H are used in which part of the high-humidity air is removed from the drying chamber and part is dried in the heat pump evaporator, which is then mixed with dry atmospheric air in the mixing chamber MC.

During the drying process, when the rate of moisture removal is reduced (final stage), bypass mode is used to reduce the heat pump capacity (Cases F and H).

For Case G, it is convenient to define another ratio to help understand the processes involved, namely, the *recirculation ratio, K*, which is defined as:

$$K = v_{3'}/v_3, \tag{4.2}$$

where $v_{3'}$ and v_3 are the volumes of air being recirculated and the total drying air used (state 3), respectively. When $K = 0$, the system reverts to a basic open system, i.e., Case C.

As an example, consider the drying process of wood using the system shown as Case F in Fig. 4.5. In the condenser, the air is heated and enters the drying chamber. The amount of air heating in the condenser depends on the type of wood being dried. The temperature differences between the inlet and outlet of the drying chamber Δt_{dc} are as follows for various kinds of wood:

- Softwoods: 2–3 °C
- Birch and beech: 1.5–2.5 °C
- Oak and larch: 1–1.5 °C.

Calculations were carried out for two different values of Δt_{dc} that reveal the optimum bypass that maximizes the COP of the heat pump. The results are displayed in Fig. 4.8 as a function of the bypass factor B, defined as:

$$B = 1/(1 - \alpha) \tag{4.3}$$

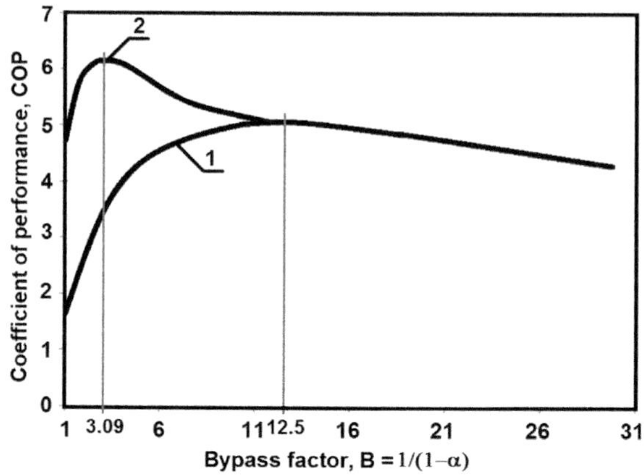

Fig. 4.8 Coefficient of performance of a heat pump drying unit (Case F) as a function of the bypass factor for $t_{mix} = 60°C$ and $\phi_{mix} = 80\%$: curve 1: $\Delta t_{dc} = 1°C$; curve 2: $\Delta t_{dc} = 3°C$.

The optimum point for curve 1 (with 1 °C temperature change) occurs at $B = 12.5$ or a bypass ratio $\alpha = 0.92$ or 92%; for curve 2 (with 3 °C temperature change) it occurs at $B = 3.09$ or $\alpha = 0.68$ or 68%. These values indicate how the system should be operated for the best heat pump energy efficiency.

Figure 4.9 shows the results of calculations done for this case. The results presented are for three kinds of wood. The efficiency of the drying plant depends on the relative volume of air recirculated, and increases with the air recirculation ratio K. The specific energy consumption, SEC (which is inversely related to the efficiency), and the relative humidity, ϕ_{mix}, of the air entering the dryer are plotted as functions of the recirculation ratio, K. It is evident that the amount of energy required per kg of moisture removed decreases linearly as K increases. The worst performance occurs when there is no recirculation, i.e., $K = 0$, where the SEC is the highest.

Typical practical values for K range from 0 to 60–70% for recirculated air with a low humidity $\phi = 20\%$. Therefore, the optimal energy efficiency using recirculation is provided at a value of $K = 60–70\%$.

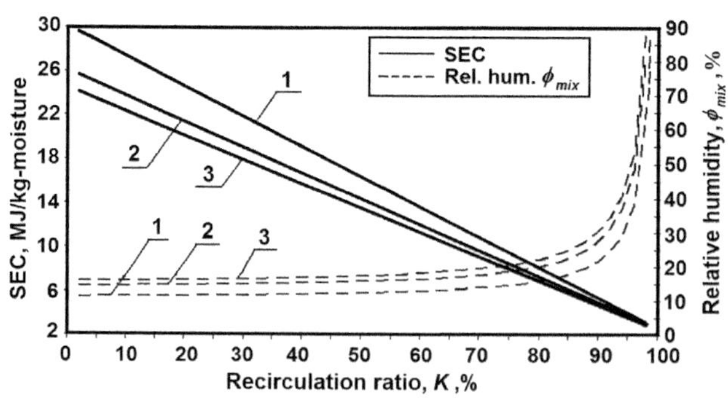

Fig. 4.9 Dependence of specific energy consumption and relative humidity of air at dryer inlet as a function of the recirculation ratio for Case G: 1 = larch; 2 = pine; 3 = oak.

In general, drying systems that incorporate a heat pump are less electricity intensive than systems with simple resistance heaters because HP systems use 2–3 times less electricity to run the compressor than is consumed in the electric heaters.

4.2 Experience with heat pumps for various drying applications

For many decades, heat pumps have been known as an effective method for increasing the potential of low-temperature and low-potential heat sources, which contributed to the growth of their widespread use for the purpose of supplying heat to public facilities. Moreover, their ability to capture the latent heat of vapor condensation of a low-potential working fluid (by means of its cooling and drying) and transfer that energy to an air flow at a high temperature level made them suitable for use in heat engineering processes of drying. Following the general trend of improving product quality, controlling the drying process and reducing energy consumption, many researchers recognized the specific features of the application of heat pumps in drying technology. This led to a rapid growth of both theoretical and applied studies of heat pump drying, as can be seen in Table 4.1 [19].

In addition to reducing the energy consumption of a drying plant, one of the main advantages of using heat pump technology includes creating well-controlled drying conditions of temperature and humidity of the drying agent. Many researchers have demonstrated the ability to create precise drying conditions for drying a wide range of products and improving their quality. A variant of the necessary optimization of components and the system design for increasing an energy efficient heat pump dryer is given in Ref. [20]. Any convection dryer may be equipped with a heat pump of the appropriate design. Heat pumps are most often used in tunnel-type chamber dryers, but it is also possible to use them in dryers of other types, e.g., rotary kilns. However, installations requiring high temperatures of a drying agent (above 80 °C), such as spraying facilities, are not suitable for the application of heat pump technology.

Figure 4.10 shows the general classification of heat pump dryers (HPDs) by mode of operation, the number of stages of drying, the number of stages of the heat pump, and types of auxiliary heat sources. Many installations from the following classification have been proposed over the last two decades.

As mentioned above, the effectiveness of the HPD under specified conditions is determined by the efficiency of the heat pump operation, which is measured by the coefficient of performance, *COP*. In relation to drying processes, researchers from some countries characterize the efficiency of the drying plant by the specific moisture evaporation coefficient, which is commonly known as SMER, defined as follows:

$$\text{SMER} = \text{Amount of moisture evaporated, kg,}$$
$$\text{per the amount of energy consumed, kWh.}$$

SMER depends on the maximum air temperature in the dryer, relative humidity of a drying agent, evaporation and condensation temperatures of the refrigerant and the

Table 4.1 Research references in the field of heat pump drying.

Researchers	Country	Application
Chou et al. (1998)	Singapore	Agrarian and marine products (mushrooms, fruits, seafood)
Chou et al. (1998), Carrington, Barneveld, et al. (1996), Carrington, Bannister, et al. (1996), Sun, et al. (1996)	New Zealand	Plank timber and wood
Prasertsan et al. (1997), Prasertsan and Saen-Saby (1998)	Thailand	Agrarian crops (bananas)
Mason and Blarcom (1993)	Australia	Nuts
Meyer and Greyvenstein (1992)	South Africa	Grain
Rossi et al. (1992)	Brazil	Vegetables (onions)
Nassikas et al. (1992)	Greece	Paper
Strommen and Kramer (1994)	Norway	Marine products (fish)
Hawlader et al. (2006, 2006)	Singapore	Ginger, guava, papaya
Alves-Filho and Ross (2006)	Norway and Ireland	General studies
Sieniutycz (2006)	Poland	General studies
Sakar et al. (2006, 2006)	India	General studies
Claussen et al. (2007)	Norway	Potatoes
Sunthonvit et al. (2007)	Australia	Nectarines
Van Der Pal et al. (2007)	New Zealand	Wood
Zhang et al. (2007)	China	Wood

efficiency of the cooling system. Figure 4.11 shows some of the SMER values for the typical range of temperatures in a HPD as a function of initial temperature of the drying agent [21,22].

The characteristic feature, according to Fig. 4.11, is that the amount of moisture removed per unit of energy consumption is sharply reduced with a decrease in the temperature of a drying agent, mainly due to the decrease in the COP of the heat pump. The theoretical maximum of SMER for convective heat drying is about 1.55 kg of moisture/kWh (based on the latent heat of moisture evaporation at 100 °C). Typical values of SMER achieved by heat pumps are about 3 kg of moisture/kWh [23]. While this is about twice as high as the ideal SMER, it is much better than traditional convection type dryers that have SMER values in the range of 0.5−1 kg of moisture/kWh. The overall values of the coefficient of moisture removal efficiency for the HPD are shown in Fig. 4.12 [24]. The heat pump and the dryer can be combined in different

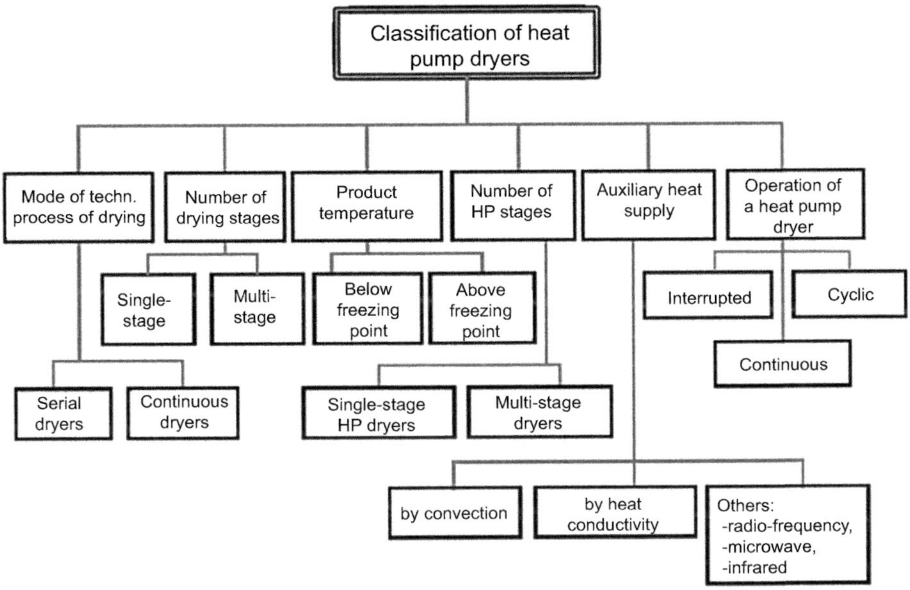

Fig. 4.10 Classification of heat pump dryers.

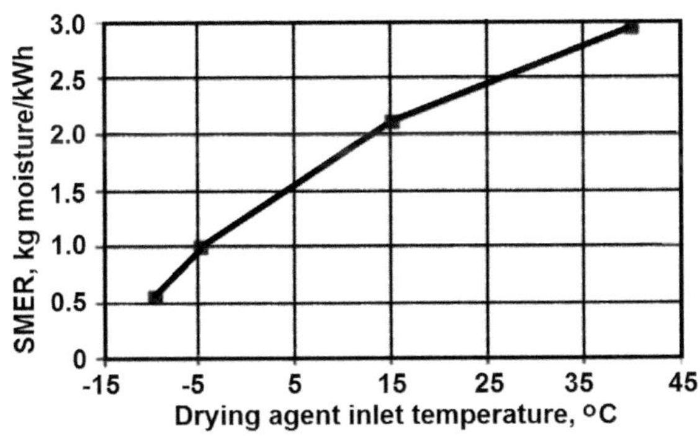

Fig. 4.11 SMER versus inlet temperature of the drying agent.

configurations, depending on the relative location of air and energy flows. An estimate of the efficiency of heat pump dryers was carried out for some arrangements in Ref. [24].

Two-stage heat pump systems are used in drying technology to achieve higher energy efficiency and improved temperature and humidity control of the drying agent (Fig. 4.13).

Figure 4.13 gives an example of a 2-stage HPD in a parallel configuration, where the evaporator of the first stage operates at a higher pressure relative to the evaporator of the second stage [25]. In terms of energy efficiency, a 2-stage heat pump can regenerate more latent heat of humid air, due to a higher *COP*. Comparative data of the *COP*

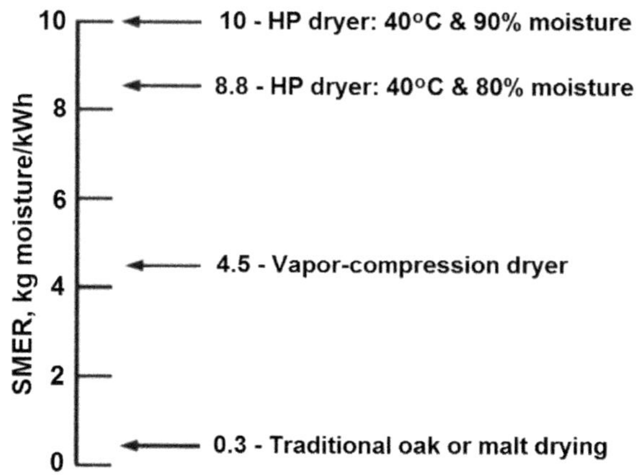

Fig. 4.12 Overall efficiency of various drying systems. Note: For the two HPDs, the 80% and 90% refer to the moisture in the outlet air.

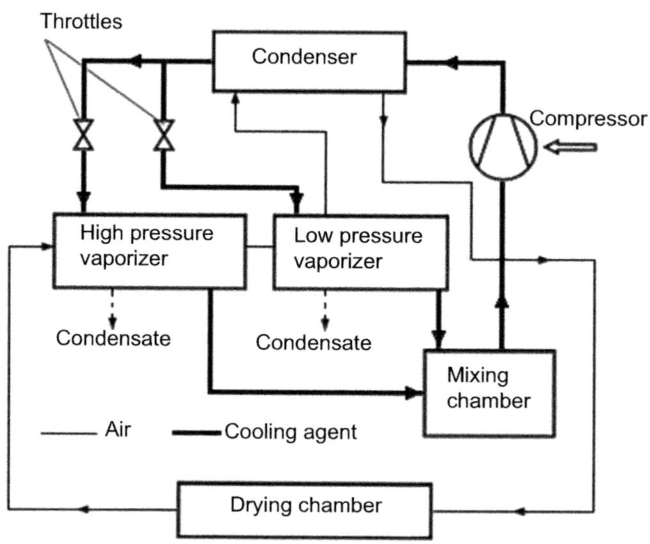

Fig. 4.13 Schematic diagram of a double-stage heat pump dryer.

for different heat pump drier arrangements are shown in Fig. 4.14. As can be seen, the efficiency of heat pump drying systems can be increased by combining it with other sources of heat, such as natural gas.

Heat pumps have significantly reduced energy consumption for drying textiles. Data presented in Table 4.2 [26] compare the energy consumption during drying with a heat pump and the original system where the heating of air is carried out by burning oil.

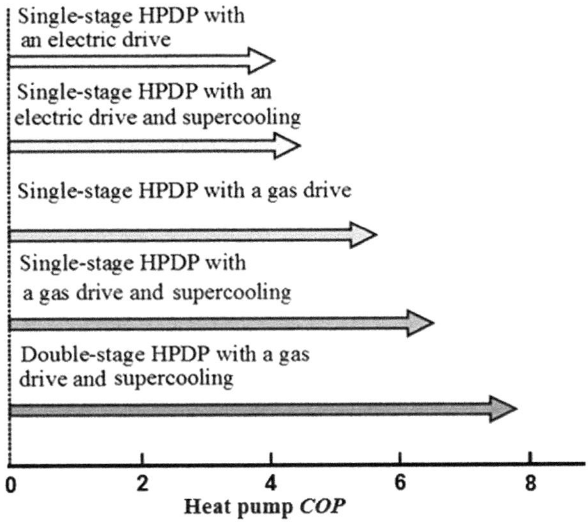

Fig. 4.14 *COP* for different types of HP system: barley drying with air at 66°C and $\varphi = 5\%$.

Table 4.2 Energy consumption when drying textiles.

Driers with oil combustion	
Oil consumption, MJ/h	599
Power consumption, kW	2.29
Drying time, h	5.75
Amount of moisture removed, kg	154
Specific energy consumption (SEC), MJ/kg	22.4
SEC, accounting for the efficiency of energy generation, MJ/kg	31.7
Heat pump dryer with electric drive	
Nominal electric power, kW	15
Average power consumed, kW	18.4
Drying time, h	5.75
Amount of moisture removed, kg	154
SEC, MJ/kg	2.02
SEC, accounting for the efficiency of energy generation, MJ/kg	7.7
Comparison	
Reduction in SEC, %	91
Reduction in SEC, accounting for the efficiency of energy generation,%	76
Reduced operating costs,%	70
Payback period, years	2.4

Table 4.3 Energy consumption when drying gypsum molds.

Driers with gas combustion	
SEC in the dryer, MJ/kg	4.6
SEC, accounting for the efficiency of energy generation, MJ/kg	7.05
Heat pump dryer	
Nominal electric power consumption of heat pump, kW	3
Mass of single loading of a moist product, kg	800
Amount of moisture removed, kg	232
Drying time, h	18
Total power consumption (electrical + thermal), kW	108
SEC, MJ/kg	1.7
SEC, accounting for the efficiency of energy generation, MJ/kg	6.2
Comparison	
Reduction in SEC, %	63
Reduction in SEC, accounting for the efficiency of energy generation,%	12

In the manufacture of clay products, one of the power-consuming operations is the drying of plaster molds [27]. Gas was burned for air heating to 50 °C in the original system. In order to reduce the energy consumption for drying, a heat pump was used that resulted in a 12% reduction in energy consumption, taking into account the efficiency of electricity generation by a gas turbine. The data to compare the initial system and the one with the heat pump are given in Table 4.3.

The use of heat pump drying technology early in its development took place only in special installations for drying ceramics and other materials that require sensitive and careful modes of heat-and-moisture treatment due to economic and technological limitations [28]. However, interest in heat pump drying of food products and biomaterials has recently increased, where it is necessary in low-temperature modes of the process, and in precise control of drying conditions for stabilizing and improving the product quality [29–32]. In most studies, the main conclusion is that the use of heat pump technology for drying, besides significantly reducing energy consumption, also significantly improves in the quality of the processed product. This is especially important in the case of heat-and-moisture treatment of food products that requires fine and precise control of technical parameters of the drying process.

The main advantage of using heat pump technology when drying thermosensitive products is the possibility of providing low-temperature conditions for the process in a temperature range from 20 °C to 60 °C [33]. The studies in Refs. [34,35] show that HPD can operate using cyclic temperature patterns, e.g., a frequency of operation (i.e., loading – drying – unloading – loading) variation, to improve the quality of various crops dried in 2-stage heat pump driers. According to the authors' data, with correctly selected technical parameters for the process and a given frequency of operation variation of temperature, the degradation of a very important quality indicator for food products, namely, color, may be reduced by 87%.

The ability to control the drying conditions quickly and efficiently is another advantage of heat pump technology for drying food products. In countries where humidity is high, a large percentage of defective products is observed during the rainy seasons due to high humidity of the air used as the drying agent. In such cases, the HPD can reduce production costs by means of controlling the moisture content of the product by regulating the removal of moisture and the latent heat of vapor formation from the air in the evaporator.

Figure 4.15 shows a tunnel HPD for drying thermosensitive food products (salted cod). According to this source, the drying temperature can be adjusted in the range of 0–70 °C, but for special purposes, the lower temperature limit can be reduced to −10 °C. The inlet relative humidity of the drying agent ranges between 30 and 40% [36].

A separate issue in the process of drying food products is the problem of material deodorization or degasification. At the same time, heat pump technology makes it much easier to solve this problem since in a closed drying system the moisture received from the material is withdrawn from the cycle in the liquid phase. Furthermore, the restoration of aromatic components or the removal of absorbed harmful substances is done much easier from the liquid phase than from a large amount of exhaust air.

Closed HPD make it possible to use N_2, CO_2 or other inert gases as drying agents to reduce the likelihood of fire or explosions, as well as to prevent the destruction of oxygen-sensitive products. But in this case it is necessary to provide for a special structure for feeding the product material to prevent the leakage of the drying agent into the environment. The efficiency of heat and mass transfer processes in dispersed systems

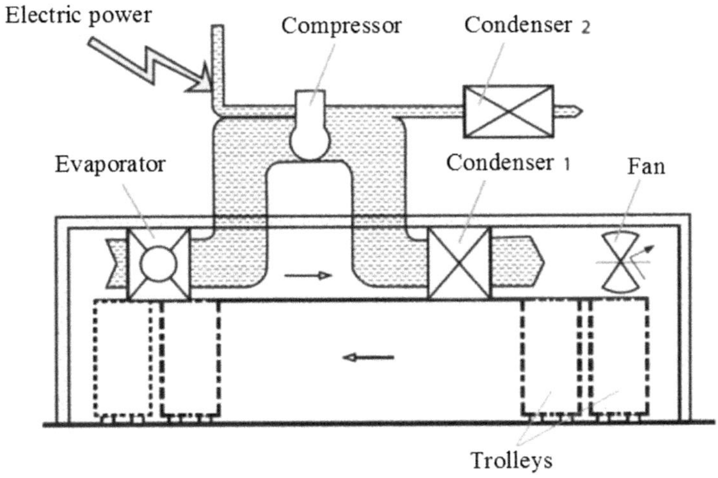

Fig. 4.15 Tunnel HPD for drying salted cod. Note: The condenser 2 dries the bypass air.

at low drying temperatures of thermosensitive food products makes it possible to use heat pump technology in combination with pneumatic, conveyor, or vibration drying plants as well as installations with a fluidized bed. Figure 4.12 shows a schematic diagram of a heat pump drier with a fluidized bed developed at the Norwegian Institute of Technology that has been completely tested for a number of food products [37].

The operating temperature, according to Ref. [37], is set at the desired level by adjusting the condenser power by using frequency control of the engine speed; this also maintains the required relative humidity of the product. This method allows the product to be dried at temperatures from 20 °C to 60 °C and relative air humidity from 20% to 90% (Fig. 4.16).

Studies conducted at the Norwegian Institute of Technology on various types of heat-sensitive materials and products of biological origin, as well as pharmaceutical preparations, medical and biotechnological products, bacteria, fruit, and vegetables have proven that heat pump drying in a fluidized bed yields a product of higher quality, although at a higher price. Therefore, this technology can be recommended for premium products. Figure 4.17 shows the scheme of a multi-stage HPD with a fluidized bed consisting of two driers placed in series and two heat pumps connected in parallel, which provides air conditioning in drying chambers. Each heat pump can operate independently under completely different conditions with different refrigerants [38].

The moist product is fed into the first dryer, where it stays until attaining certain humidity. The semi-dry product is then fed to the second dryer, where it can be treated at a higher temperature before the drying process is completed. Research shows that consistent drying conditions allow the moisture content to be reduced at low temperatures to preserve product quality, followed by finishing drying at higher temperatures in order to increase the overall efficiency of the heat pump drying process.

References [39–41] analyze the use of heat pump technology in the food industry in Australia. Based on their own experience, the authors rated more than 30 units of heat pump driers used at Australian food processing enterprises. The conclusions

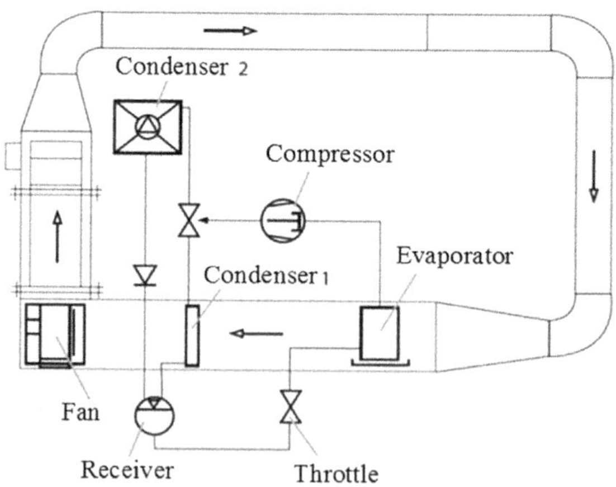

Fig. 4.16 HPD with a fluidized bed for drying food products.

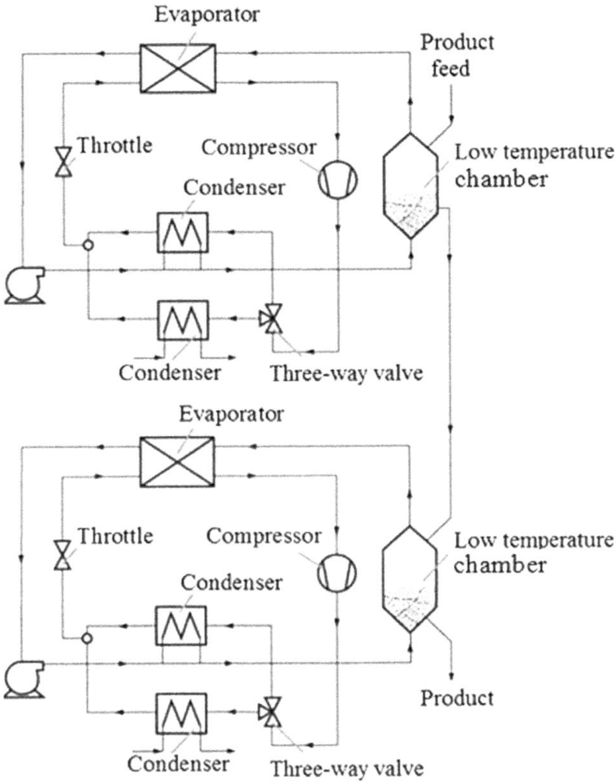

Fig. 4.17 Multi-stage HPD with a fluidized bed.

reached were that the high efficiency these technologies led to an increase in producers' confidence in heat pump drying technologies and to a significant increase in the number of installed HPD with a specific moisture removal coefficient of 500–1000 kg per day.

In the country of Georgia, a heat pump plant was put into operation at the Sumtretsk tea-drying plant [42]; a schematic of the installation is shown in Fig. 4.18.

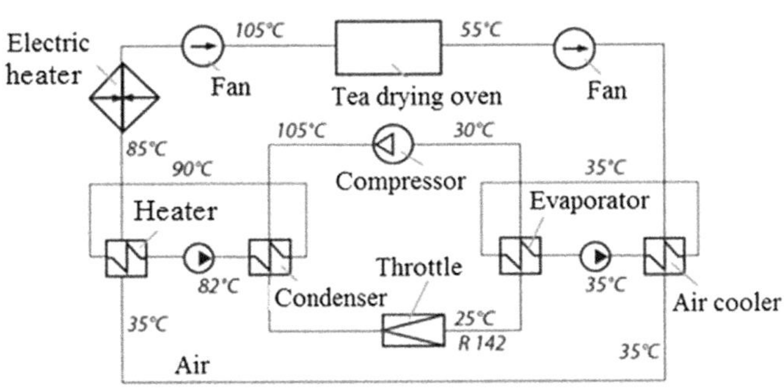

Fig. 4.18 HPD process flow diagram for heat supply in tea leaf drying.

The exhaust air of the dryer is fed into the dehumidifier where it is cooled with water to below the dew point. Water from the dehumidifier enters the evaporator of the heat pump. The dried air is fed to a heater where it is heated by water coming from the HP condenser. Additional heating is carried out in an electric heater. According to the authors, the implementation of heat pump plants at all tea factories in Georgia will save 100,000 tons of high-quality fuel oil annually.

Research carried out by scientists from many countries studying combinations of HPD with additional sources of heat, namely, microwave energy, solar energy, energy of infrared rays, radio frequency pulses, etc. deserve particular attention.

In References [34,35] studies of HPD in combination with microwave drying mechanisms were performed. The main conclusion is that the use of microwave energy during the heat pump drying of food products and some of building materials can significantly increase the productivity of the dryer and the quality of the products. However, the cost of energy to operate microwave generators lowers the cost-effectiveness of HPD to the level of traditional convection drying plants.

HPD systems using solar collectors as an auxiliary source of thermal energy are also used. One particular such combination is illustrated in Fig. 4.19. References [43,44] present the analysis of the performance of solar HPD operation, the main conclusion of which is that, owing to the intermittent nature of sunlight, systems having a tank-accumulator are the most effective option for using solar energy as an additional heat source (Fig. 4.20). The operational effectiveness of such systems depends significantly on the duration of solar daytime activity.

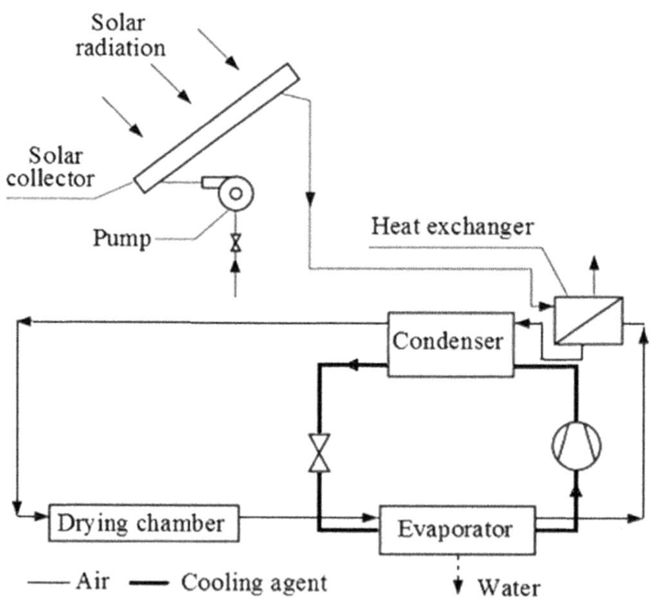

Fig. 4.19 HPD scheme with solar collector.

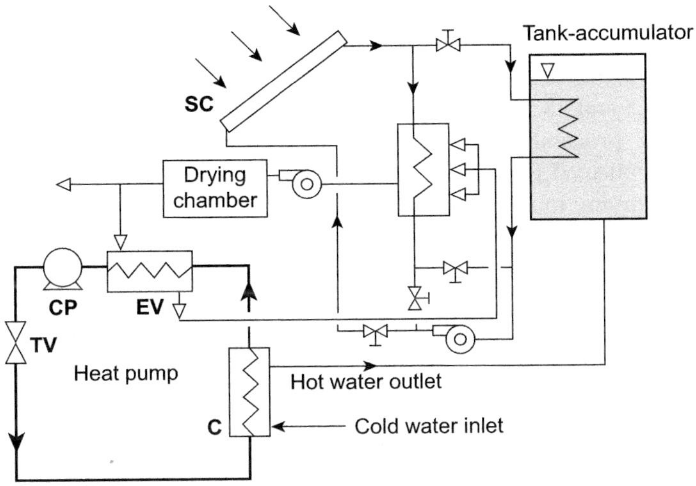

Fig. 4.20 HPD scheme with a solar collector and a tank-accumulator.

The use of solar collectors can improve the performance of an installation, and the economic effect depends on the type of material to be dried and its cost. Also, in numerous references one can find a large number of different schemes for combining a heat pump with other heat energy sources in order to create the most efficient drying plant, adapted to specific conditions to dry various types of products.

It should be noted that in all the cases mentioned above, the thermal efficiency of the heat pump system, which is measured by the heat pump *COP*, is greater than 1. However, the cost of electrical energy required to drive the HP can be higher than the cost of recovered heat. Thus, the profitability of the whole system depends essentially on the relative cost of electric and thermal energy. However, besides taking into account the payback period of the HPD, it is also necessary to consider both the higher quality of the product being manufactured and, in some cases, the substantial increase in the productivity of the installation.

4.3 Grain drying with heat pumps

Grain production is one of the main components of the agrarian sector of an economy, and is traditionally the main agricultural product of many countries, in particular, Canada, China, Mexico, Russia, South Africa, Ukraine and United States. The US, China, Russia, the EU and Ukraine were the largest grain producers according to the data of 2011. The area devoted to raising grain crops in different countries ranges from 20 to 60% of planted lands, and the gross grain harvest reaches many millions of tons. While about 15–20% of grain crops are exported, the main part, i.e., 80–85% remains for the domestic needs of the producing countries, which means that a large number of processing and storage complexes for grains and cereals must be available.

Ukraine is among the top ten producers of wheat and corn, where 60% of the steppe region of agricultural land is taken up with the cultivation of grain crops, one of the largest amounts in the world. For the year 2018, the wheat and corn harvest in Ukraine amounted to 24.5 and 35.5 million tons, respectively. The harvested crops are subject to technological processing, storage for local use, and exportation. The moisture content of the collected grain is 10–30% higher than the equilibrium value, which necessitates its drying to improve the quality and shelf life. Readers unfamiliar with crop drying processes, may refer to Ref. [45] which presents the fundamentals, but without any discussion of the use of heat pumps, which is our main purpose in this section.

Furthermore, 20–90% of the total cost of agricultural grain output resides in the cost of processing the harvest, which mainly involves the most power-consuming process of heat-and-moisture treatment; up to 80% of the harvest is subject to this form of processing. Also, the constant growth of prices for traditional energy sources together with the increase in grain production volumes leads manufacturers (1) to increase the energy efficiency of technological processes of drying, (2) to use secondary energy resources, and (3) to develop optimal schemes of heat-and-moisture treatment using alternative energy sources.

Performing the drying process in a timely and proper manner not only raises the quality of the grain during storage, but also improves its food characteristics. By maintaining optimal drying methods, further ripeness of grain is achieved, the degree of maturity is aligned, uniform distribution of moisture is established, and the appearance and other technical characteristics are improved. The process of heat-and-moisture treatment leads to the destruction of pests and microorganisms, and in some cases, improves the technical properties of defective grain. Grain drying technology for foodstuffs determines the grade of flour being obtained, the quality of the cereals, and the "shelf life" for their storage. Seed and feed grain drying has its own characteristics. Processes for seed grain are done at low temperature and require special technological parameters of the drying processes, which depend on their intended purpose.

The use of grain as a ready-made agricultural product necessitates its heat-and-moisture treatment to a certain level of humidity. When wheat grain is used in the domestic market it is enough to dry it to 16–17% moisture. However, when grains are exported or used as seed material for long-term storage, the final humidity should be 11–14%, assuming an initial humidity of 19–35%, depending on the grade and subsequent special purpose. The main methods of drying grain are by convection, conduction, electricity, sorption, and infrared rays. The selection of the most effective method depends on the purpose of grain, its volume, and technical parameters of the heat-and-moisture treatment process.

The drying processes make up 90% of the energy consumption for converting the harvested grain into a finished product. The most cost effective and, consequently, the main method of drying is by convection. Convective drying has its specific features and is characterized as either high temperature (drying agent $\geq$ 80 °C) or low temperature (25–80 °C). Since the 1980s, high-temperature heat-and-moisture processing has been mainly used for feed grain and low-grade grain. In most cases, low-temperature drying methods are used for food types of grain. This is because grain loses some of its biological and food characteristics as a result of being treated with a high temperature drying agent.

Current grain drying facilities combine drying and cooling technology. In this case, the heat acquired by the air as it cools the grain is captured and used for further grain drying. One important feature of the technical process of drying grain crops is their gradual cooling after heat-and-moisture treatment to retain the characteristics imparted by the process. The main purpose is to prevent cracking the grain surface from too rapid a temperature drop during abrupt cooling. Up to 80% of all dried grain is subject to gradual cooling since only about 20% of all grain is feed grain where surface cracking is not important.

In traditional heat supply systems for grain drying, non-renewable fuels are used to heat the drying agent. In this case, drying installations can be both open type and closed recirculation type; see Sect. 4.1. Recycling leads to higher efficiency of a drying plant [17], as well as to obtaining the requisite air relative humidity, which is particularly important for drying seed grain, and contributes to the reduction of mechanical stresses in the material.

Therefore, recirculation is used for drying grain and other materials for which the production quality is largely determined by the drying regime. Figure 4.21 shows schematic diagrams of the operation of traditional periodic open-type grain dryers (Fig. 4.21A) and systems with partial recirculation of a waste drying agent (Fig. 4.21B). The energy costs of traditional drying schemes based on natural gas heating are on average about 46.4 MJ per tonne of grain with a 1% decrease in moisture content, and the consumption of 0.7–19 m^3 of natural gas, depending on the cereals, indicating the low energy efficiency of traditional systems.

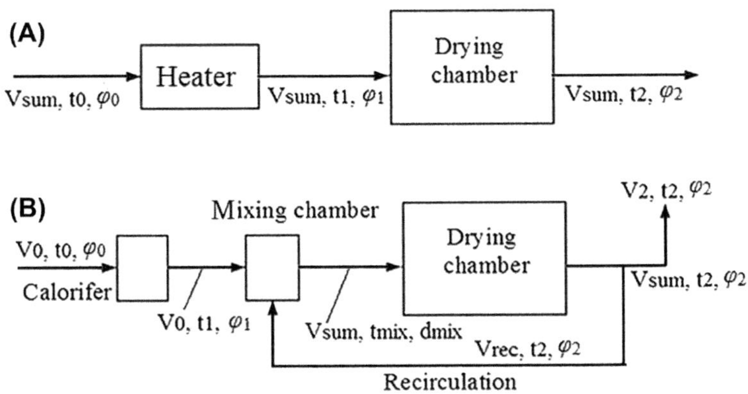

Fig. 4.21 Schematics of traditional batch-processing grain drying plants: (A) a traditional dryer with a once-through use of a drying agent; (B) a dryer with partial recirculation of a drying agent: V_{sum} – the total required air flow; V_0 – air flow at the inlet to the heater; V_{rec} – recirculated air flow rate; V_2 – exhaust air flow rate; t_0, φ_0 – temperature and relative humidity at the inlet to the heater; t_1, φ_1 – temperature and relative humidity at the entrance to the drying chamber; t_2, φ_2 – temperature and relative humidity at the outlet from the drying chamber; t_{mix}, φ_{mix} – temperature and relative humidity at the outlet of the mixing chamber.

Problems of coping with increasing grain production, rising prices, and energy-intensive and expensive methods of processing the grains have led to a search for more efficient methods of drying, such as heat pumps for heat-and-moisture treatment. Starting from the 1950s, leading manufacturers of dryers launched large-scale research campaigns into the possibility of more effective use of heat pump units in these processes. In 1952 the first patents were obtained for technical systems using condensation-type heat pumps without recirculation (Fig. 4.22A) and with partial recirculation of a drying agent (Fig. 4.22B). It will be seen that the heat pump in Fig. 4.22 essentially replaces the heater in Fig. 4.21.

The heat pump unit included in the convection drying scheme performs two functions: (1) dehumidification for a wet drying agent by cooling the air below the dew point in the evaporator, and (2) heating of the drying agent. The drying agent that passes through the heat pump evaporator serves as a low-temperature source of heat. Reference [46] shows that using a heat pump as a heat recovery unit for a drying agent leads to a significant increase in the thermodynamic and energy efficiency of the drying system. Fig. 4.23 shows the thermodynamic efficiency of a drying plant operation as a

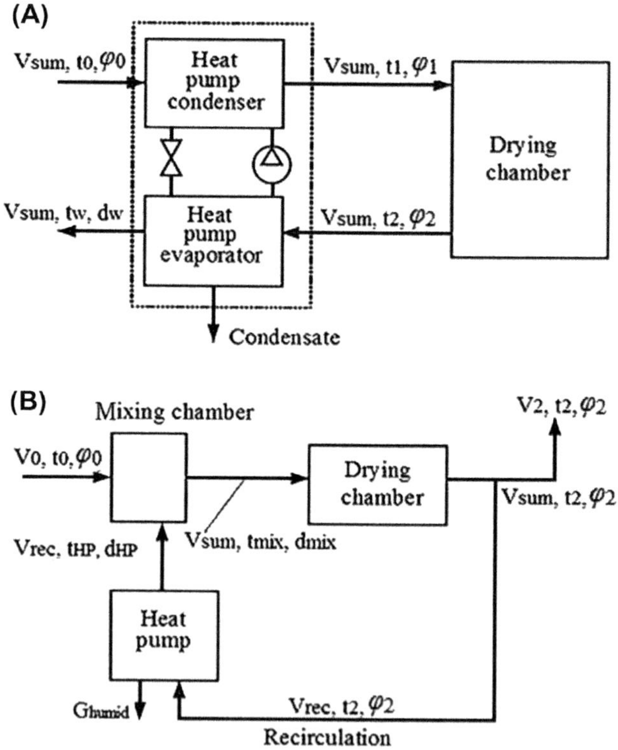

Fig. 4.22 Schematic diagrams of heat pump driers for grain: (A) HPD with once-through use of a drying agent; (B) HPD with substantial recirculation: t_w, φ_w — air parameters after the evaporator; t_{HP}, φ_{HP} — air parameters after the heat pump; G_{humid} — moisture removed.

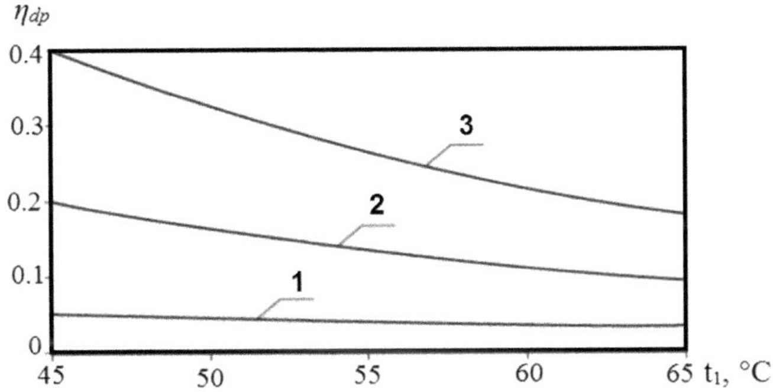

Fig. 4.23 Efficiency of energy usage in the dryer as a function of the temperature of the drying agent: 1 – traditional dryer; 2 – actual HPD; 3 – ideal Carnot HPD.

function of the drying agent temperature for a traditional drying plant with a once-through drying agent (Fig. 4.21A) and a similar heat pump system (Fig. 4.22A). Also included for comparison is the ideal Carnot heat pump. The authors of [46] defined the operational efficiency, η_{dp}, as the thermodynamic efficiency of using external energy for moisture evaporation; namely, it is the ratio of the heat rate directly used for drying in the drying chamber to either: (1) the energy rate supplied in the heater for a traditional drying plant or (2) the power to drive the heat pump compressor for a HPD.

As one can see from Fig. 4.23, the efficiency of the ideal Carnot HPD and a real HPD are five and three times higher, respectively, than a traditional dryer for a given ambient air temperature. However, in this case, the heat pump performs only the role of a heat supplier; if a heat recuperator was included, the efficiency would be higher. So adding a heat pump or one with a recuperator increases the capital cost but lowers the operating cost, and therefore a feasibility study is needed to define the optimal arrangement.

A significant increase in the efficiency of the dryer plant operation can be achieved by including a heat pump in the recirculation line of the drying agent (Fig. 4.22B). In the case of a traditional dryer plant, a significant amount of the spent drying agent is discharged along with the evaporated moisture withdrawn from the cycle, which in turn requires adding fresh incoming air and more energy to heat it to the requisite temperature. When the HPD is operated, the spent dryer agent, which is also a low-potential heat source for the heat pump, practically does not leave the dryer plant cycle, and evaporated moisture from the material is removed in the heat pump evaporator in the form of condensate. Thus, the necessity to add fresh incoming air and energy consumption for its heating almost disappears. Insignificant discharge of the dryer agent (~3–14 %) is due to the need for the energy extraction from the heat pump drive, which is superfluous in the drying cycle, according to the energy balance of the system. Under partial recirculation, part of the air needs to be replaced by fresh air at reduced humidity. Therefore, the air must be partially removed.

The results of a numerical analysis of the operational efficiency, η_{dp}, of a traditional and a heat pump drying plant with recirculation of the exhaust dryer agent are

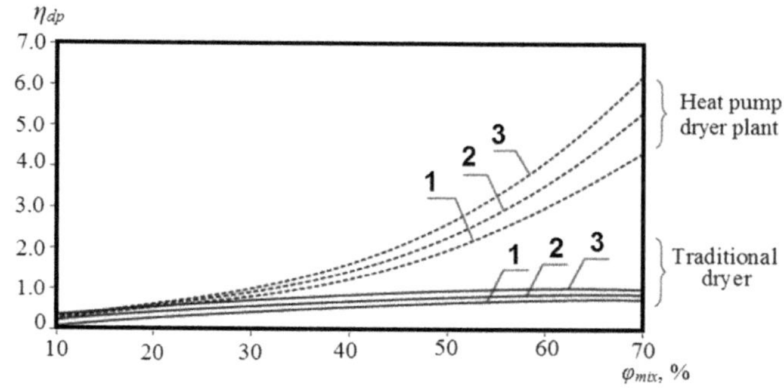

Fig. 4.24 Operational efficiency of a heat pump and traditional dryer as a function of the outlet relative humidity of a drying agent, φ_{mix}: *1* – t_{mix} = *65°C; 2* – t_{mix} = *55°C; 3* – t_{mix} = *45°C.*

presented in Figs. 4.24 and 4.25 [46]. The graphs pertain to the theoretical convective low-temperature process of drying wheat grain in the first drying period.

As can be seen from Fig. 4.24, an increase in the regulated relative humidity of the drying agent leads to a significant increase in the operational efficiency of the recirculating heat pump dryer plant. In the range of variation $\varphi_{mix} = 30$–70%, η_{dp} increases by 1.5–5 times for a HPD compared to a traditional dryer plant.

It should be noted that at $\varphi_{mix} < \sim 30\%$, the η_{dp} of the HPD does not differ much from that of the traditional system. That is due to the low moisture content in the exhaust dryer agent which is a source of low-potential heat for the heat pump dryer plant.

The efficiency of the HPD increases somewhat with a decrease in the dryer agent temperature (Fig. 4.25); this comes about because of the shrinking of the temperature limits of the cycle and, consequently, an increase in the COP. Therefore, it is in the low-temperature processes of grain drying that the use of the HPD can be recommended [47]. In general, the low-temperature drying process using heat-pumps as the main source of heat can include auxiliary heat sources for use during start-up to warm up

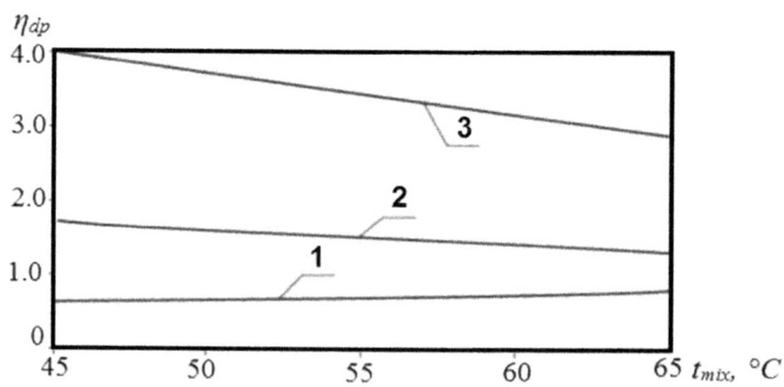

Fig. 4.25 Operational efficiency of a heat pump and traditional recirculating dryers as a function of the temperature of a drying agent t_{mix} at $\varphi_{mix} = 40\%$: 1 – traditional dryer; 2 – real cycle heat pump; 3 ideal Carnot heat pump.

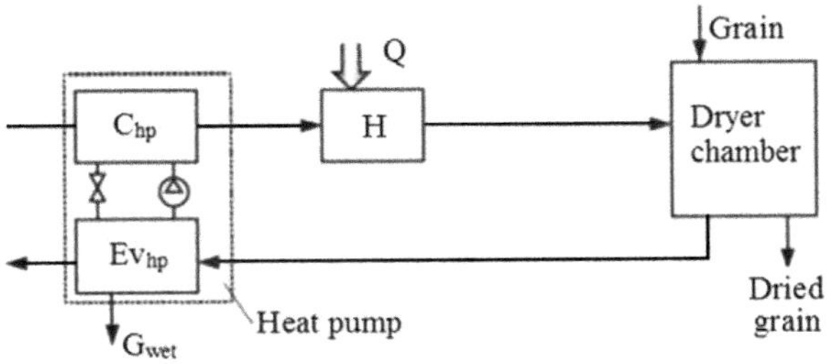

Fig. 4.26 HPD with an auxiliary heater. C_{hp}, heat pump condenser; E_{vhp}, heat pump evaporator; H, auxiliary heater.

empty hoppers or at times of heavy load on the compressor of the heat pump. One of the options for implementing such a technology is presented in Fig. 4.26.

Figure 4.27 presents a schematic diagram with recirculation of the exhaust drying agent into a mixing chamber after the primary air flow has passed through the auxiliary heating device, H (option (A) in the figure). This allows for adjusting the humidity and cooling the heated drying agent to the required parameters. Auxiliary heat source dryers can also be made with recirculation before the auxiliary heating device (option (B) in the figure).

Using auxiliary heat sources in drying units in addition to heat pumps, using the heat of an exhaust drying agent, and the latent heat of condensation of evaporated moisture, all lead to stabilization of the dryer operation, improvement in controlling dryer operation, equalization of load on the heat pump, and, consequently, to reducing external energy for moisture removal from the grain.

Figure 4.28 illustrates a grain HPD with "warm" and "cold" drying chambers, and a regenerative heat exchanger installed between the HP condenser and the electric heater [67]. According to the authors, this system significantly reduces energy costs by using the heat of the exhaust drying agent. Exhaust air is directed to a regenerative heat exchanger, where its temperature decreases, thereby reducing the thermal load on the cooler-evaporator in a regeneration cycle and consequently decreasing the energy consumption. Using the exhaust air heat provides an increase in the temperature of fresh air entering the electric heater, while at the same time reducing the energy costs

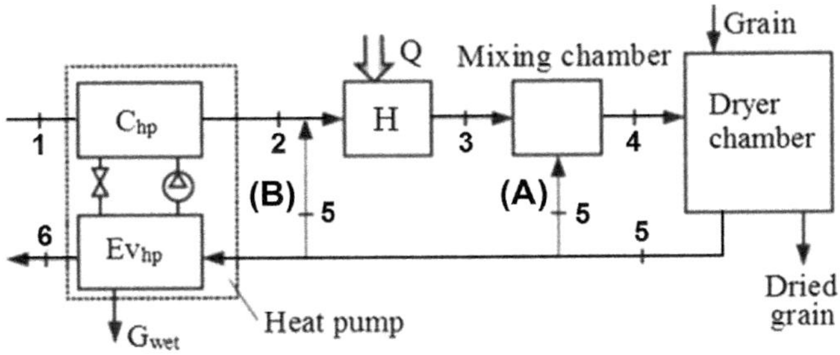

Fig. 4.27 Heat pump drying plant with an auxiliary heater, H, and two different modes of recirculation, options (A) and (B).

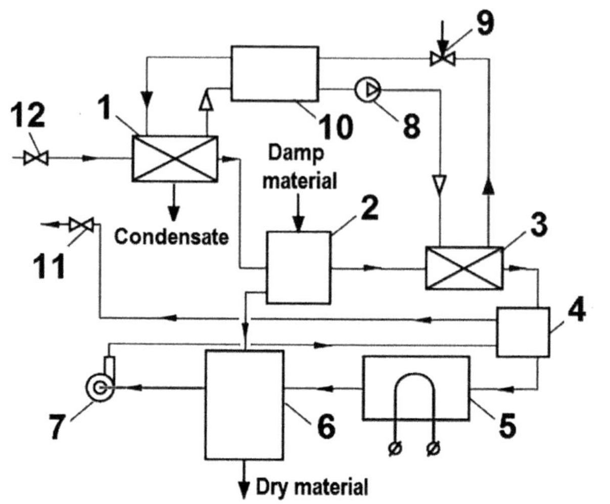

Fig. 4.28 Scheme of a grain HPD with heat regeneration: 1 – HP evaporator; 2 –"cold" drying chamber; 3 – condenser; 4 – regenerative heat exchanger; 5 – electric heater of a DA; 6 – "warm" drying chamber; 7 – fan; 8 – HP compressor; 9 – throttle; 10 – regenerative heat exchanger; 11, 12 –vent and supply air valves.

for heating the drying agent (DA) to the specified temperature entering the "warm" drying chamber.

The product to be dried is successively supplied by means of a conveyor into the drying chambers 2 and 6. In the circulation circuit a DA is fed by means of the fan 7 through the valve 12. The drying agent enters the cooler-evaporator 1, where it is dried and cooled by condensing the moisture on the cold surface of the cooler-evaporator. Then the DA enters the "cold" drying chamber 2, where the dry cold DA assimilates a portion of moisture contained in the dried product. From the "cold" drying chamber, the DA is sent to the heater-condenser 3, where it is heated by the heat of condensation of hot refrigerant vapors circulating in the HP circuit. Passing the regenerative heat exchanger 4 and the heater 5 in sequence, the DA is heated to the maximum temperature permitted by the technology and enters the "warm" drying chamber 6. The heated drying agent absorbs the residual moisture from the product and enters the regenerative heat exchanger 4, heating the DA after the heater-condenser 3. Finally, the DA is discharged into the atmosphere through the valve 11.

Raising the temperature of the DA entering heater 5 from the exhaust air heat leads to a 25–30% reduction in the energy consumption to achieve the specified temperature entering the "warm" drying chamber [67]. Combining the HPD with the cooling chambers that provide final drying and improving grain quality allows a significant amount of cooling air heat to be utilized (Fig. 4.29).

Thus, as shown in this section, heat pump technology for heat supply took a step beyond the needs of public heat consumers long ago and now finds applications in various heat engineering processes, especially in industrial drying processes, particularly grain drying. However, one of the most promising directions for the use of heat

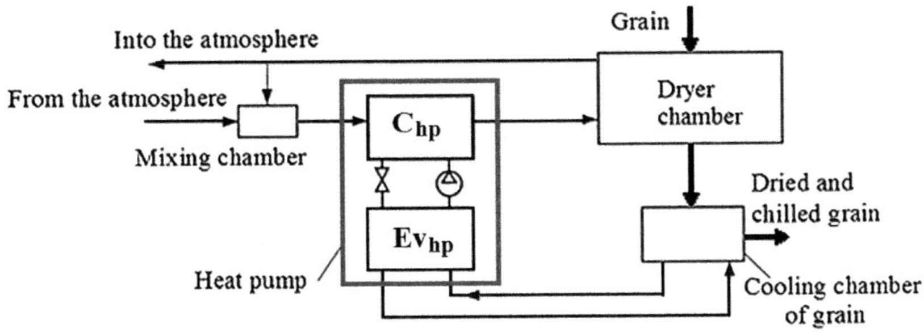

Fig. 4.29 Diagram of HPD with a grain cooling module. C_{hp}, HP condenser; E_{vhp}, HP evaporator.

pumps at present is their use for heat supply in wood and timber drying processes that will be discussed below.

4.4 Wood drying with heat pumps

The worldwide total round wood timber production reached more than $3{,}797 \times 10^6\ m^3$ in 2017 (Table 4.4), of which about $485.1 \times 10^6\ m^3$ was sawn wood from both conifers and non-conifers (Table 4.5). Round wood production has been growing at roughly $43 \times 10^6\ m^3$ per year for the last decade. About 58% of the sawn wood planking is made in the top five countries, China, U.S., Canada, Russia and Germany [44].

Table 4.4 Round wood production in the top ten countries in 2017 and the worldwide trend from 2008 to 2017 [44].

Country	$10^6\ m^3$	% Total	Year	$10^6\ m^3$
	Year 2017		World totals	
USA	419.6	11.05	2008	3463.7
India	354.0	9.32	2009	3359.5
China	327.7	8.63	2010	3544.4
Brazil	256.8	6.76	2011	3609.7
Russia	212.4	5.59	2012	3638.4
Canada	155.1	4.08	2013	3686.5
Indonesia	118.3	3.12	2014	3720.6
Ethiopia	113.6	2.99	2015	3711.8
DRC	89.2	2.35	2016	3776.1
Nigeria	75.9	2.00	2017	3797.1

Table 4.5 Sawn wood production (conifer and non-conifer) in the top ten countries in 2017 [44].

Country	10^6 m^3	% Total
Year 2017		
China	86.03	17.7
USA	80.37	16.6
Canada	49.50	10.2
Russia	40.58	8.4
Germany	23.17	4.8
Sweden	18.41	3.8
Brazil	14.60	3.0
India	6.89	1.4
Viet Nam	6.00	1.2
Indonesia	4.17	0.9
World total	485.13	100

Sawn wood is the most common form of construction lumber and is aimed at satisfying the needs of three categories of usage: furniture manufacture, construction, and industry. In order to achieve high throughput and rapid return of capital, manufacturers in the last two categories dry the raw product at high temperatures, which allows the wood to dry in less than 2–4 days, reducing the moisture content from 100–140% moisture to 10–20%. Rapid treatment of this type is feasible since the final quality and appearance of the wood is not critical in construction and industry. However, this mode of drying is unacceptable for the furniture manufacturing industry owing to the occurrence of various types of material defects such as cracks, guttering, loss of color, etc. Planking for this application is subjected to heat and moisture treatment under milder temperature regimes and longer drying times to obtain high-quality material suitable for furniture.

In dryers with traditional heat sources, increasing the drying time inevitably leads to an increased usage of non-renewable primary fuel, and, as a consequence, to an increase in the production cost. The presence of atmospheric pollutants from the combustion emissions of traditional heat generators of drying plants and the concern about the ecological aspects of vital activities of humanity, which are particularly sharply formulated in recent times, demand that we pay attention to alternative methods of heating for the technological processes of convective drying of wood. Thus, energy-saving heat-pump systems were recognized and have long been applied.

Wood drying was one of the first commercial-scale applications of heat pumps for drying purposes that require fine temperature and humidity control. The pioneer in this area was WESTAIR, which has been manufacturing such equipment for over 10 years. The schematic diagram of the operation of a HP wood dryer is shown in Fig. 4.30. Air dried in the evaporator after heating in a condenser is fed to a stack of boards, shown edge-wise in a dozen groups. Prior to operating the compressor, the temperature in the chamber is raised to 20–25 °C by means of an electric heater.

To obtain a higher HP COP, the temperature of the hot air was decreased from 60 °C to 45 °C in the original design. Although this doubled the drying time from 3 to 6 weeks, energy savings amounted to more than 20%.

Table 4.6 presents data from an example of a high-temperature timber drying plant in England with a HP. Data pertain to 32 mm thick boards that have their moisture reduced to 12%. The reduction in the specific energy consumption SEC is 71% for the HPD itself, and 18% if the efficiency of the external process that generates the thermal or electrical energy is accounted for.

In those countries where the woodworking industry is a strong component of the economy, many scientists and engineers are engaged in research and development aimed improving the performance of HP systems for drying lumber. Thus, data on the use of HPD for pine drying may be found in Refs. [48–52]. Table 4.7 compares the power consumption for three types of heat generators used for drying pine. The total primary energy savings in this case is 1.3 GJ/m^3 for the HPD compared to the drier with a gas heat generator and 1.5 GJ/m^3 compared to a drier using solid fuels. Similar savings are also possible when drying hardwood.

Significant successes in heat supply of convective drying chambers for drying wood with the help of heat pumps was achieved by western European woodworking

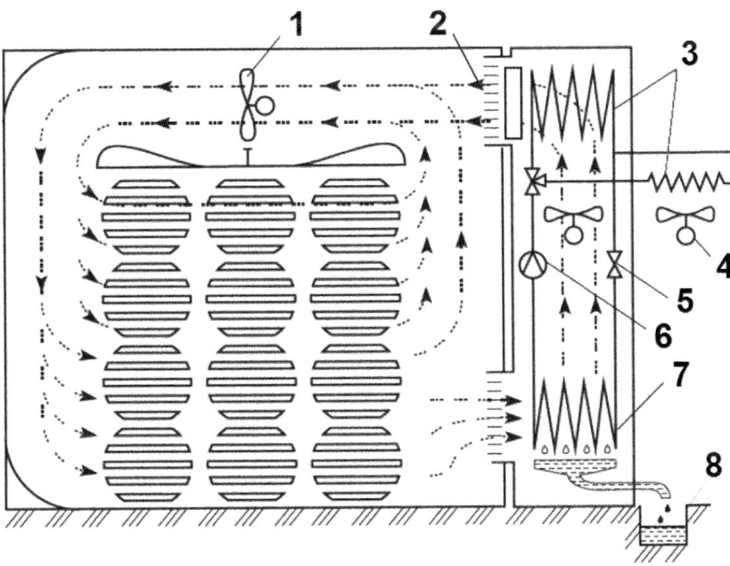

Fig. 4.30 Schematic of a wood dryer with a heat pump: 1 – fan for circulating air; 2 – direction of heat flow; 3 – condenser; 4 – cooling fan; 5 – throttle valve; 6 – compressor; 7 – evaporator; 8 – condensate collector.

Table 4.6 Energy consumption to dry boards of 32 mm thickness to 12% moisture.

Driers with oil burning	
Volume of downloaded product, m^3	21
Energy consumption, kWh/m^3	29.3
Amount of moisture removed, kg/m^3	91.1
SEC, kJ/kg	12.3
SEC, accounting for the efficiency of combustion energy generation, kJ/kg	15.8
Dryer with high temperature heat pump	
Volume of downloaded product, m^3	11.5
Energy consumption, kWh/m^3	88
Amount of moisture removed, kg/m^3	91.1
SEC, kJ/kg	3.5
SEC, accounting for the efficiency of electrical energy generation, kJ/kg	12.9
Comparison	
Reduction in SEC, %	71
Reduction in SEC, accounting for the efficiency of energy generation, %	18

Table 4.7 Energy consumption when drying pine, a softwood, using three methods.

Energy consumption	Traditional solid fuel dryer	Traditional gas dryer	Heat pump dryer
Fuel, GJ/m^3	2.4	2.2	–
Energy, GJ/m^3	0.3	0.3	0.6
Total energy, GJ/m^3	2.7	2.5	0.6
Total primary energy, GJ/m^3	3.3	3.1	1.8

enterprises. The first research in this area was carried out by the Spanish firm EBAC; its heat pump driers operate on most wood-processing enterprises in Spain [50]. One type of HPD circuit design is presented in Fig. 4.31. This scheme does not provide for use of excess heat coming from the thermal equipment, which simplification in turn reduces the cost of the heat pump itself [53]. Excessive heat along with part of the moisture is discharged through discharge valve 8, and incoming fresh air enters the chamber through air inlet valves 7. One disadvantage of such an installation is that during the winter, when operating with negative Celsius ambient temperatures, the temperature control in the drying chamber can become unstable since some portion of heat will be spent on heating the incoming air to the design temperature.

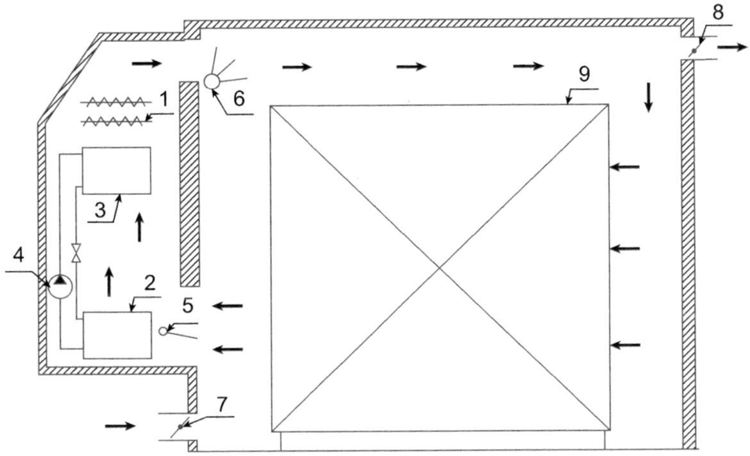

Fig. 4.31 EBAC heat-pump dryer with air intake and outlet on the side of the drying line: 1 – electric heater; 2 – HP evaporator; 3 – HP condenser; 4 – HP compressor; 5 – control latch; 6 –humidifier; 7 – intake valve/channel for fresh air; 8 – exhaust valve/ channel; 9 – wood stack.

Particular attention is paid to the issues of environmental safety of HPD operation. In Ref. [54] an estimate was made of the effect of greenhouse gas (GHG) emissions for a specific HPD of plank timber in comparison with traditional heat supply methods (Fig. 4.32). The horizontal axis shows the drying plant using various fuels. However, the electricity that is used in the dryers may be received from power plants that burn solid fuel (left half of the figure), or that burn gaseous fuel (right side). Therefore,

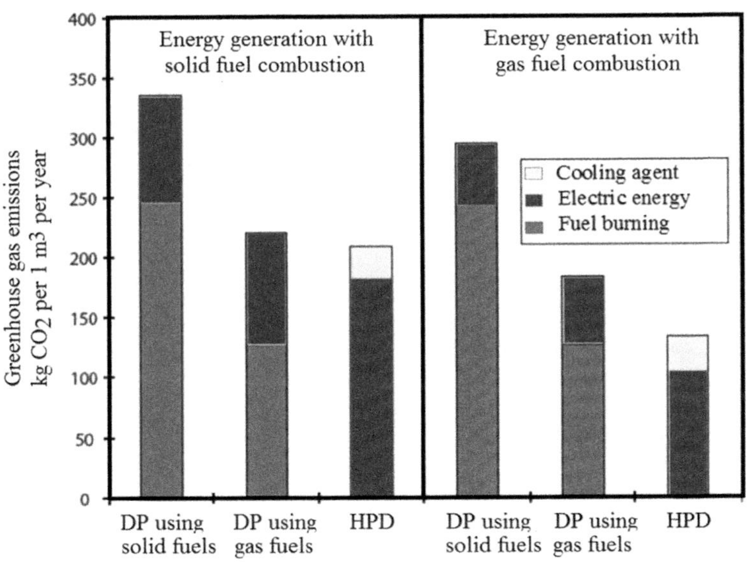

Fig. 4.32 GHG emissions for three different types of drying plants [55].

overall CO_2 emissions differ depending on the type of fuel that is used to generate the dryer's electricity.

Based on the data shown in the figure, one can conclude that by deploying heat pump technology for drying wood compared to solid fuel plants, the emission of greenhouse gases can be reduced by about 125–160 kg of CO_2/m^3 of the drying chamber per year. As for plants using natural gas as primary fuel, this indicator ranges from 10 to 50 kg of CO_2/m^3 of the drying chamber per year. When this environmental benefit is combined with superior thermodynamic energy performance, it is clear that HPD demonstrate important advantages over traditional dryers.

According to the data from Ref. [21], the worldwide production of heat pump wood dryers for the last few years is about 800–1000 units per year with a growth trend indicated. This may indicate an increase in the recognition of heat pump technologies for drying wood.

4.5 Summary

Eight typical drying systems are presented. These include two basic open air systems with supplementary heaters and recirculation of air, the drying agent. Six arrangements include a heat pump: two systems are open and may use a supplementary heater; the other four are closed loop and may include recirculation or bypass. Thus a wide assortment of dryers is available for a variety of drying applications. A survey is included of worldwide research being carried out in drying with heat pumps, along with a hierarchy that classifies the many uses of heat pumps in this industry.

The systems are described, analyzed, and the results presented with the aid of the psychrometric chart, in tables and in graphs. Optimizations are performed. Comparisons are drawn among the systems to show which are preferred in certain situations. Care is taken to define the alternative fuel/energy use and the mode of drive for the heat pump compressor in making comparisons, since many possible arrangements exist in practice. When proper accounting is done to include the energy required for alternatives to heat pumps in drying grains and wood, heat pumps are the most effective on the basis of overall energy consumption per unit of dried product.

Nomenclature

B	bypass factor
COP	coefficient of performance
$E_{cooling}$	cooling coefficient
h	enthalpy
K	recirculation ratio
m	flow rate
P	pressure
Q	capacity
q	heat load
t	temperature
V	air flow

Greek letters

α	bypass ratio
η	operational efficiency
η_{ex}	exergy COP
ν	volume
ξ	concentration
φ	relative humidity

Subscripts

DA	drying agent
dc	drying chamber
DP	dryer plant
GHG	greenhouses gas
HP	heat pump
HPDP	heat pump dryer plant
MC	mixing chamber
PC	psychrometric chart
SEC	specific energy consumption
SMER	amount of moisture evaporated

Review questions

1. Describe the types of heat pump dryers according to the scheme of implementation of the drying process.
2. Name the conditions for effective usage of the heat pump in a dryer plant.
3. Give the classification of heat pump dryers.
4. Describe the criteria of efficiency for using heat pumps in drying plants.
5. Discuss the main reasons for using heat pumps in drying technology in the 20th century.
6. Name the auxiliary sources of heat energy when operating heat pump dryers.
7. Name the prerequisites for the use of heat pumps in the processes of drying grains.
8. Name the functions of the heat pump in the grain drying plants of different systems.
9. (A) What determines the efficiency of heat pump grain dryers and (B) what is the effectiveness of their use compared to traditional systems.
10. In what modes of drying wood it is expedient to use a heat pump?
11. Describe the types and designs of heat pump dryers of wood.

Example

Determine the parameters at the characteristic points of the process and construct the process in the enthalpy-concentration (h, ξ) diagram. Find the heat loads of the apparatus, the specific energy consumption, the refrigerating coefficient, and the exergy efficiency of the plant (Fig. 4.E1 and Tables 4.E1 and 4.E2).

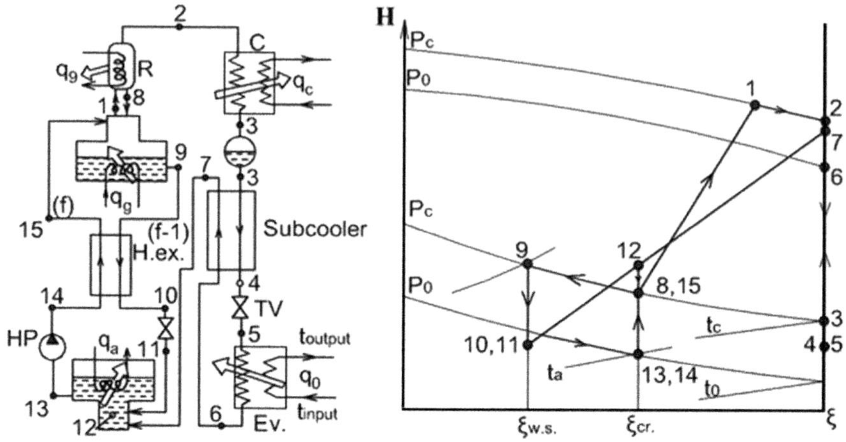

Fig. 4.E1 Scheme and process in the H, ξ-diagram of the water-ammonia refrigeration unit.

Table 4.E1 Initial data.

Parameters	**Case A**	**Case B [55]**
Cooling capacity, Q	1000 kW	1000 kW
Brine temperature at evaporator inlet, $t_{b.inlet}$	−10 °C	−20 °C
Brine temperature at evaporator outlet, $t_{b.outlet}$	−20 °C	−30 °C
Cooling water inlet temperature, $t_{c.w.inlet}$	20 °C	20 °C
Cooling water outlet temperature, $t_{c.w.outlet}$	25 °C	25 °C
Heating steam temperature, t_s	100 °C	130 °C
Pressure, P	100 kPa	275 kPa
Working fluid: ammonia		
Absorbent: water		
Accepted temperature differences, °C:		
Evaporator Δt_{ev}	5	3
Condenser Δt_{cond}	5	5
Absorber Δt_{abs}	5	5
Generator Δt_{gen}	10	6
Cooler $\Delta t_{3\text{-}7}$	10	10
Heat of brine $\Delta t_{14\text{-}10}$	10	10
Reflux condenser $\Delta t_{r.c.}$	15	15

Table 4.E2 Solution.

Parameters	Formula	Case A	Case B [55]
Cooling agent evaporation temperature, C	$t_0 = t_{b.outlet} - \Delta t_{ev}$	−25	−33
Ammonia evaporation pressure, MPa	by TS-diagram	0.154	0.1
Condensation temperature, C	$t_{cond} = t_{c.w.outlet} - \Delta t_{cond}$	30	30
Condenser pressure, MPa	by TS-diagram	0.858	1.2
Weak solution temperature at generator output, C	$t_9 = t_s - \Delta t_{gen}$	90	124
Weak solution parameters, kJ/kg	$\xi_{w.s.}; h_9$	0.30; 180	0.225; 462
Strong solution temperature at absorber outlet, C	$t_{13} = t_{input.app.} - \Delta t_{ab}$	25	25
Strong solution parameters, kJ/kg	$\xi_{s.s.}; h_{13}$	0.39; 210	0.315; 0
Solution circulation ratios	$f = G_{s.s.}/D = G_{15}/G_2 = (\xi_2 - \xi_{w.s.})/(\xi_{s.s.} - \xi_{w.s.})$	11.2	8.62
Steam parameters after reflux condenser, C; kJ/kg	$t_2 = t_{c.w.outlet} + \Delta_{r.c.};\ \xi_2 = 1$	40; 1350	40; 1660
Equilibrium vapor parameters at generator outlet:	–	$P_1 = 0.858MPa$ $t_1 = 90°C$ $\xi_1 = 0.954$ $h_1 = 1820kJ/kg$	$P_1 = 1.2MPa$ $t_1 = 105°C$ $\xi_1 = 0.925$ $h_1 = 1890kJ/kg$
Mixture concentration	$\xi_8 = \xi_{13}$	0.39	0.315
The enthalpy of the solution at point 8	–	$h_8 = 150$ kJ/kg	$h_8 = 377$ kJ/kg

Continued

Table 4.E2 Solution.—cont'd

Parameters	Formula	Case A	Case B [55]
Specific withdrawal of reflux from reflux condenser (reflux ratio)	$\varphi = G_8/G_2 = (\xi_2 - \xi_1)/(\xi_1 - \xi_8)$	0.081	0.123
Specific heat load of reflux condenser, kJ/kg	$q_{r.c.} = (h_1 - h_2) + \varphi(h_1 - h_8)$	605.3	416
Weak solution parameters after heat exchanger, C; kJ/kg	**a)** $t_{10} = t_{14} + \Delta t_{h.ex.} = t_{14} + \Delta t_{14-10}$ **b)** h_{10}	**a)** 35 **b)** 100	**a)** 35 **b)** 62
Strong solution enthalpy at generator inlet, kJ/kg	$h_{15} = h_{14} + \frac{f-1}{f}(h_9 - h_{10})$	102	352
Specific heat load of heat exchanger, kJ/kg	$q_{h.ex} = (f-1)(h_9 - h_{10})$	2856	3030
Specific heat load of condenser, kJ/kg	$q_c = h_2 - h_3$	1200	1160
Cooling agent vapor temperature after cooler, C	$t_7 = t_3 - \Delta t_{cooler} = t_c - \Delta t_{3-7}$	20	20
Specific heat load of cooler, kJ/kg	$q_c = h_7 - h_6$	50	111
Throttle inlet enthalpy	$h_4 = h_3 - q_c$	100	389
Specific cooling capacity of plant, kJ/kg	$q_0 = h_6 - h_4$	1150	1185
Heat released on absorption, kJ/kg	$q_a = (h_7 - h_{10}) + f(h_{10} - h_{13})$	1960	2160
Generator heat load of, kJ/kg	$q_g = (h_1 - h_9) + f(h_9 - h_{15}) + \varphi(h_1 - h_8)$	2646	2550

Thermal balance of plant, kJ/kg	$q_{letdown} = q_{alloted}$ $q_{letdown} = q_g + q_0$ $q_{alloted} = q_a + q_c + q_{r.c.}$	3826 3765	3735 3736
Working fluid (ammonia) mass flow rate, kg/s	$m_{NH_3} = \frac{Q_0}{q_0}$	0.87	0.85
Thermal loads, kW: **a)** generator **b)** absorber **c)** cooler **d)** condenser **e)** reflux condenser	**a)** $Q_g = m \cdot q_g$; **b)** $Q_a = m \cdot q_a$; **c)** $Q_c = m \cdot q_c$; **d)** $Q_{cond} = m \cdot q_{cond}$; **e)** $Q_r = m \cdot q_r$;	**a)** 2305 **b)** 1705 **c)** 43.5 **d)** 1044 **e)** 526	**a)** 2170 **b)** 1835 **c)** 94.5 **d)** 985 **e)** 354
Specific heat consumption	$q_{sp} = Q_g/Q_0$	2.3	2.17
Cooling coefficient	$E_{cooling} = Q_0/Q_g$	0.43	0.463
Exergy COP of the plant	$\eta_{ex} = \frac{q_0(\tau_q)_0}{q_0(\tau_q)_s} = E_{cooling}\frac{(\tau_q)_0}{(\tau_q)_s}$	0.37	0.373
	$(\tau_q)_0 = (T_0 - T_{env})/T_0$	−0.18	−0.22
	$(\tau_q)_s = (T_s - T_{env})/T_s$	0.21	0.273

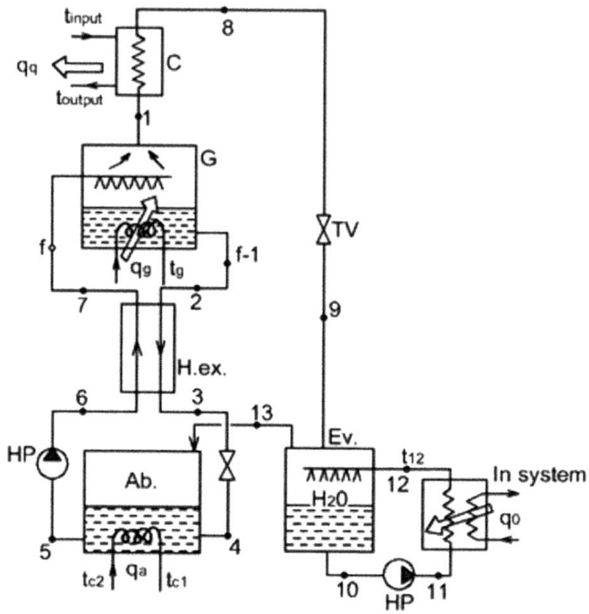

Fig. 4.P1 Flow diagram of a lithium-bromide absorption refrigeration unit.

Exercise

Analyze the scheme of a lithium-bromide absorption unit with a cooling capacity $Q = 1000$ kW (Fig. 4P1). Initial data: the temperature of cold water at the output of the evaporator $t_{output\ ev.} = 5$ °C, the temperature of heating water $t_{h.\ w.} = 80$ °C, the temperature of supply water at the input to the apparatus of $t_{input} = 20$ °C and at the output of $t_{output} = 25$ °C. Finite temperature differences: in the condenser $t_c = 5$°C, in the absorber $t_{ab} = 8$°C, in the consumption system $t_{c.s.} = 3$°C, in the generator $t_g = 10$°C, in the heat exchanger $t_{h.e.} = 10$°C.

Answers. $Q_g = 1220$ kW; $Q_{cond} = 1020$ kW; $Q_{h.ex.} = 21.6$ kW; $Q_{ab} = 1200$ kW; $\Delta Q = 1.23$; $E_{cool.} = 0.82$; $\eta_e = 0.233$.

References

[1] Heat pumps: vision vs. reality. Energy recovery. Thomas Nowak; November 04, 2016. https://www.ee-ip.org/articles/detailed/bd3568515e538ad350faf063915c6141/heatpumps-vision-vs-reality/.

[2] Large scale heat pumps in Europe: 16 examples of realized and successful projects. European Heat Pump Association [undated brochure].

[3] Ray D. Heat pumps. Transl. Eng. Energoizdat; 1982. 224 pp. [in Russian]

[4] Mellor JD. Fundamentals of freeze-drying. London: Academic press; 1978. 380 p.

[5] Oetjen G-W. Freeze-drying. 2nd ed. Weinheim: Wiley; 2004. 395 p.

[6] Keey RB. Drying. Principles and practice. Oxford: Pergamon; 1975. 376 p.
[7] Strumillo C, Kudra T. Drying: principles, applications, and design. New York: Gordon and Breach; 1986. 448 p.
[8] Keey RB. Drying of loose and particulate materials. New York: Hemisphere; 1992. 504 p.
[9] Mujumdar AS, editor. Handbook of industrial drying. 3rd ed. New York: Taylor & Francis; 2007. 1280 p.
[10] Alves-Filho O. Heat pump drying of fruits and roots. The influence of heat and mass transfer on dryer characteristics [Ph.D. thesis]. Trondheim, Norway: Norwegian University of Science and Technology; 1996. 120 p.
[11] Sosle V. A heat pump dehumidifier assisted dryer for agrifoods [Ph.D. thesis]. Montreal, Canada: McGill University; 2002. 120 p.
[12] Hodgett DL. Efficient drying using heat pumps. Chemical Engineer 1976:510–2.
[13] International symposium on the industrial application of heat pump, 1982.
[14] Sylla R, Abas S, Tai K. The potential for heat pump in drying and dehumidification System. International Journal of Energy Research 1982;6(4).
[15] Ray D. Energy saving in industry: reference manual for engineering and technical workers. Transl. Eng. Energoizdat; 1983. 208 pp. [in Russian]
[16] Lykov MV. Drying in chemical industry. Chemistry 1970. 432 pp. [in Russian].
[17] Romanov EV, Orlov AY. The possibility of using heat pumps in the drying process. Dept. of Chemical Engineering, GOU VPO "TGTU"; 2008 [in Russian].
[18] Rubinstein H. Drying with heat pumps. Energy Technology 1983;(2).
[19] Kudra T, Mujumdar S. Heat pump drying: advanced drying technologies. 2nd ed. 2008. New York.
[20] MacArthur JW. International Journal of Refrigeration 1984;7(2):123.
[21] Song X. Low temperature fluidized bed drying with temperature program [Ph.D. thesis]. Trondheim, Norway: The Norwegian Institute of Technology, Division of Refrigeration Engineering; 1990 [in Norwegian].
[22] Strommen I, Kramer K. New applications of heat pumps in drying process. Drying Technology 1994;12(4):889–901.
[23] Prasertsan S, Saensaby P. Heat pump drying of agricultural materials. Drying Technology 1998;16(1 and 2):235–50.
[24] Saensabai P, Prasertsan S. Effect of component arrangement and ambient and drying conditions on the performance of heat pump dryers. Drying Technology 2003;21(1): 103–27.
[25] Perry EJ. Drying by cascaded heat pumps. Proc. Institute of Refrigeration Management; 1981. p. 1–8.
[26] Matsuo K, Senshu T, Hayashi M, Kokuba H. How can heat pumps be made more cost competitive?. In: Proc. Int. Energy agency heat pump conference. May, Graz, Austria. Boras, Sweden: IEA Heat Pump Centre; 1984. p. 87–96.
[27] Geeraert B. Energy saving in drying application with special emphasis on the use of heat pumps. Revue Energy. Primery 1983;18(1).
[28] Van der Pal M, Sun Q, Simpson I, Scharpf E, Carrington G. Influence of kiln venting on brown stain in Pinus Radiata sapwood. Drying Technology 2007;25(6):1105–10.
[29] Sunthonvit N, Srzednicki G, Craske J. Effects of drying treatments on the composition of volatile compounds in dried nectarines. Drying Technology 2007;25(5):877–81.
[30] Hawlader MNA, Perera CO, Tian M. Comparison of the retention of 6-gingerol in drying of ginger under modified atmosphere heat pump drying and other drying methods. Drying Technology 2006;24(1):51–6.

[31] Hawlader MNA, Perera CO, Tian M, Yeo KL. Drying of guava and papaya: impact of different drying methods. Drying Technology 2006;24(1):77−87.

[32] Claussen IC, Strommen I, Egelandsdal B, Strætkvern KO. Effect of drying methods on functionality of a native potato protein concentrate. Drying Technology 2007;25(6): 1091−8.

[33] Prasertsan S, Saen-saby P, Prateepchaikul G, Ngamsritrakul P. Heat pump dryer. Part 3: experiment verification of the simulation. Int J Energy Res 1997;21:1−20.

[34] Chou SK, Chua KJ. Heat pump drying systems. In: Mujumdar AS, editor. Handbook of industrial drying. 3rd ed. Boca Raton, FL: Taylor & Francis; 2007. p. 1103−31.

[35] Chua KJ, Chou SK, Hawlader MNA, Ho JC. A two-stage heat pump dryer for better heat recovery and product quality. Journal of The Institution of Engineers, Singapore 1998; 38(6):8−14.

[36] Magnussen OM, Strømmen I. Heat pump drying of heavily salted codfish. In: Nordic refrigeration meeting, copenhagen, May 21−23; 1981 [in Norwegian].

[37] Strommen I, Jonassen O. In: Strumillo C, Pakowski Z, editors. Performance tests of a new 2-stage countercurrent heat pump fluidized bed dryer, Drying'96. Lodz: Lodz Technical University; 1996. p. 563−8.

[38] Alves-Filho O, Strømmen I. In: Strumillo C, Pakowski Z, editors. Performance and improvements in heat pump dryers, Drying'96. Lodz: Lodz Technical University; 1996. p. 405−16.

[39] Chen G, McHuge J, Bannister P, Carrington CG, Sun ZF. Design and applications of controlled-atmosphere dehumidifier fruit dryers. IPENZ Transactions, Institute of Professional Engineers New Zealand 2000;27(1):31−4.

[40] O'Neill M, Rahman MS, Perera CO, Smith B, Melton LD. Colour and density of apple cubes dried in air and modified atmospheres. International Journal of Food Properties 1998;1(3):197−205.

[41] Chen G, Bannister P, Carrington CG, Sun ZF. Economic performance of enhanced dehumidifier kilns. Proceedings of the IPENZ annual conference, vol. 2. Wellington, NZ: Institute of Professional Engineers New Zealand; 1997. p. 144−8.

[42] Gomelauri WI, Vezirishvili OS. Experience of development and application of heat-pump installations. Thermal Power Engineering 1978;(4) [in Russian].

[43] Hawlader MNA, Chou SK, Jahangeer KA, Rahman SMA, Lau E. Solar assisted heat pump dryer and water heater. Applied Energy 2003;74(1−2):85−193.

[44] FAOSTAT, Forestry Production and Trade, Food and Agriculture Org. of the U.N. http://www.fao.org/faostat/en/#data/FO/visualize.

[45] Geankoplis CJ. Transport processes and unit operations [Chapter 7]. Boston: Allyn and Bacon; 1978.

[46] Bezrodnyi MK, Kudrya DS, Vovk VV. Thermodynamic analysis of heat pump drying installations for grain drying," technical thermal physics and industrial heat power engineering. Dnepropetrovsk 2012;4:27−40 [in Ukrainian].

[47] Bannister P, Chen G, Carrington CG, Sun ZF. Guidelines for operating dehumidifier timber kilns. 2002. Energy Group Limited Research Report EGL-RR-02. Available at: http://www.eglnet.com.

[48] Carrington CG, Sun ZF, Bannister P, Chen G. Optimising efficiency and productivity of a dehumidifier batch dryer − Part 1: capacity and airflow. International Journal of Energy Research 2000;24:187−204.

[49] Carrington CG, Sun ZF, Bannister P. Dehumidifier batch drying, Effect of heat losses and air-leakage. International Journal of Energy Research 2000;24:205−14.

[50] Carrington CG, Wells CM, Sun ZF, Chen G. Use of dynamic modelling for the design of batch-mode dehumidifier driers. Drying Technology; 2002.
[51] Chen G, Bannister P, Carrington CG. A survey on dehumidifier wood drying operation in New Zealand. Research Report. Energy Group Limited; 2002.
[52] Ebac industrial products. [Electronic resource], Access mode: www.ebacusa.com.
[53] Ebac lumber dryer. Owner's manual. Kiln construction guide. Troubleshooting guide; 2004. p. 82.
[54] Bannister P, Chen G, Grey A, Carrington CG, Sun ZF. Economic reduction of greenhouse gas emissions through enhanced dehumidifier timber drying. In: Proceedings, 19th international congress of refrigeration. Linz, Austria: International Institute of Refrigeration; 1997. p. 241–9.
[55] Martynov AV. Installations for the transformation of heat and cooling. Collection of tasks. Moscow: Energoatomizdat; 1989. p. 194.

Heating with geothermal systems

5

5.1 Geothermal direct heat usage: International experience

Direct heat applications of geothermal energy can be found anywhere in the world, unlike electric power generation which is limited to moderate-to-high temperature resources. There is a wide spectrum of uses from geofluid temperatures as low as 25−30 °C (aquaculture) and as high as 180−190 °C (absorption refrigeration). The range of applications as a function of temperature is shown in Fig. 5.1 which is taken from Ref. [1] and based on the Lindal diagram. Heat pumps are included since earth-coupled heat pump systems can be used for both heating and cooling anywhere in the world.

The latest information on worldwide geothermal direct usage was presented at the 2015 quinquennial congress of the geothermal industry [2]. The following discussion is based on that work. Geothermal direct heat usage has been growing steadily over the last two decades. Starting from 1995, the installed thermal capacity has increased by a factor of 7.8 as of 2015, while the heat utilization has increased by a factor of 5.3; see Fig. 5.2.

By far the most popular application is geothermal (or earth-coupled) heat pumps. The top five countries in terms of heat pump installed capacity are the United States,

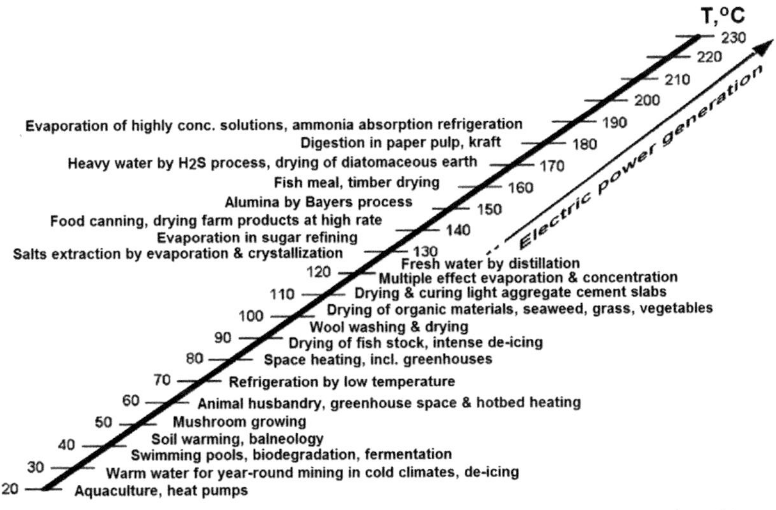

Fig. 5.1 Geothermal direct heat applications as a function of temperature, after [1].

Low-Temperature Energy Systems with Applications of Renewable Energy
https://doi.org/10.1016/B978-0-12-816249-1.00005-4

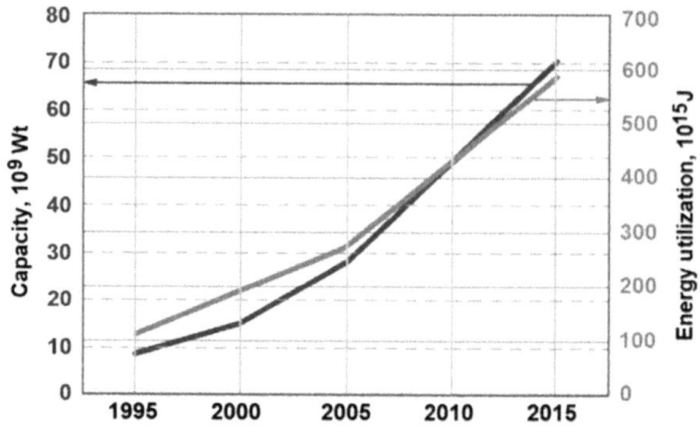

Fig. 5.2 Growth of geothermal direct heat usage since 1995, redrawn from [2].

China, Sweden, Germany and France. Data for all applications in 2015 are shown in Table 5.1. A total of over 70 GWt of thermal capacity was installed in some 82 countries, producing 588×10^{12} kJ of heat energy utilization. The capacity factor is found

Table 5.1 Geothermal direct heat usage by application for 2015 [2].

Application	Thermal capacity, MWt	Percent of total, %	Heat utilization, TJ	Percent of total, %	Capacity factor, %
Geothermal heat pumps	49,898	70.95	325,028	55.30	20.7
Bathing and swimming	9140	13.00	119,381	20.31	41.4
Space heating	7556	10.74	88,222	15.01	37.0
Greenhouse heating	1830	2.60	26,662	4.54	46.2
Aquaculture pond heating	695	0.99	11,958	2.03	54.6
Industrial uses	610	0.87	10,453	1.78	54.3
Cooling/Snow melting	360	0.51	2600	0.44	22.9
Agricultural drying	161	0.23	2030	0.35	40.0
Others	79	0.11	1452	0.25	58.3
Totals	70,329	100	587,786	100	26.5

by dividing the actual heat utilization by the maximum possible utilization if the installed capacity operated for a full year.

In many cases, geothermal plays a significant part in a country's thermal energy supply system or energy infrastructure. A few of these are described below.

- Iceland: Geothermal hot water from numerous fields is piped to cities and towns to cover 90% of the space heating needs of buildings in the entire country.
- Japan: There is a thriving balneology industry in Japan owing to the national culture of visiting hot spring spas (onsen, in Japanese). The statistics are impressive: 2000 onsens, 5000 public baths, and 1500 hotels serving 15 million guests per year.
- Sweden: As a cold-weather country, Sweden heats 20% of all buildings using geothermal heat pumps.
- Switzerland: A European leader in geothermal heat pumps, Switzerland has 90,000 units installed or about 2.25 units per square kilometer of total area, a remarkable figure given that only 6.8% of the country's land area is covered with buildings and residences. Thus there is a rate of 33 units per inhabited square kilometer.
- Turkey: Favored with many geothermal resources, most of low-to-moderate temperature, Turkey has 90,000 apartment residences geothermally heated in 16 cities, nearly 30% of the total units in the country. Turkey is also the world leader in geothermal greenhouses with 300 ha in operation.
- United States: There are 1.4 million geothermal heat pumps providing full comfort control in residences and buildings, and the growth rate is strong at 7.0% per year.

Table 5.2 lists the top ten countries by installed capacity, heat utilization, and capacity factor. Not surprisingly the same ten countries appear in the first two categories. Ten relatively small countries rank as the best in capacity factor. Also notice that the top ten countries account for roughly 80% of all the world's capacity and utilization of geothermal heat that is distributed among 82 countries.

5.2 Modern technology, geothermal field development, and geothermal heating systems

Geothermal energy has been recognized for millennia by the inhabitants of planet Earth, who both feared its power and devised means to capture it for useful purposes [3]. In this section we discuss ways to find exploitable geothermal resources using modern technology, drilling techniques that reach several kilometers into the earth to unlock the potential, and systems for distributing that energy to consumers.

5.2.1 Exploration

In early times one need not have advanced degrees or highly trained personnel to recognize geothermal resources. Nature revealed these subterranean caches of energy through surface thermal manifestations such as hot springs, steam vents or fumaroles, mineral deposits such as sulfur, obsidian or calcite, spectacular geysers, and patches of altered hot ground. Early peoples were attracted to places with such characteristics and

Table 5.2 Top 10 countries in geothermal direct heat usage ranked by installed capacity, heat utilization and capacity factor.

Rank	Country	Installed capacity, MWt	Country	Heat utilization, TJ	Country	Capacity factor, %
1	China	17,870	China	174,352	Algeria	98.6
2	United States	17,416	United States	75,862	Mexico	84.9
3	Sweden	5600	Sweden	51,920	Madagascar	85.3
4	Turkey	2886	Turkey	45,126	Caribbean	88.2
5	Germany	2849	Iceland	26,717	Israel	84.4
6	France	2347	Japan	26,130	Nepal	77.5
7	Japan	2186	Germany	19,531	Guatemala	77.5
8	Iceland	2040	Finland	18,000	Belarus	76.1
9	Switzerland	1733	France	15,867	Honduras	73.9
10	Finland	1560	Switzerland	11,837	Costa Rica	66.6
	Total	56,487	Total	465,342	Average	81.3

benefitted from healing baths, the warmth of the earth, and tools fashioned from the volcanic rocks. Although weapons were made from hard, sharp obsidian rocks and used to attack enemies, geothermal places were usually declared neutral and often sacred sites where all people, friend and foe alike, could gather socially and recuperate.

While there are many places marked by obvious surface geothermal activity, there are many more that are not, so-called "blind resources" that can only be discovered with sensitive instrumentation, painstakingly deployed, whose signals can be interpreted to reveal structures and activity deep below the surface beyond the human senses.

It was once thought that geothermal activity was solely associated with the movement of the gigantic tectonic plates that comprise the cracked crust of the earth. Subduction zones are created where one plate slides beneath an adjacent one, giving rise to deep-seated melting of the crust and the rising of magma, melting its way upward toward the surface. Volcanoes are thus brought into life - an event witnessed by ancient people time and again. Legends and myths reveal that they were in awe of the destructive power of volcanoes. Residual heat left in the near-surface crust radiated and conducted its way through the soils and gave rise to the aforementioned manifestations and peaceful uses of this natural energy.

5.2.1.1 Geological observations

An understanding of geology and the history of geologic events is the starting point for exploration. Age-dating of rocks tells how young an area might be in geologic time. Places with no current geothermal surface activity may yet show traces of extinct ancient behavior that hint at anomalous formations now hidden. Even with no surface evidence at all, there may still be commercial geothermal resources lurking beneath the surface. Several techniques are now available to help uncover them.

5.2.1.2 Geochemical techniques

If any fluids are present in the area, they can be subjected to chemical analysis to determine their origins. Warm or hot springs are obvious examples that are suitable for study. The presence of these may be associated with faults or fractures that serve as conduits for geothermal fluids. It is possible to estimate the temperature of the deep fluids from an analysis of surface waters. The concentration of various minerals, such as silica, dissolved in the fluids can be correlated to deep temperatures using the solubility-temperature characteristics of the fluids. The pH of the fluids plays an important role in this assessment.

5.2.1.3 Geophysical assessment

The spatial subsurface distribution of physical properties reveals discontinuities that indicate the presence of faults. The density of the rocks affects the local gravitational acceleration, and extremely sensitive instrumentation may be used to detect these slight variations. Thus, gravity highs and gravity lows can be discovered which may indicate zones with different porosity or permeability, which are very important in regard to storage and transport of geothermal fluids. The electrical conductivity is another important physical property that is easily measured. The magnetic field in a prospective area is of great interest and can be measured and interpreted to provide a fuller picture of the subsurface structure.

5.2.1.4 Synthesis of findings

The most important lesson during exploration is that no one individual method is dispositive about the value of a particular site; all of the information gathered from exploration tools of various types must be used to piece together a full assessment of the field's potential. The data must be synthesized in a comprehensive manner, lest one measurement be given undue excessive weight and influence the final outcome. Ultimately, even a most promising area based on exploration studies must be proven by drilling deep wells.

5.2.2 Drilling

The best use of the exploration studies is to determine targets for the first deep wells. Since well-drilling is an expensive and risky operation, one needs to target the most attractive site for the first well, along with sites for the next ones if the first well proves successful. In the early days of geothermal exploration, obvious sites were targeted and shallow wells were sufficient to tap into reservoirs that had existed for millennia without exploitation. These wells were used to supply fluids for mineral recovery, as at Larderello in Italy where boric acid was produced for several centuries until the mid-1900s, or to generate electricity from geothermal steam, as was done also at Larderello from about 1904 to the present day.

Nowadays, places far from tectonic plate boundaries in areas not known for conventional geothermal resources, such as Western Europe, are being drilled for thermal waters to heat buildings, and other direct heat applications. Since the fluid temperature in such places is controlled mainly by the natural geothermal gradient of about 30−33 °C per kilometer, it is necessary to drill quite deep, typically 3000−4000 m, to reach fluids with usable temperatures. Drilling rigs have been developed for such deep operation, and some even are automated to the extent that crews on the drilling platform are not needed; all operations are controlled remotely from the rig cabin, including adding drill pipe and screwing together the drill string.

Even though acceptable temperature may be found, a well still needs adequate permeability to allow sufficient flow rate of geothermal fluid. A poor drilling operation can damage the formation, plugging narrow fractures that serve as fluid conduits, and leave an unproductive hole in the ground. Some damage can be repaired by chemical treatment to clean out the plugged fractures or by thermal shock from injecting cold water and thereby restore permeability and allow the well to be salvaged for commercial use.

In some basins in Europe, a deep sedimentary formation with good permeability exists at drillable depths. In such places two wells can be drilled into the same layer but with widely spaced points of penetration that permit fluid to be withdrawn from one well, used for various heat applications, and then reinjected into the same formation. Done properly using directional drilling, the two wellheads may be close to each other on the surface, say 10 m apart, but kilometers apart where they intercept the deep sedimentary formation. This minimizes interference between the cold injectate and the hot producing fluid.

5.2.3 Development

Once a resource has been identified and characterized, the next step is to match it to the requirements of particular applications. Naturally, certain applications will be in mind when the project is first conceptualized, but only when the resource becomes known and characterized can specific, appropriate applications be brought into consideration.

For example, in Europe where low-to-moderate temperature geothermal fluids can be accessed at depths of 2500 m and deeper, applications such as district heating, horticulture, aquaculture, recreation, and balneology are typical appropriate applications. The return on investment for various facilities will be weighed against the cost of development, construction, installation, operation, and maintenance to determine feasibility of any project, even before the resource is verified. This involves risks for investors and requires some form of incentive or insurance to mitigate the risk of failure to find a commercial resource. Governments and municipalities often provide such guarantees for a limited time period in order to stimulate sustainable and environmentally friendly projects. The drilling phase is the most risky, especially when depths of 4000–5000 m are needed to reach adequate temperatures, because sufficient permeability must be achieved in addition to good temperatures. The costs and risks escalate when the reservoir must be stimulated (e.g., hydrofractured) to improve the permeability and thereby enhance the productivity and injectivity of the wells.

5.3 Geothermal district heating systems

Many examples exist of geothermal district heating systems. Only two will be presented in this section; interested readers can find many others in the literature. The first one is a very old system in Europe that dates back about 700 years and the other is a fairly modern one in a city in the western United States.

5.3.1 Chaudes-Aigues, France

One of the earliest district heating systems in historic time occurred in a small French town, Chaudes-Aigues. Set in the valley of the Remontalou River in south-central France, the community of fewer than 900 residents hosts about 30 natural hot springs with waters ranging in temperature from 45 °C to as hot as 82 °C; see Fig. 5.3. The Remontalou flows through the elongated town from generally south to north; the elevation varies from about 800 masl in the south to about 755 m in the north.

Certainly the springs were known to the early inhabitants of the area dating back to at least 450 AD. Excavations in the area have unearthed artifacts and thermal baths in caves that date back to the Romans. By 1332 a district heating system had been

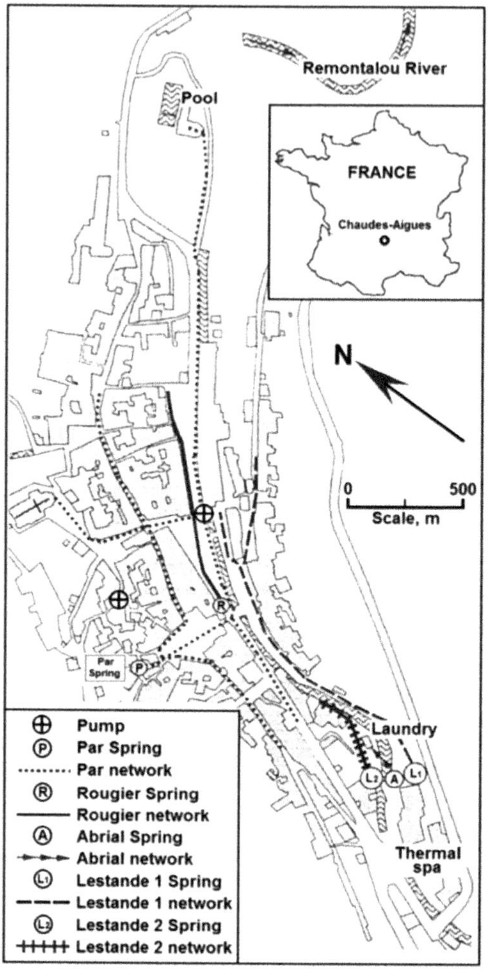

Fig. 5.3 Chaudes-Aigues, France, with recent district heating networks [4].

established for some of the houses in the town [4]. Residents could divert hot water from a municipal channel running past their house. The diverted water flowed under the first floor of the house thereby heating the interior; see Fig. 5.4. About 40 houses received the hot waters free-of-charge, but the owners were obligated to maintain their part of the system which involved removal of scale formation in the conduits. This arrangement was the forerunner of today's radiant floor heating systems.

The main supply was the Par Spring (Fig. 5.5) which was situated in the heart of the town. In the 1400s widespread use of the hot water took hold not solely for heating but many other industrial and commercial uses, as well as for health and therapeutic benefits.

The temperature of the Par Spring is 82 °C, said to be the hottest spring in Europe, and the mass flow rate is about 5 kg/s; the total flow from all 30 springs is about 11 kg/s [5]. The town church was included as part of the district heating system. A smaller

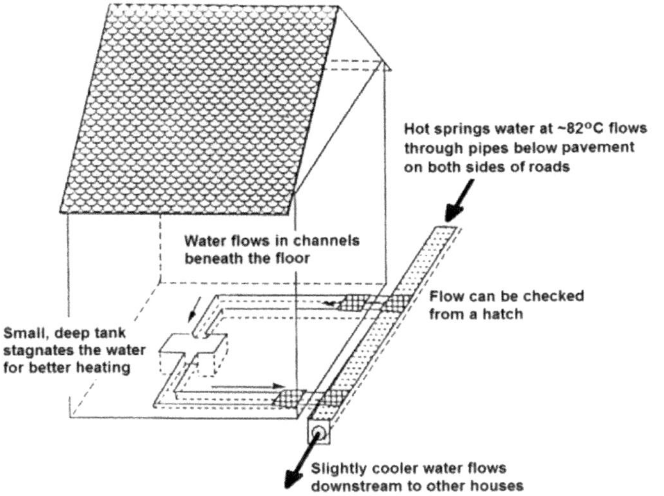

Fig. 5.4 Typical house in Chaudes-Aigues hot water district heating network [4].

spring, the Abrial, was used exclusively to supply a communal laundry room (lavoir) which is still in place (Fig. 5.6).

This pattern of usage continued for many centuries. However, in 2009 the town council decided that the district heating system would be taken out of service in order to supply hot water to a proposed new, large thermal spa that would cater to tourists; see Fig. 5.7. Only those houses fortunate enough to be situated over a private spring were able to continue to heat their homes from geothermal water. However, the church continues to be heated in the winter by the waters from the Par Spring [5].

Fig. 5.5 Par Spring, Chaudes-Aigues, France, hottest spring in Europe at 82 °C [5].

Fig. 5.6 (L) The Lavoir (small building with dark roof, lower center) in Chaudes-Aigues, 2013; Google Earth Street View, Sept. 8, 2013; (R) Lavoir interior, hot water inlet pipe at rear of the tub; http://au-detour-d-un-chemin.over-blog.com/article-les-sources-de-chaudes-aigues-cantal-116257053.html.

Fig. 5.7 Caleden thermal spa in Chaudes-Aigues [6].

5.3.2 Boise, Idaho, USA

Downtown Boise, Idaho, is home to the oldest and largest geothermal heating district in the United States. Before recorded history, the land where Boise sits today was occupied by Native Americans, notably the Nez Perce, going back at least 12,000 years. They were drawn to the place by the natural hot springs that emanated from the west face of a massive granite outcrop. The harsh winters were made more bearable by the warmth afforded by the springs. In fact the area became known later as "Peace Valley" in accordance with Native-American culture that recognized natural hot springs as sacred places at which all were welcome to come, socialize, and enjoy the comfort of the warm waters [7].

In modern history, the use of the geothermal energy goes back to the early 1890s when a pair of private entrepreneurs dug wells in an area where they had noticed snow did not build up during the winter, an intuitive exploration technique. The waters that flowed spontaneously from their wells were hot enough to heat homes in the area, and led to the construction of a natatorium housed in a newly-built, magnificent, Moorish-style building that made direct use of the geothermal waters [8]; see

Fig. 5.8 Vintage postcard of the Natatorium, geothermal swimming pool, Boise, Idaho, USA, c.1900 [8].

Figure 5.8. The Natatorium pool water came straight from the nearby wells, making it a popular gathering place for residents. The activity it generated spurred the development of the city until the 1930s when part of the roof collapsed into the pool from deterioration caused by the geothermal steam. That marked the end of the Natatorium; it was condemned and dismantled in 1934. Today at the same spot a large outdoor swimming pool remains, but it is not filled with geothermal water. Since 2015, however, the pool water is heated by geothermal fluid via a heat exchanger [9,10]; see Fig. 5.9.

Starting in the 1970s and driven partly by the worldwide energy crises, a more concerted effort got underway to exploit the geothermal resource. Several entities began drilling wells along the fault system lying northeast of the city. Nearly all wells were successful and produced prolific flow rates of hot water at around 79 °C. Aggressive competition among the companies resulted in the present-day situation in which four separate systems of district heating operate side-by-side and intermingled in the eastern part of Boise [2,7]. The four lines that comprise the district heating system are: (1) the Boise Warm Springs Water District (BWSWD); (2) State of Idaho Capitol Mall (SICM); (3) City of Boise, including the extension to Boise State University (CB) (see Fig. 5.10), and (4) U.S. Veteran's Administration (VA).

Fig. 5.9 The Natatorium, Boise, Idaho, USA, 2018: TripAdvisor image.

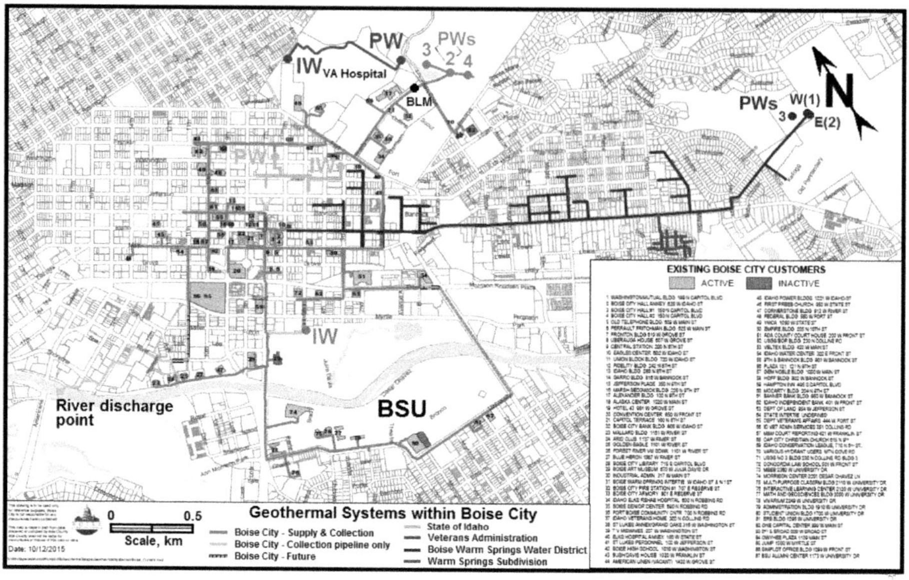

Fig. 5.10 City of Boise geothermal district heating networks. Note: a major fault runs linearly across the upper part of the map just above a line connecting the wells shown there, modified map supplied courtesy of [17].

As of 2018, Boise had a population of 217,000, making it the 99th largest city in the United States. The Idaho State Capitol building is the only capitol in the U.S. that is totally heated by geothermal energy.

As can be seen from the geologic map and cross-section in Figs. 5.11 and 5.12, respectively, the northeast section of Boise is bounded by a major fault, the Foothills Fault, and several smaller associated ones, forming a highly fractured network. This fault marks a small portion of the northeastern boundary of the Snake River plain that coincides with the trace of the magma intrusion that now lies beneath Yellowstone National Park.

All of the productive wells lie to the south of the main fault and many of them apparently intersect its many permeable zones. Excellent permeability is found where linears cross the Foothills Fault. The wells closest to the Foothills Fault produce from shallow depths of 182–244 m; those farther away need to reach about 1000 m to find permeability. Many of the early wells flowed spontaneously (artesian flow), but as more were drilled and the fluid production increased to meet growing demand, it was necessary to install downhole pumps to maintain and control the flows. A moratorium was even imposed on well drilling to limit the drawdown of water in the producing wells. Currently there is a cap in place on the total fluid withdrawal and the entire system is monitored for sustainability by the Idaho Water Resources Board. The IWRB is responsible for the formulation and implementation of a state water plan, financing of water projects, and the operation of programs that support sustainable management of Idaho's water resources.

Each of the four district heating lines will now be described in some detail.

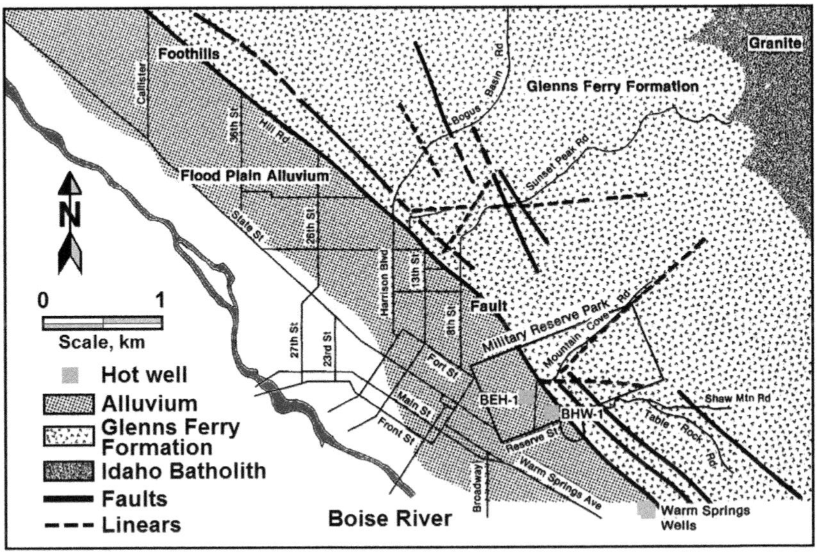

Fig. 5.11 Geological map of Boise, Idaho showing the Foothills Fault and related linears, after [11].

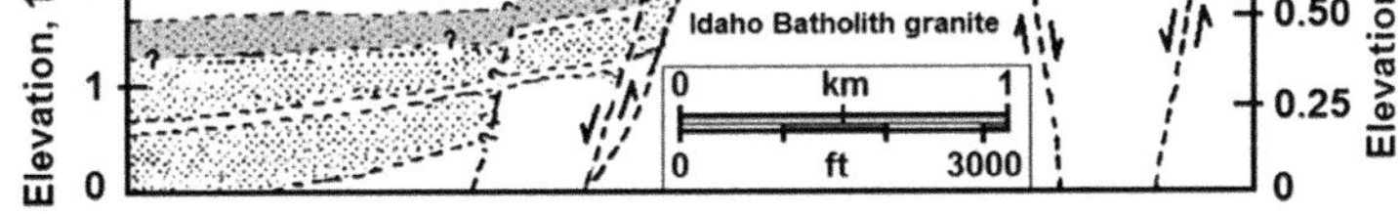

Fig. 5.12 Cross-section looking NW through the well BHW-1 in Fig. 5.11, modified from [12].

5.3.2.1 Boise Warm Springs Water District (BWSWD)

The original Boise Warm Springs geothermal district heating system began in 1892 and was the first such company in the United States. The current Boise Warm Springs Water District was established in 1987 as a private company in the state of Idaho. This company serves about 300 users, nearly all private residences, with hot water from its two production wells in the eastern part of the city; see Fig. 5.13. The wells are located just south of the Foothills Fault, penetrate the hanging wall, and intercept the fault at a shallow depth. The water is at a temperature of about 79 °C, a value that has stayed remarkably constant over more than a century of exploitation. Usually only one of the two wells is needed to meet the demand.

An interesting feature of the BWSWD is that it is an open system. The hot geothermal fluid is sent out via a single-pipe distribution line with users tapping

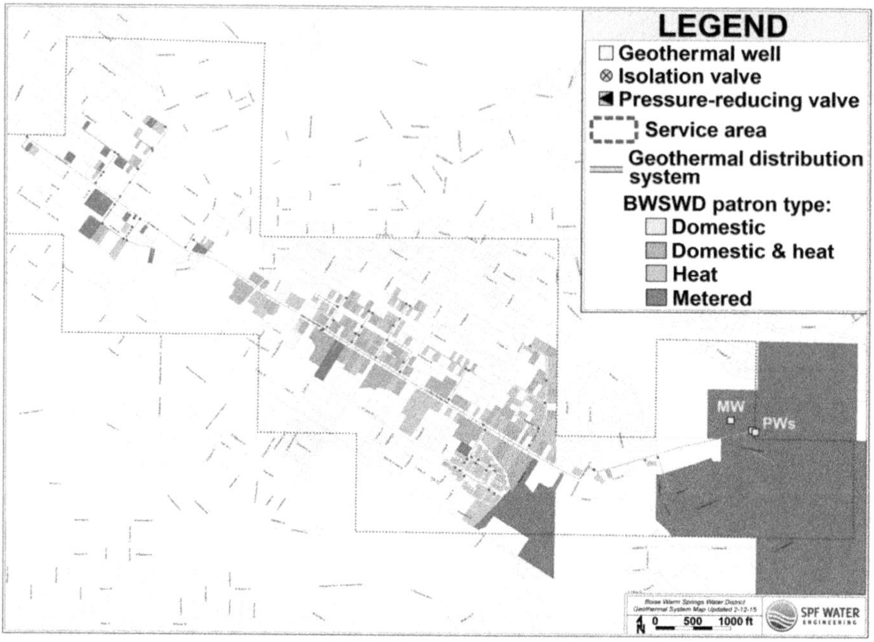

Fig. 5.13 Boise Warm Springs Water District service map. *MW*, monitoring well; *PW*, production wells (2 of them) [13].

into the line at their residence. They use it for domestic hot water, e.g., baths and showers, and/or to heat their homes. Domestic use results in the water being disposed of via a sewer, but the cooled heating water is disposed of as the residents wish. Some use it in swimming pools, while others may send it into shallow wells on their property. There is no organized reinjection and the water is not returned to the BWSWD.

The annual hot water usage in the BWSWD since 2008 is shown in Fig. 5.14. For example in 2016–17, the average volumetric flow rate was 1.77 m^3/min (468.0 GPM) or an average mass flow rate of 28.2 kg/s. The two lower consumption years corresponded to warmer winters than normal in 2008–09 and 2014–15; the highest consumption occurred in colder winters, 2011–12 and 2016–17.

5.3.2.2 State of Idaho Capitol Mall (SICM)

The State of Idaho began to exploit the Boise geothermal resource in 1981 to serve state government buildings. Two wells were drilled: an injection well, CM#1, 656 m (2152 ft) deep; 12-in casing to 533 m (1750 ft); open hole to total depth; and a production well, CM#2, 923 m (3030 ft) deep; 12-in casing, solid to 383 m (1258 ft), perforated to 777 m (2550 ft); open hole to depth. The producer is capable of delivering more than 3.4 m^3/min (900 GPM) or 55.5 kg/s, initially at a temperature of 71 °C; the current temperature is 68 °C. Neither well intercepts the fault system associated with the Foothills Fault because they are completed into the down-

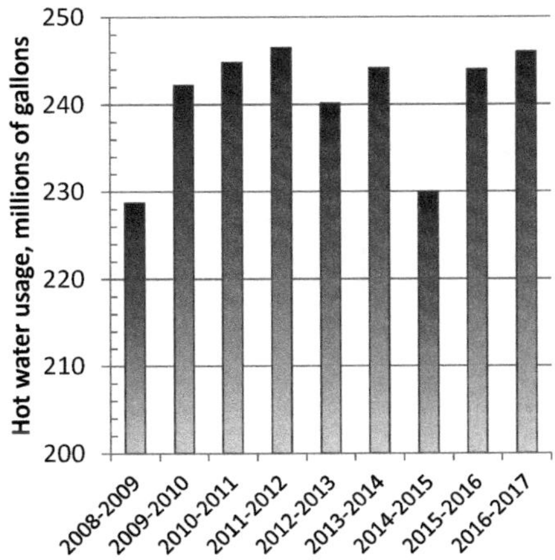

Fig. 5.14 Annual hot water usage in BWSWD [14]; Pers. Comm. Del Eytchison. Note: 1000 gallons = 3.786 m^3.

thrown layers of rhyolite that are characterized by good permeability; see Fig. 5.15. The wells are separated by about 377 m at the surface but end up in different layers of rhyolite at different depths.

The SICM system serves the State Capitol and 10 other buildings (see Fig. 5.16). The total gross heated floor area amounts to over 8.16 ha. The system is a double-pipe arrangement where the produced geofluid in delivered in one pipeline and after use is collected in a return pipeline and reinjected. When the system began operations in 1982, only seven buildings were served. The DPW building was then a shop, but joined the system in 1984. The Alexander House was added in 1992, but only uses collected cooled water just prior to its being reinjected. The Parking Garage #2 was built in 2014 (around and over the wellhead of the production well) and added to the system; however, it uses a very small amount. The Capitol Annex (formerly the Old Ada County Courthouse) was added in 2015 when its renovation was completed.

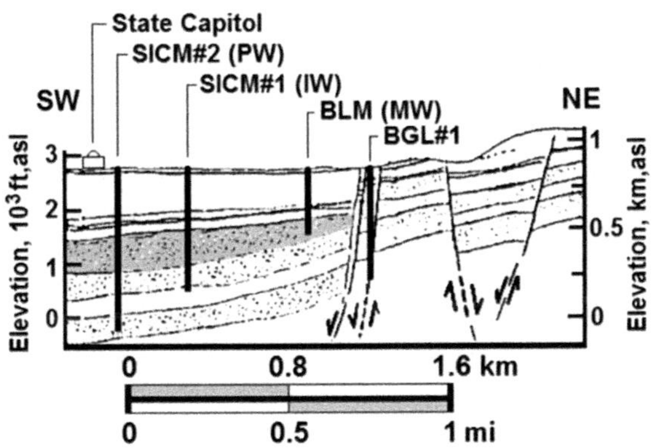

Fig. 5.15 Cross-section looking NNW through the SICM, BLM monitor, and CB (BGL) wells; see Fig. 5.10, modified from [15]. See Fig. 5.12 for rock types.

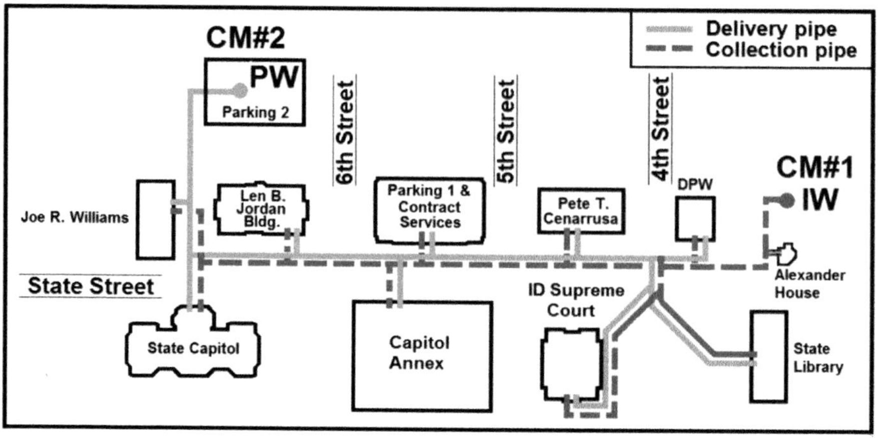

Fig. 5.16 State of Idaho Capitol Mall geothermal water distribution system. Notes: Parking garage 2 uses very little hot water, only in stairwells, and returns the used water to the supply line with insignificant drop in temperature; Alexander House receives no hot water from a supply line but only takes some of the cooled geofluid and returns it to the collector after use, modified from [16]; Ric Johnston pers. comm.

As was mentioned earlier, it is noteworthy that the Idaho State Capitol is the only U.S. capitol building fully heated by geothermal energy. Furthermore, all the buildings shown in Fig. 5.16 rely solely on geothermal and have no back-up boilers. This is feasible because of an interconnection between the SICM and the CB systems, whereby if one system is out of commission for maintenance, the other system supplies the needed hot water. This has happened three times over the last eight years. The total heat supplied annually to the 11 buildings amounts to 26.4–33.8 TJ, depending on the severity of the winter season. The running cost for this heat is only for the electricity and maintenance of the production and injection pumps.

5.3.2.3 *City of Boise (CB)*

The City of Boise began their district heating operations in 1983 when they bought the three wells BGL-2, -3, and -4 that had been drilled in 1981 by Boise Geothermal Ltd. Close by was the BLM well that serves as a monitoring well; see Fig. 5.10. In 1988 the distribution system was fully operational. The system grew to serve 92 buildings with a total heated area of 58.5 ha. Initially the cooled water was disposed of in the Boise River, but that ended in 2001. Table 5.3 gives some details on the CB wells.

The distribution system is a double-pipe arrangement: one line supplies hot water, the other collects the cooled water from each user. The water is used only as a heating medium, and each user employs a heat exchanger appropriate to their needs. The temperature and pressure in the supply line are 71–79 °C and 345–448 kPa g, while in the collection line they are 43–49 °C and 172–241 kPa g. Note that the return temperature is near ideal for balneology, but no spas are planned for this water, and it is being reinjected directly. However, there are plans to use the disposal line as a heat source for heat pumps during winter in the section of the city closest to the injection well.

Table 5.3 Production well information for CB system [17].

	BGL-2	BGL-3	BGL-4	Inj. well
Year Drilled	1981	1981	1981	1999
Depth, m	268	579	335	1097
Temperature, °C	80.6	72.2	80.6	46.1[a]
Pump power, kW	93	149	37	N.A.
Max. flow rate, kg/s	73.6	122.6	49	219
Max. flow rate, GPM	1200	2000	800	3500

[a]Injectate average temperature.

There are some cross-over connections between the supply and collection lines that can be used to maintain the water velocity. A few users actually use the lower-temperature water for specific purposes. Disposal is via an injection well about 370 m from the Boise River. It is about 1.8 km southwest of the production wells, but is significantly deeper than the production wells, namely, about 1100 m deep. Being ~2 km from the Foothills Fault, the injection well may have found a permeable layer within a rhyolite formation or perhaps at the interface of the rhyolite and the granitic formations; see Fig. 5.12.

The water level in the producing wells has recovered to essentially its preproduction levels after reinjection began. Prior to using the injection well, the water level had been drawn down by about 10 m below the original level, but as soon as injection began in 1999, the level began to rise and now (2018) has stabilized at about 0.6–0.9 m below the undisturbed level. This is crucial because the pumps must be set below the water level to function. Based on the pressure recovery observed in the monitoring wells that are close to the Foothills Fault, there is hydraulic connection between the production wells and the injection well, but it is not too strong since the production temperature has so far been unaffected.

5.3.2.4 Veteran's Administration (VA)

The United States Veterans Administration (VA) owns two wells, one producer (VA#1) and one injector (VA#2), that allow the VA to heat their complex of buildings with geothermal water. The wells were drilled in 1983 and have the following dimensions: VA#1: 508 m (1666 ft) deep, 16-in casing to 43 m (140 ft), 12-in to 244 m (802 ft), 8-in perforated to 457 m (1500 ft), open hole to depth; VA#2: 561 m (1840 ft) deep, 10-in casing to 40 m (130 ft), 7-in perforated to 451 m (1480 ft), open hole to depth. The wells are separated by about 538 m [18].

The production well appears to penetrate the fracture zone associated with the Foothills Fault, the same area feeding the nearby City of Boise wells, and possibly a portion of the upper rhyolite formation used by the SICM injection well [18]; see Figs. 5.10 and 5.15.

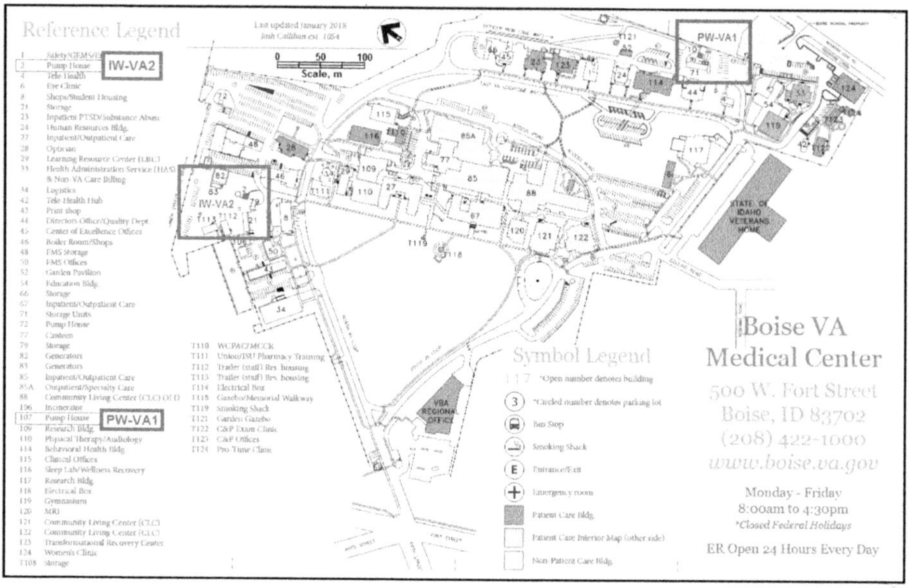

Fig. 5.17 Boise Veterans Administration Hospital: Geothermal wells [19].

The heating system was put into operation in 1988. Approximately 3.72 ha spread over 22 buildings on the hospital grounds receive heat in this way; see Fig. 5.17 [19]. From 1997 to 2002, the VA produced on average 1.34 m^3/min (354 GPM) or 21.8 kg/s. The geofluid temperature has held fairly steady at 71 °C throughout the operation of the system. All produced geofluid is returned to the reservoir [20].

5.4 Geothermal heated greenhouses

5.4.1 Basic concepts and engineering

As is true of all greenhouses, geothermally-heated greenhouses allow growing on a year-round basis in all climates, particularly cold ones with a limited outdoor growing season. Low-to-moderate geofluid temperatures can be used effectively to control the soil and air temperatures in the greenhouse. These geofluids would not be sufficiently energetic for use in a geothermal power plant but are well-matched to agricultural and horticultural applications. A comprehensive manual of geothermal greenhouse design and operation is available in Ref. [21].

Over the last 25 years there has been steady growth in geothermal greenhouse heating, both in terms of installed thermal power capacity and in energy utilization; see Table 5.4. The same has not been true for geothermal aquaculture (discussed in the next section) which retracted and then has grown modestly for the last ten years.

Many important crops thrive under temperature conditions easily achieved with geofluids. For example, Fig. 5.18 shows the growing curves for three common crops

Table 5.4 World geothermal energy usage for greenhouses and aquaculture [2].

Year	Greenhouse	Aquaculture	Greenhouse	Aquaculture
	Capacity, MWt		Energy, TJ	
1995	1085	1097	15,742	13,493
2000	1246	605	17,864	11,733
2005	1404	616	20,661	10,976
2010	1544	653	23,264	11,521
2015	1830	695	26,662	11,958

as functions of temperature. Lettuce does best at ~13–14 °C, tomatoes at ~20 °C, and cucumbers at ~27–28 °C. On large farms, each crop can be confined to its own building where ideal conditions can be maintained. Table 5.5 gives recommended temperatures for a variety of crops.

Natural gas, where available, may be used to provide heat to greenhouses in cold climates, but at the expense of the environment owing to the emissions of carbon dioxide, ironically called a "greenhouse gas." Geothermal greenhouses often rely on backup natural gas-fired systems in times of extreme cold weather. Carbon dioxide, which typically is found dissolved in geofluids, is often separated from the geofluid and used within the greenhouses to enhance crop growth.

Engineering design of a greenhouse begins with the application of the principles of thermodynamics and heat transfer. Once the site and materials of construction (glass, plastic, polyethylene, fiberglass, wood or aluminum framing) are decided, the analysis of heat loss can get started. Heat losses are related to conduction, convection and radiation, lumped together as transmission losses. Additional heating is needed to warm outside air that is used for ventilation in accordance with requirements for air-change.

It is common to select as the design outdoor conditions not the coldest temperature on record, but one that applies for all but about 22 days of the heating season [23]. This

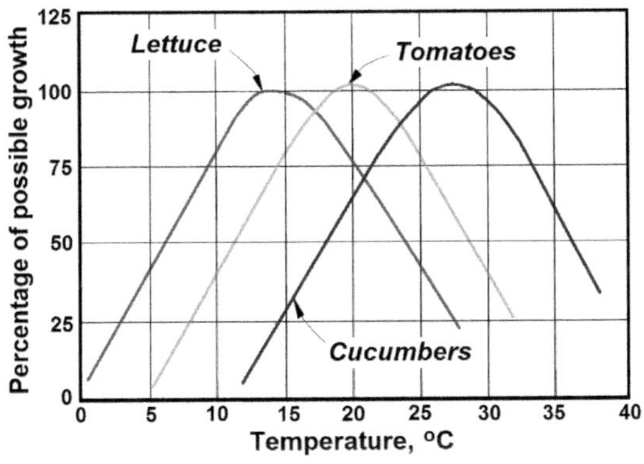

Fig. 5.18 Growth curves for three crops versus temperature [22].

Table 5.5 Temperature requirements for typical greenhouse crops, °C [23].

	Day	Night
Vegetables		
Peppers	18–29	16–18
Tomato	21–24	17–18
Cucumber	24–25	21
Lettuce (hydroponic)	24	18
Flowers		
Roses	16–17	17
Poinsettias	21–27	18–22
Easter Lilies	16	n.a.
Carnations	24	10
Geraniums	21–27 (max)	n.a.
Fuchsia	21 (min)	18 (min)

means that some auxiliary heating may be needed for a small period of time but prevents an over-designed system for most times. A similar selection may be made for winds as they are a strong determinant of convective losses. For example, a wind of 13 m/s creates about 50% higher heat loss than still air.

Transmission losses are proportional to the exposed surface area, whereas the ventilation losses are proportional to the volume of the building. Generally the heating is done by means of floor coils and fan coils; the former may be used to heat the soil for the crops, while the latter is used to maintain the air temperature at desired levels. One problem with buried heating coils in soil is that a temperature gradient is set up whereby some plants see soil hotter than the optimum while others see lower temperatures, and only some see the ideal temperature. This can create uneven growth conditions.

A generalized schematic of the geothermal heating system is shown in Fig. 5.19. The temperatures in the circuit will be tailored to the conditions in the geothermal reservoir and the needs of the crops. In most cases disposal is by means of an injection well.

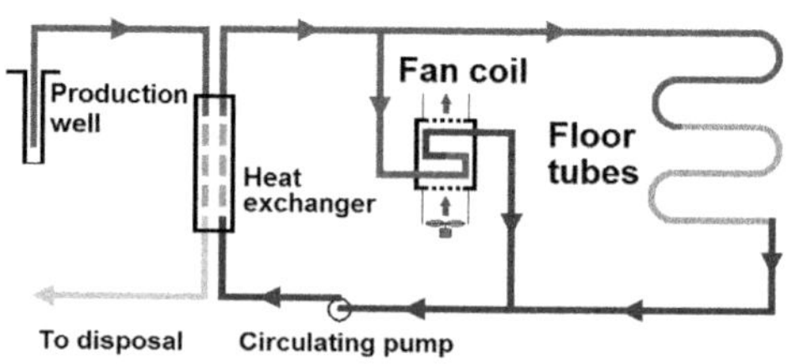

Fig. 5.19 Simplified generalized arrangement for a geothermal greenhouse.

5.4.2 Case study: Cuckoo Polder, Netherlands

The Netherlands, a country of roughly 17 million people and covering 4.2 million hectares, is beginning to exploit its significant geothermal potential. The source of the energy is not the traditional volcanic- or tectonic-related variety, but deep hydrothermal fluids found in sediments below 2500 m depth. Fig. 5.20 maps the areas with good and possible resources; note that two large cities, The Hague and Rotterdam, lie within the good areas.

The example chosen for this section is the greenhouse development known as “Aardwarmteproject Koekoekspolder” or “Geothermal energy project at Cuckoo Polder.” It is located in IJsselmuiden where extensive horticultural development is taking place. The current complex of greenhouses is served by one doublet, i.e., one production and one reinjection well. There are plans to enlarge the system by adding up to six more doublets. The concept involves a geothermal cluster in which crops with differing temperature requirements are served in parallel and in series to guarantee appropriate growing conditions for each crop. The various greenhouses are owned by different entities that cooperate to allow the optimum flow and growing conditions to be met continuously even under changing conditions. All flows are computer controlled using inputs from the users.

Cuckoo Polder is shown in Fig. 5.21 within the roughly rectangular tilted area bounded by roads. The doublet is marked by the star. It is within this area, roughly 1800 m × 2290 m or 412 ha, where up to six more doublets are expected to be drilled.

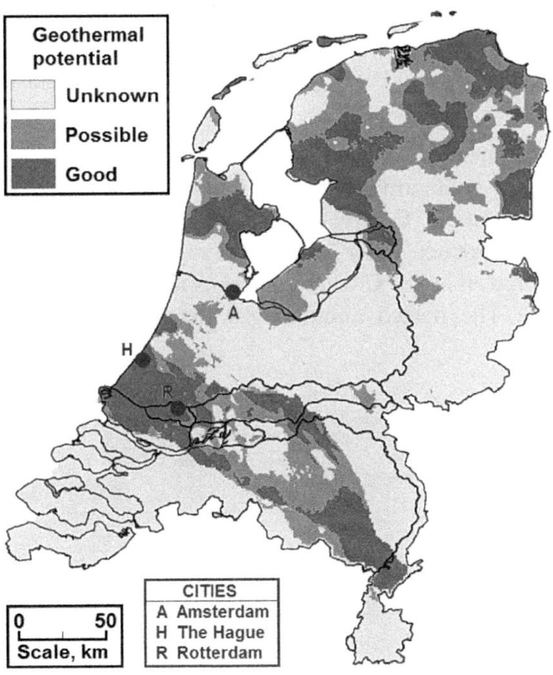

Fig. 5.20 Areas with geothermal potential in the Netherlands, modified from [24].

Fig. 5.21 Satellite view of Cuckoo Polder greenhouses. Google Earth image, Sept. 24, 2016.

The cluster arrangement is depicted in Fig. 5.22. A doublet of wells, separated by about 10 m, services a geothermal hot water plant. The geofluid is pumped from a depth of about 2500 m at 74 °C. It passes through a trio of plate-and-frame heat exchangers where it heats treated water from 26 °C to about 72 °C. The cooled geofluid is sent to the injection well at 28 °C where it returns to the reservoir about 1.5 km from the production spot. The heated circulating water travels through a sequence of

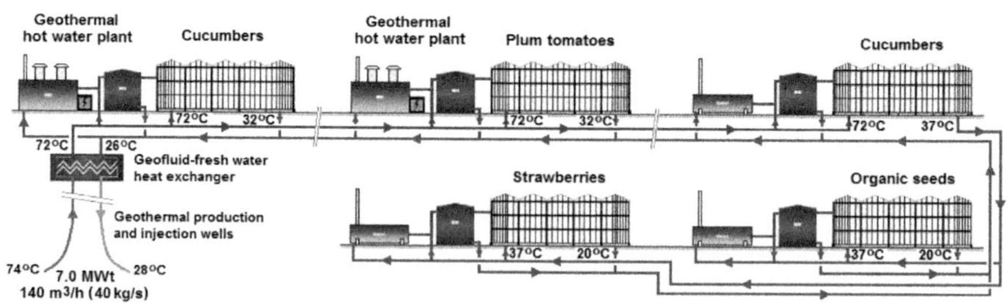

Fig. 5.22 Greenhouse arrangement at Cuckoo Polder, after [24].

Fig. 5.23 Piping in the hot water plant at Cuckoo Polder [25].

greenhouses via a double-pipe system wherein the hot water feeds each greenhouse off the supply line and the cooled water is collected in the return line. The hottest water is used to grow cucumbers and plum tomatoes. The water is reduced to about 37 °C after these greenhouses, and is further used for low-temperature growing of organic seeds and strawberries. By then the water temperature has dropped to essentially 20 °C where it is returned to the collector line, mixing with the water at 32 °C being discharged from the first two greenhouses, and arrives at the heat exchanger for reheating. Some of the piping in the hot water plant is shown in Fig. 5.23.

There are three plate-and-frame (P&F) heat exchangers in the hot water plant. P&F units are particularly appropriate for this application owing to their flexible design and tolerance of impure geofluids. Fig. 5.24 is a schematic of a P&F heat exchanger showing how the hot fluid and the cold fluid pass on opposite sides of corrugated, thin plates. Gaskets prevent mixing of the fluids and keep them in their proper channels. The units are very

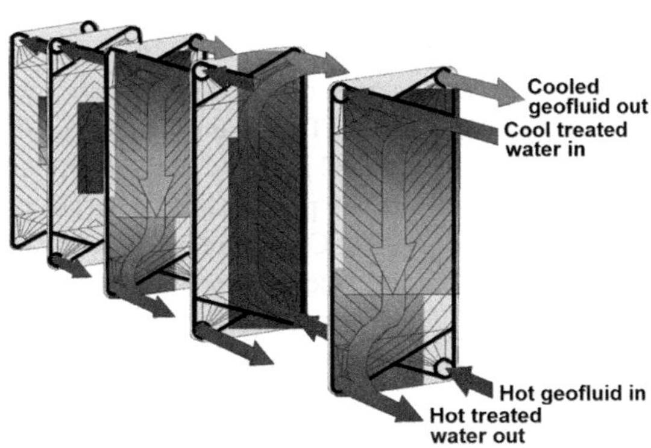

Fig. 5.24 Plate-and-frame heat exchangers as used in Cuckoo Polder greenhouses, after [22].

efficient and compact. Owing to their flexible arrangement, i.e., plates can be removed or added as needed, these units can be adjusted depending on the geofluid temperature or heating requirements. They are also easy to disassemble for periodic cleaning.

For more information about greenhouses in the Netherlands, particularly the history, policies, and regulation of the industry, the reader is referred to Ref. [26].

5.5 Geothermal aquaculture

5.5.1 Introduction

Aquaculture is the practice of raising aquatic species under controlled conditions to enhance growth and minimize disease. Geothermal aquaculture uses geofluids, whether taken directly from purpose-drilled wells or tail waters from power production or direct heating, to maintain uniform temperature and fluid chemistry for optimum growing conditions.

A guide to the use of geothermal water in the raising of aquatic species may be found in Ref. [27]; only a summary can be given here. The geothermal fluid must be of the proper temperature to match the growing needs of the desired species. Where the geofluid is too hot, as is often the case for waste fluids from power plants, fresh water is mixed with the geofluid to achieve the desired temperature for the particular species being raised.

Tilapia is one of the most popular fish for geothermal aquaculture owing to their love of warm water and tolerance for brackish conditions. Other species that match up well with geothermal fluids include shrimp, giant prawns, carp, eels, catfish, sea bass, striped bass, sturgeon, arctic char, salmon, and even crocodiles and alligators. Oregon in the U.S. is home to several commercial aquaculture operations including a company called "Gone Fishing" that raises tilapia for sale in large cities [28]. The geofluid is obtained by pumping from wells; the fluid temperature is 82 °C.

5.5.2 Case studies: prawn, salmon and arctic char farming

Giant Malaysian prawns grow best in water at 27–30 °C and reach maturity in 5–7 months. Geofluids that have been used for space heating are often suitable in temperature to be further used in aquaculture. This is the case at the Oregon Institute of Technology (OIT) in the U.S. where 56 °C geofluid leaves the campus district heating system and flows into ponds where the prawns are raised. A temperature control system maintains the pond water at 27 °C despite very cold winters [29]. Many practical suggestions based on experience can be found in this reference. The most important fact is that such facilities could not be contemplated in cold climates without the availability of naturally warm or hot geofluids owing to the economics involved.

Iceland is one of the leading countries in geothermal aquaculture owing to the abundance of geothermal resources. Figure 5.25 is a schematic flow diagram for one such

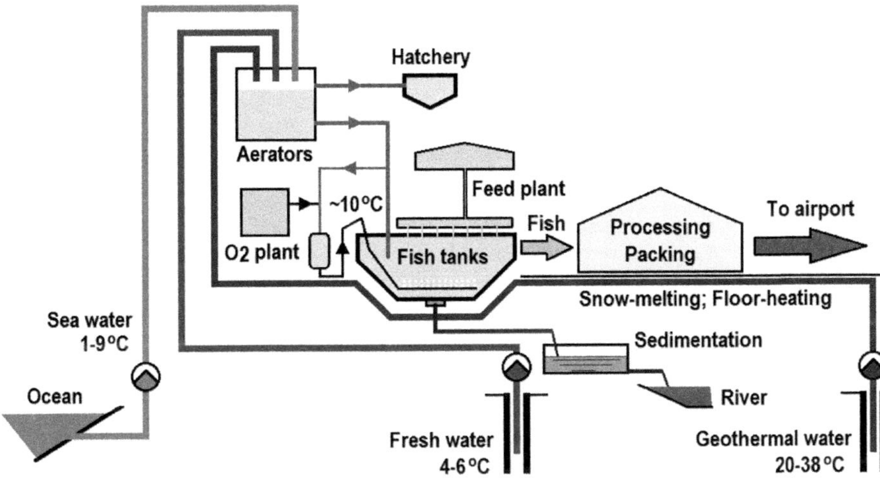

Fig. 5.25 An Icelandic geothermal fish farm flow diagram, redrawn from [30]. The emergency back-up system is not shown.

plant. It is located in northern Iceland where there is easy access to cold ocean water, cold fresh water, and geothermal fluids at various temperatures from very shallow wells. By careful mixing, the ideal conditions are created for raising salmon and arctic char. Given that it can be fatal to the fish if there is a failure anywhere in the fluid supply system, it is necessary to provide an emergency back-up system that can be a significant initial capital expense.

An interesting application of geothermal aquaculture is associated with the Wairakei geothermal power station in New Zealand [1]. A fish farm raises giant Malaysian river prawns with the aid of the hot effluent from the power plant. Located close to Taupo, the facility opened in 1987 and currently is the only prawn farm in New Zealand. Since the waste brine from the power plant is not suitable for direct use in the growing ponds (it is laden with silica and too hot), about 70 kg/s (in summer) and 110 kg/s (in winter) of 90 °C brine is used in heat exchangers to maintain the right temperature for the prawns. Fresh water from the Waikato River adjacent to the site is heated to 27–31 °C and used in the ponds; see Fig. 5.26. The farm has its own hatchery on site. The prawns reach maturity in eight months, and about 7.8 tonnes of prawns can be produced each year from the 19 ponds covering 2.75 ha on the farm; two of the ponds are dedicated to prawn fishing by the public [31,32].

The unique aspect is that the farm is simultaneously a tourist attraction. The Huka Prawn Park is a family venue where the prawns may be seen in their ponds and patrons may fish for them from the banks of the ponds. There is also a restaurant on-site where the products of the farm may be enjoyed [33]. In this way, the technical venture and the tourist attraction form a combined operation that not only enhances the owner's profits, but also demonstrates the attractiveness of geothermal energy to the general public.

Fig. 5.26 Huka prawn farm and Wairakei geothermal power station, New Zealand; enhanced Google Earth image of January 10, 2008.

5.5.3 Case study: alligator and crocodile farming

Many of the 23 species of crocodiles and alligators are critically endangered or vulnerable [34]. Because protection for these creatures prohibits their taking for leather, meat, or other commercial products, there is motivation to raise them under controlled farm conditions. Furthermore, since they thrive in warm waters, using geothermal fluids is a natural means of providing a suitable environment. As with other forms of geothermal aquaculture, geofluids that may not be suitable for direct use may then be used as a heating medium to create the desired temperature conditions to produce a thriving population of crocodilians.

There is a synergy available when combining fish farming with crocodilian farming, namely, the use of the waste products from the former to provide feed for the latter. Instead of dumping the waste from fish production and harvesting into rivers or landfills, crocodiles and alligators will eat that material. One example is the Colorado Gators reptile park in south-central Colorado, near Hooper, in the U.S. The operation started as a geothermal tilapia farming facility, until the owners realized that alligators would solve their fish waste disposal problem [35]. In 1987 they bought 100 baby alligators and thus began a combined tilapia-alligator farm. Furthermore, like the Huka Prawn Park, they later added a public recreation area that now is a popular tourist attraction.

For a comprehensive exposition on raising crocodiles in farms, the reader is referred to Ref. [36].

5.6 Heating circuits with single- and two-stage heat pump systems

In this section we will present in technical detail how heat pumps can be designed to work with low-to-moderate geothermal waters to provide heating for various applications, including the ones described in the previous sections.

Low-temperature geothermal resources are by far the most prevalent. About 68–72% of geothermal energy reserves have temperatures less than 130 °C; see Fig. 5.27. Currently, of the geothermal resources most widely used for binary plants, more than 77% have a water temperature of 100–200 °C.

The economic and thermodynamic efficiency of geothermal systems is determined in part by the temperature of the geothermal fluid after use by the consumer (electric power or heat supply system). Existing geothermal power systems discharge waste geofluid with a final temperature of 80–100 °C and above, which makes them less energy efficient. Lowering the temperature below 30–35 °C is possible with the use of heat transfer technologies in combined heat supply systems.

In contrast to high-temperature geothermal systems using expensive wells with a depth of more than 1500–2000 m, the use of low-temperature thermal waters is economically more efficient. Significant reserves of fresh and slightly mineralized waters with a temperature of 20–50 °C occur at shallow depths, but these waters are not used for electricity generation. However, in comparison with medium- and high-temperature thermal waters, low-temperature waters have a number of advantages: low capital costs for their production (wells with a depth of 50–500 m); low mineralization, which reduces scaling and corrosion of metal pipes; and the possibility of implementing an artesian (self) flow mode or pumping with low energy costs.

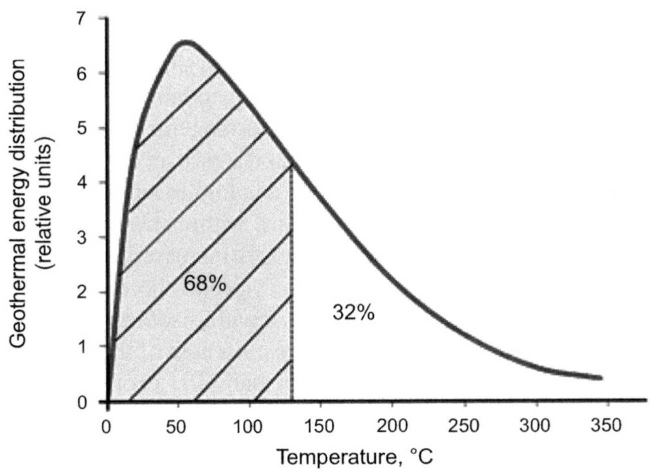

Fig. 5.27 Distribution of world geothermal energy reserves as a function of resource temperature.

5.6.1 Single-stage heat pumps

Using geothermal waters in the temperature range of 30–60 °C as low-energy heat sources for heat pump units (HPU) allows raising the HPU condensing temperature to 90–95 °C, leading to a high coefficient of performance, *COP*. According to the thermophysical properties of heat pump working fluids, as the condensing temperature in the cycle increases, the latent heat of condensation decreases and the inner irreversible losses due to throttling increase, which causes a reduction in *COP*. It is possible to provide a high temperature for heat transfer (85–90 °C) using a single-stage heat pump system by choosing high-temperature working fluids as refrigerants. At the same time, the total heat output may decrease, which would result in smaller heat exchangers and compressors for the HPU. In any case, the selection of the working fluid for the HPU, subject to predetermined temperature limits of a thermodynamic cycle, will always involve a compromise. This poses a challenge for the system designer to achieve both high performance and high temperatures for heating applications.

In this connection, by using a numerical simulation of the thermodynamic characteristics of a geothermal heat pump with a given technological scheme we can find a working fluid which will provide effective performance under high-temperature conditions. The influence of the particular technological scheme of the HPU elements on the heat transfer efficiency will be considered.

The efficiency of using a low-potential geothermal fluid in a HPU also depends on its final temperature. This is achieved by the use of one HPU or two or more HPUs connected in series. With the successive movement of thermal water through the evaporators, the HPU allows the process of evaporation of the working fluid at different temperature levels, which leads to an increase in the total COP and saving energy on the compressor drive.

Depending on the parameters of the geothermal fluid (flow rate and temperature) and the customer's requirements for the final temperature, up to three HPUs can be included in the system. Cascade or multi-stage HPUs can decrease the thermodynamic losses while raising the complexity and initial cost of the system. We will examine this generalized case, and later specialize it to the simplest single-stage system.

It is possible to expand the working range of geothermal heating systems with HPUs using several units in a series counterflow arrangement; see Fig. 5.28A. Water heating in a condenser and cooling of geothermal water in an evaporator are realized in stages, in which each successive cycle operates at a higher temperature level for evaporation and condensation of the refrigerant. Thus, within specified temperature limits of the basic working cycle, as the number of stages increases, the performance of the HPU approaches the ideal limit of the Carnot cycle. In the unattainable limit of an infinite number of stages, an effective cycle having non-isothermal processes of evaporation and condensation would be reached. The closeness of the approximation to the Carnot cycle is measured by the ratio of the actual HPU coefficient of performance COP_{HPU} to that of the theoretical Carnot HPU cycle COP_{Carnot} [37–41].

Fig. 5.28B shows a single HPU having a multi-stage compressor which can be configured with interstage cooling to reduce the overall work of compression [42–45]. Fig. 5.28C shows an open geothermal heating system, supplying both hot

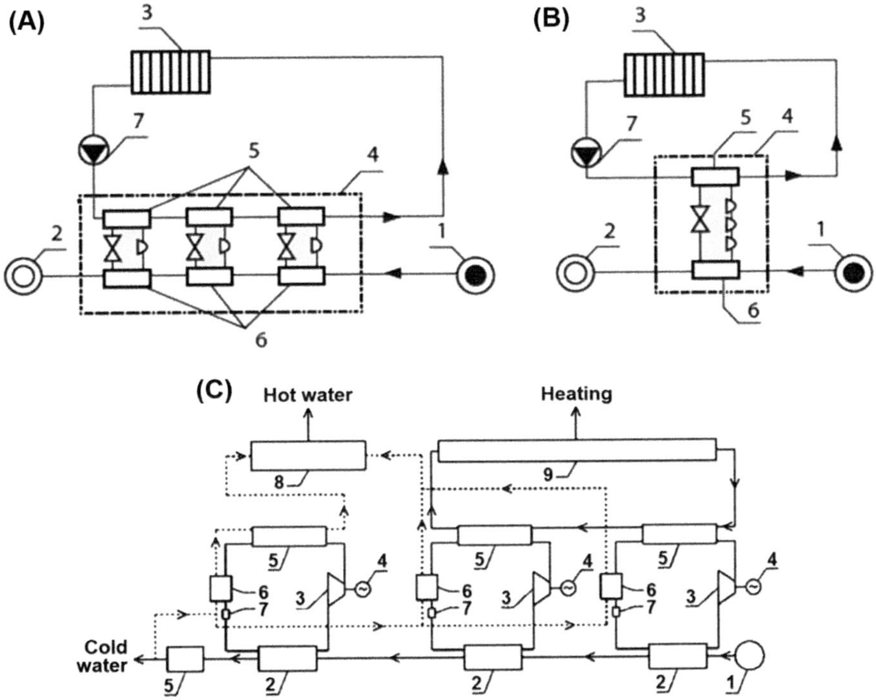

Fig. 5.28 Principal multi-stage HPU configurations for geothermal heating: (A) Series, countercurrent, cascade system; (B) Single HPU with multi-stage compression. *1*, geothermal production well; *2*, injection well; *3*, heat consumer; *4*, HPU; *5*, condenser; *6*, evaporator; *7*, circulation pump. (C) Open geothermal heating system, hot and cold water supply (see text for equipment identification).

and cold water. Fresh or slightly mineralized water is supplied from well 1. In the evaporators 2 the water is cooled; it is cleared of chemical and mechanical impurities in block 5, and supplied to the consumer. A part of the purified water is supplied to the coolers 6 and the condenser 5 of one of the HPUs where it is heated and supplied to the consumer 8. The other two HPUs provide water heating to a higher temperature which supplies the heating system 9. Each HPU consists of an evaporator 2, a compressor 3, electric motor 4, condenser 5, cooler 6 and throttle 7.

To improve the efficiency of simple, single-stage heat pumps, the following equipment is installed: a receiver after the condenser ensures a uniform supply of the working substance to the throttle valve and maintains the pressure in the condenser; an aftercooler following the condenser provides additional cooling of the working fluid with an external cooling medium. The thermal performance of a heat pump with a subcooler SC (Fig. 5.29) is determined by the properties of the working fluid and the actual operating conditions [46–50].

At higher working fluid temperatures, the irreversible throttling losses increase (cf. processes 4-1′ and 5−1). To reduce these losses, a subcooler (SC) is included in the

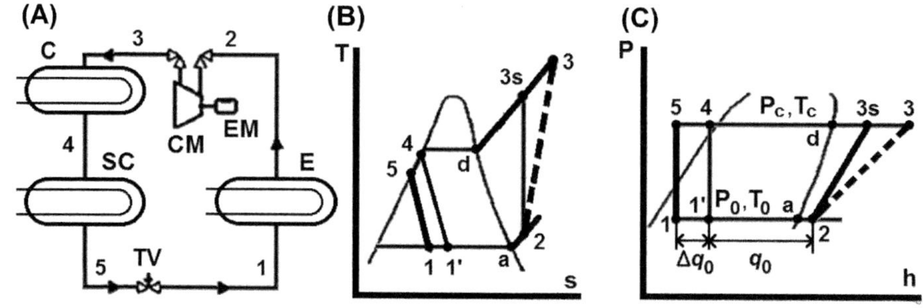

Fig. 5.29 Refrigerator or heat pump with subcooler SC: (A) system; (B) *T-s* cycle; (C) *P-h* cycle.

heat pump system (Fig. 5.29). Subcooling the working fluid below the condensation temperature increases the heat exchanged in the evaporator from the heat source to the working fluid (per unit of working fluid) and reduces the specific electric power consumption in the compressor per unit of delivered heat. Furthermore, the higher the condensation temperature, the deeper the cooling of liquid condensate. The throttling process $4-1'$ (Figs. 5.29 and 5.30) is isenthalpic, i.e., $h_4 = h_{1'}$ and $h_5 = h_1$. The temperature drop of the liquid working fluid in the condensate cooler (process 4–5) from T_c to T_5 results in an increase in the specific heat output for the value $h_4 - h_5$.

The (T,s) and (P,h) diagrams show the process of subcooling the working fluid (4–5). The COP for a refrigerator with a subcooler is determined by the formula:

$$COP = \frac{q_0 + \Delta q_0}{w} \tag{5.1}$$

$$\Delta q_0 = h_4 - h_5 \tag{5.2}$$

If technical and economic calculations show that the value of the efficiency improvement is greater than the cost of the subcooler, then installing a subcooler is feasible.

A regenerative heat exchanger may be installed on the liquid line of the refrigerant after the condenser and on the vapor line after the evaporator (Fig. 5.30) to provide subcooling of the liquid before entering the throttle valve which results in increased specific cooling capacity.

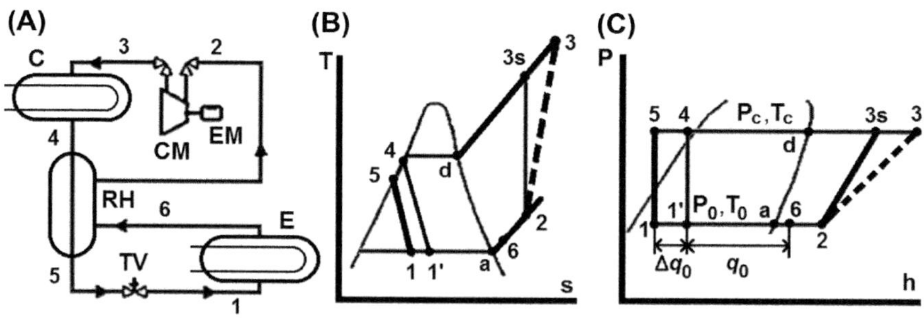

Fig. 5.30 Refrigerator or heat pump with regenerative heat exchanger RH: (A) system; (B) ***T-s*** cycle; (C) ***P-h*** cycle.

Systems using working fluids with high heat capacity of saturated liquid are more effective. The installation of the regenerative heat exchanger ensures "dry running" of the compressor, i.e., superheated vapor at the compressor inlet, and an increase in the specific cooling capacity. The COP for a refrigerator with a regenerative heat exchanger is determined by the formula:

$$COP = \frac{q_0 + \Delta q_0}{w + \Delta w} \tag{5.3}$$

where, $\Delta w = w_2 - w_1$, is the increment of the compressor power in the scheme with the regenerator (w_2), and without the regenerator (w_1); Δw arises from the divergence of isobars as the temperature is raised.

If $\Delta q_0 < \Delta w$, then the cycle with the regenerative heat exchanger is less efficient. Even if the COP is not improved, the regenerative heat exchanger provides for a dry compression process which is more effective and practical than a wet one.

An economizer EC installed after the condenser, also provides liquid subcooling before throttling (Fig. 5.31). However, the economizer does not guarantee a dry compression process. The subcooling in the economizer, process (4—6), before throttling (6—1) is carried out at the expense of part of the useful cooling capacity.

The specific cooling capacity q_0* obtained in the process 5-2″ is internal to the cycle, whereas the specific cooling capacity q_0 in the process 1-2′ is the beneficial effect of the refrigerator. The maximum increase in specific cooling capacity by subcooling the working substance before throttling is $\Delta q_0 = h_4 - h_6$.

The heat balance of the economizer EC is determined by the formula:

$$(\dot{m}_4 - \dot{m}_5)(h_4 - h_6) = \dot{m}_5(h_{2''} - h_5) \tag{5.4}$$

where $\dot{m}_5$ is the liquid flow rate bled from state 4 and sent to the first throttle valve TV1. If the mass flow through the compressor is defined as unity, i.e., $\dot{m}_4 \equiv 1$, then we find

$$h_6 - h_4 = \left[\frac{\dot{m}_5}{1 - \dot{m}_5}\right](h_5 - h_{2''}) \tag{5.4a}$$

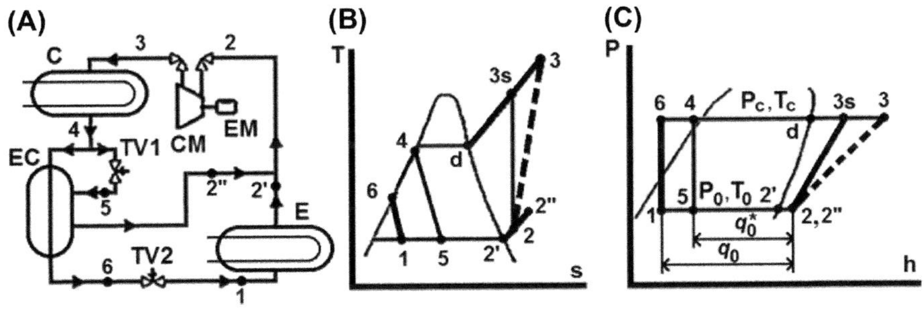

Fig. 5.31 Refrigerator or heat pump with economizer: (A) system; (B) *T-s* cycle; (C) *P-h* cycle.

The inlet enthalpy to the compressor is found from the mixing equation as follows:

$$h_2 = \dot{m}_5 h_{2''} + (1 - \dot{m}_5) h_{2'} \tag{5.4b}$$

The subcooling temperature at point 6 is determined by the formula:

$$T_6 = T_5 + \Delta T \tag{5.5}$$

$$\Delta T = 2 - 5^\circ\text{C (typ.)} \tag{5.6}$$

The COP for a refrigerator with an economizer is determined by the formula:

$$COP = \frac{q_0}{w} = \frac{(1 - \dot{m}_5)(h_{2'} - h_1)}{h_3 - h_2} \tag{5.7}$$

With reference to the single-stage system with a subcooler shown in Fig. 5.29, the actual coefficient of performance *COP* of the heat pump cycle (a-2-3-d-4-5-1-a) is determined by the formula:

$$COP = \eta_i \eta_{el} (h_3 - h_4)/w \tag{5.8}$$

where η_i is the isentropic efficiency of the compressor, η_{el} is the electromechanical efficiency of the compressor, h_3 and h_4 are the enthalpy of the working fluid, and w is the specific isentropic compression work for an ideal compressor, the process 2—3s.

The specific isentropic compression work w in the compressor is obtained from the equation:

$$w = \frac{k}{k-1} P_e v_2 \left[\left(\frac{P_{3s}}{P_2} \right)^{\frac{k-1}{k}} - 1 \right] \tag{5.9}$$

where k is the isentropic index; P_2 and P_{3s} are the evaporation and condensation pressures, respectively, kPa; and v_2 is the compressor inlet specific volume, m^3/kg.

The *COP* for heat pump system with a subcooler, regenerative heat exchanger, or economizer is given by:

$$COP = \frac{\eta_i \eta_{el} [(h_3 - h_4) + (h_4 - h_5)]}{w} \tag{5.10}$$

The working fluid mass flow rate is calculated in the following way:

- For a HPU without a condensate cooler:

$$\dot{m}_1 = \frac{\dot{m}_w c_{pw} (T_{input} - T_{output})}{h_2 - h_4} \tag{5.11}$$

- For a HPU with a condensate cooler:

$$\dot{m}_2 = \frac{\dot{m}_w c_{pw}(T_{input} - T_{output})}{(h_2 - h_4) + (h_4 - h_5)} \tag{5.12}$$

where $\dot{m}_w$ is the mass water flow rate, kg/s; c_{pw} is the water specific heat capacity, kJ/(kg K); T_{input} and T_{output} are the HPU evaporator input and output water temperature, °C. The power to drive the HPU compressor drive is found from:

- For a HPU without a condensate cooler:

$$\dot{W}_1 = \frac{\dot{m}_1(h_2 - h_3)}{\eta_{em}} \tag{5.13}$$

- For a HPU with a condensate cooler:

$$\dot{W}_2 = \frac{\dot{m}_2(h_2 - h_3)}{\eta_{em}} \tag{5.14}$$

One can determine quantitatively the advantage of any HPU system for given conditions by adopting the ratios $\dot{W}_1/\dot{W}_2$ and $\dot{Q}_1/\dot{Q}_2$, where:

$$\frac{\dot{W}_1}{\dot{W}_2} = \frac{\dot{m}_1}{\dot{m}_2} = \left(1 + \frac{h_4 - h_5}{h_2 - h_4}\right) \tag{5.15}$$

- For a HPU without a condensate cooler:

$$\dot{Q}_1 = \dot{m}_1(h_3 - h_4) \tag{5.16}$$

- For a HPU with a condensate cooler:

$$\dot{Q}_2 = \dot{m}_2(h_3 - h_5) \tag{5.17}$$

$$\frac{\dot{Q}_1}{\dot{Q}_2} = \left(1 + \frac{h_4 - h_5}{h_2 - h_4}\right)\left(\frac{h_3 - h_4}{h_3 - h_5}\right) \tag{5.18}$$

With addition of a condensate cooler to the HPU, other conditions being equal, the compressor power requirement and heat delivery capacity are reduced. However, the percent reduction in power is higher than that of heat delivery, which indicates that the COP is higher for HPUs with a condensate subcooler.

5.6.2 Two-stage and multi-stage heat pumps

Preliminary calculations carried out for a 3-cycle system using the refrigerants R114, R245fa, R141b showed that increasing the number of cycles beyond three produces a negligible increase in COP_{tri}/COP_{Carnot}; see Fig. 5.28C. It is possible to select several sets of refrigerants for each HPU in the cascade system.

A peculiar feature of cascaded counterflow schemes is strong sensitivity of the cycles to changes of input and output parameters of the heat carrier and geothermal water. Any deviations from specified operating conditions for one HPU lead to changes of operating parameters of all the other units and consequently of the whole system. Multistage HPUs, which include one common condenser and evaporator, and several compressors connected in series with each other are less sensitive to changes in input parameters; see Fig. 5.28B.

The COP for the entire 3-cycle counterflow cascade system is equal to:

$$COP = \frac{\sum\limits_{n=3} \dot{Q}_c}{\sum\limits_{n=3} \dot{W}_{el} + \dot{W}_{aux}} \tag{5.19}$$

where $\dot{W}_{el}$ is electric power for the compressor drive, and $\dot{W}_{aux}$ is the power spent on auxiliary equipment.

An increase in the efficiency of 2-stage refrigerators and heat pumps is provided by increasing the specific cooling capacity by subcooling of the liquid after the condenser or step-by-step throttling and reducing the specific adiabatic compression work of the top-stage compressor by using two stages with inter-stage cooling. See Fig. 5.36 in the worked example at the end of Section 5.6.3.

Exergy analysis may be used to assess the approach to thermodynamic perfection of a HPU by applying the concept of exergy losses for all system components [39,44,51–55]; see Section 5.6.3. Generally the relative value of exergy destruction in any component is:

$$\chi = \dot{E}_{Dk} \Big/ \sum \dot{E}_{Dk} \tag{5.20}$$

where $\dot{E}_{Dk}$ is exergy destruction in the element under study and $\sum \dot{E}_{Dk}$ is total value of exergy destruction in the HPU.

The exergy destruction applied to the HPU is found from:

$$\delta = \frac{\dot{E}_{Dk}}{\dot{W}_{el} + \dot{Q}_{ev}\tau_{ev}} \tag{5.21}$$

where $\tau_{ev} = \frac{T_{ev} - T_{env}}{T_{ev}}$ is the Carnot factor for the evaporator.

Exergic weight of the element is:

$$\xi = \frac{\dot{E}_{Dk}}{\dot{Q}_c \tau_c} \tag{5.22}$$

where $\tau_c = \frac{T_c - T_{env}}{T_c}$ is the Carnot factor of the condenser.

Exergy efficiency of cycle is:

$$\eta_{ex} = \frac{\dot{Q}_c \tau_c}{\dot{W}_{el} + \dot{Q}_{ev} \tau_{ev}} \tag{5.23}$$

The characteristics of a multistage HPU are determined by the characteristics of its separate elements. On the one hand, the volumetric capacity of all the stages must be equal to the refrigerant flow rate appropriate to heat load on the evaporator. On the other hand, the compressor outlet pressure must correspond to the condensation pressure which goes with the condenser heating efficiency and the heat carrier parameters of a heating system.

Despite the equality of mass flow rate through all the compressors in a multistage scheme, the values of specific vapor volumes at suction in each compressor are different. Therefore, it is necessary to coordinate the work of each stage considering intermediate pressure, and in so doing try to minimize the total work input.

The results of system simulations show that, under equal conditions, the sensitivity of the COP for a three-stage HPU heating system to changes for input parameters of geothermal water and heat carrier is less than for a cascaded, counterflow system. The average value of COP for a cascaded, counterflow system and also the cooling value of geothermal water in HPU evaporators are higher. These factors favor the use of a cascaded, counterflow HPU system for projects with a low geofluid mass flow rates.

One particular feature of operation for all geothermal designs comprising doublet-well systems (one production well and one injection well) is the strongly unsteady character of the hydraulic processes in the initial period of operation. It is known that before entering quasi-steady operation, the system at first exhibits rapid pressure drop at the production wellhead, together with a pressure rise in the injection well, while the mass flow rate of the water changes considerably. Concerning a HPU, changes of geothermal water flow will lead to changes in most operation parameters, especially in a cascaded, counterflow system. Thus, it is reasonable to seek configurations of HPU systems which could compensate for this deficiency.

The use of high-temperature refrigerants [45,48,51,56,57], such as R245fa, R236fa, R142b, R114, R123/R290 in a 1-stage HPU can, in general, lead to a decrease in the specific unit heat production and, consequently, to an increase in its mass and overall dimension characteristics. In addition, in this case, it becomes problematic to find appropriate sources of low-potential heat with a temperature of 30–40 °C, since to obtain high condensation temperatures with a compression ratio of less than 6 in the HPU cycle, it is necessary to have a sufficiently high evaporation temperature. Otherwise, if the choice of the temperature limits of the cycle is such that the compression ratio is greater than 6, then it is advisable to use a system with 2-stage compression. It should be noted that unless the volumetric performance of a 1-stage compressor is at least 10% higher than the overall performance with a 2-stage compressor, then using a 1-stage scheme becomes economically infeasible, regardless of its relative simplicity, since both energy consumption and service cost increase.

Two-stage schemes are formed by the inclusion of auxiliary elements and additional lines of pipelines in the circulation circuit of a cooling agent. All the complications of 2-stage schemes are aimed at reducing internal losses from irreversibility in the cycle, namely, reducing the vapor temperature at the end of the compression process and increasing the liquid subcooling before throttling.

5.6.3 Cascade heating circuit installation

Since two-stage HPU systems with one working fluid do not provide high thermodynamic efficiency, cascade schemes with several working liquids are used [37,38,45,57]. The flow diagram of a cascade heat pump system is shown in Fig. 5.32; the cycle processes are shown in Fig. 5.33 in log *P*-*h* coordinates.

The geothermal fluid is cooled in process 9–10 from a temperature of 35°C to 26°C. The HPU transfers heat from the condenser in process 11–12 to the heating-system water in the temperature range of 70/90°C. The product of mechanical efficiency and electrical efficiency of the generator is $\eta_{me} = 0.95$. As a result of the simulation and optimization process, the compressor isentropic efficiency turns out to be 60% for the R407c loop and 74% for R134a loop (Table 5.6).

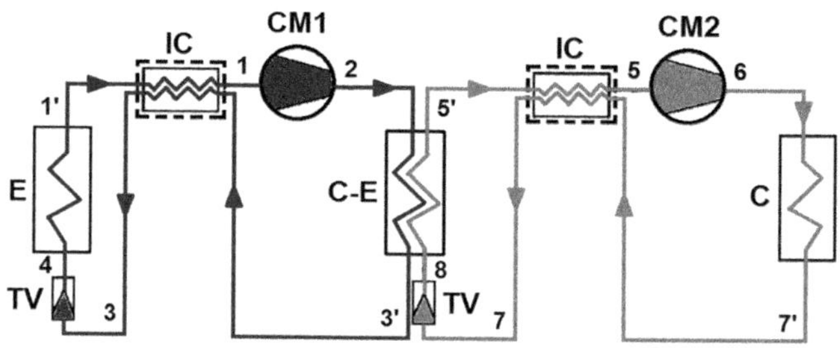

Fig. 5.32 Flow diagram of a cascade heat pump system. *C*, Condenser; *CM1, CM2*, compressors; *E*, evaporator, *IC*, intercoolers; *TV*, throttle valves.

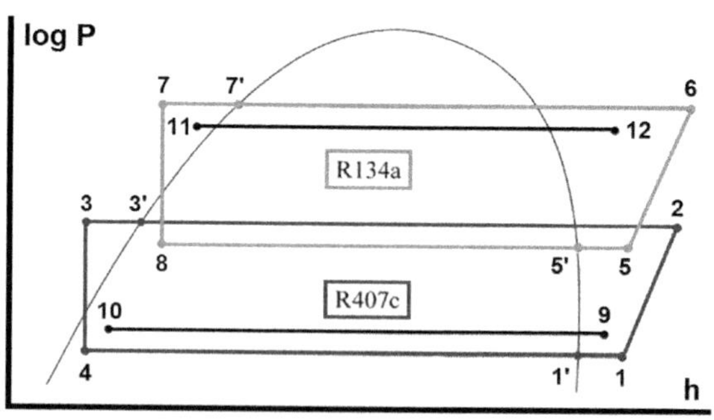

Fig. 5.33 Cycles of the 2-stage cascade HPU.

Table 5.6 Thermodynamic state-point properties for the 2-stage HPU in Fig. 5.32.

Cycle points	*P*, kPa	*T*, °C	*v*, m³/kg	*h*, kJ/kg	*s*, kJ/kg·°C	*e*, kJ/kg
Working fluid = R407*c*						
1	1025.0	25	0.0221	414.30	1.7284	62.02
2	2441.9	72.4	0.0099	446.92	1.7669	83.16
3	2441.9	54.1	0.0010	284.44	1.2755	67.19
4	1025.0	21.1	0.0072	284.44	1.2900	62.86
Working fluid = R134a						
5	1311.1	67.4	0.0171	443.96	1.7695	58.53
6	2670.2	104.2	0.0084	465.34	1.7845	75.44
7	2670.2	80	0.0011	322.28	1.3832	52.02
8	1311.1	49.8	0.0057	322.28	1.3943	48.71

Note: Dead state is taken at $T_0 = 25$ °C and $P_0 = 0.1013$ MPa.

A simulation of the 2-stage system shown in Fig. 5.32, with the intercooler in place, was carried out for ranges of values of the engineering parameters, and the results showed that the optimal refrigerant for lower cycle was R407c and for the upper cycle it was R134a. The vaporization temperature and condensation temperature considerably affect the value of the COP, because when $T_e = 25$°C and $T_c = 95$ °C, the COP = 2.58, and for $T_c = 80$ °C; COP = 3.44.

The calculated results of the simulation, optimized for COP in heating mode according to Figs. 5.32, 5.34 and 5.35 are as follows:

- evaporation temperature in the lower cycle – 25°C;
- evaporation temperature in the upper cycle – 67.4°C;

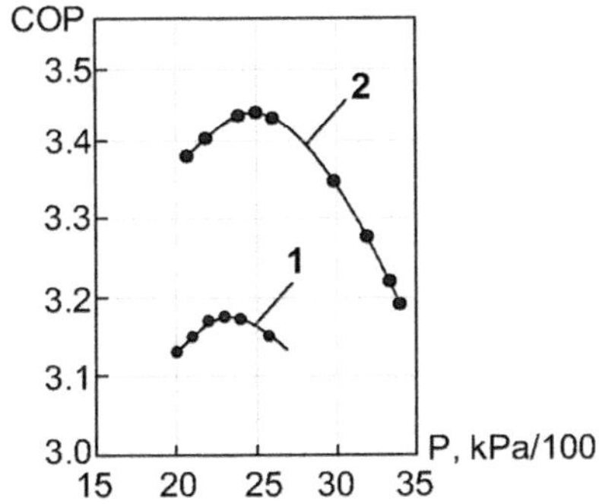

Fig. 5.34 For the system in Fig. 5.32, COP versus outlet pressure first-stage compressor: Curve 1 for R404a in the first stage and R134a in the second stage; Curve 2 for R407c in the first stage and R134a in the second stage. For both cases, $T_c = 80$°C, $T_e = 25$°C.

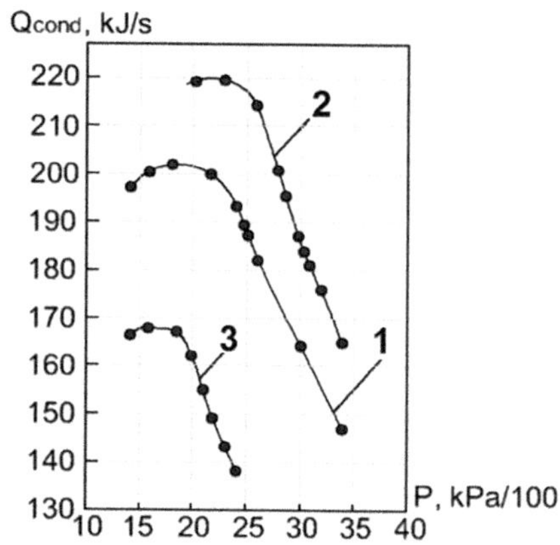

Fig. 5.35 For the system in Fig. 5.32, condenser thermal power versus the outlet pressure at the first-stage compressor: Curve 1 for R407c in the first stage ($T_c = 80°C$, $T_e = 25°C$) and R134a in the second stage; Curve 2 for R407c in the first stage ($T_c = 95°C$, $T_e = 25°C$) and R134a in the second stage; Curve 3 for R404a in the first stage ($T_c = 80°C$, $T_e = 25°C$) and R134a in the second stage.

- condensation temperature in the lower cycle – 72.4°C;
- condensation temperature in the upper cycle – 104.2°C;
- cooling capacity in the lower cycle – 133.5 kW
- heat productivity in the lower cycle – 159.2 kW
- cooling capacity in the upper cycle – 159.2 kW
- heat productivity in the upper cycle – 188.3 kW
- compressor power in the lower cycle – 25.7 kW
- compressor power in the upper cycle – 29.1 kW.

Example. This worked example is for a 2-stage system with an economizer EC incorporated between the two stages of compression; see Fig. 5.36. The working fluid was selected as R123 owing to its effectiveness over the chosen temperature range between the evaporator and the condenser (Tables 5.7–5.9).

A thermodynamic exergy assessment is given below for the 2-stage HPU shown in Fig. 5.36 and involves calculating the exergy losses in the elements of the HPU that includes an economizer, parallel throttling and subcooling of the working fluid. The working equations are presented and the results are shown graphically in Figs. 5.37 and 5.38.

Exergy loss in the evaporator is:

$$\dot{E}_{Dk}^{EV} = \dot{m}_{EV}[h_{10} - h_{11} - T_{env}(s_{10} - s_{11})] + \dot{Q}_0\tau_0 \tag{5.24}$$

where h, s are enthalpy and entropy; T_{env} is ambient temperature; $\dot{m}_{EV}$ is mass flow rate of the coolant at the low pressure level; $\dot{Q}_0$ is cooling performance of the HPU; τ_0 is exergy temperature Carnot function of low-potential source heat.

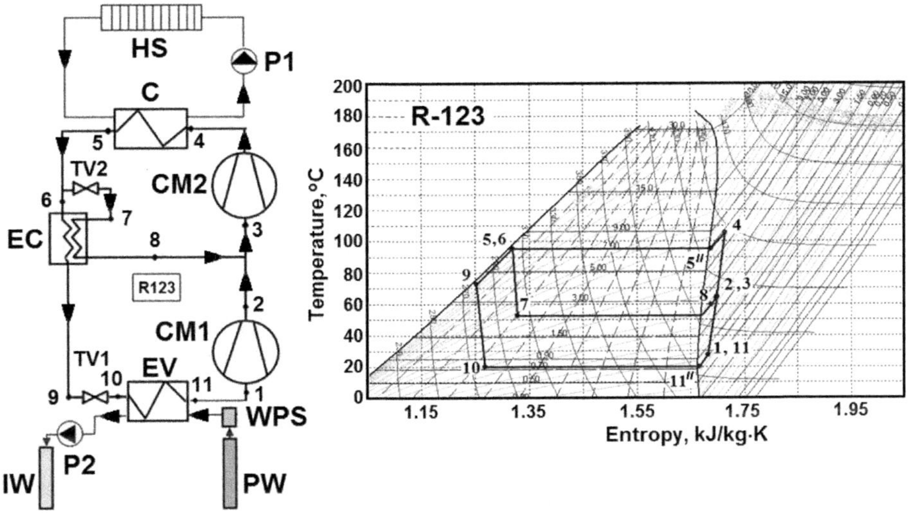

Fig. 5.36 Flow diagram of 2-stage HPU with an economizer EC. *C*, condenser; *CM1, CM2*, compressors; *EC*, economizer; *EV*, evaporator; *HS*, consumer's heating system; *P1*, circulation pump; *P2*, well pump; *PW, IW*, production and injection wells, respectively; *TV1, TV2*, throttles; *WP*, water purification system.

Table 5.7 Thermodynamic parameters of 2-stage HPU at $T_c = 95°C$. Cooling capacity = 66 kWt; Thermal power = 95.1 kWt; CM1 = 12.8 kW; CM2 = 3.25 kW; CM1 eff. = CM2 eff. = 76.3%; EC = 16 kWt; R123 flow rate in EV, $\dot{m}_{EV} = 0.542$ kg/s, in C, $\dot{m}_C = 0.655$ kg/s; COP 2.27.

	Parameters						
Points	***P*, MPa**	***T*, °C**	***V*, dm³/kg**	***h*, kJ/kg**	***S*, kJ/(kg K)**	***x***	$\dot{m}_i/\dot{m}$
1	0.076	27.00	206.26	399.58	1.6831		0.828
2″	0.231	57.18	71.69	417.55	1.6831		0.828
2	0.231	64.39	73.78	423.14	1.6999		0.828
3	0.231	63.56	73.55	422.50	1.6980		1.000
4″	0.702	98.69	24.25	441.50	1.6980		1.000
4	0.702	105.31	25.03	447.40	1.7137		1.000
5″	0.702	95.00	23.79	438.19	1.6890		1.000
5	0.702	95.00	0.79	302.27	1.3198		1.000
6	0.702	95.00	0.79	302.27	1.3198		0.828
7	0.231	52.58	20.94	302.27	1.3293	0.290	0.172
8	0.231	59.58	72.39	419.41	1.6887		0.172
9	0.702	72.58	0.75	277.91	1.2522		0.828
10	0.076	20.00	65.73	277.91	1.2683	0.325	0.828
11″	0.076	20.00	200.68	394.67	1.6666		0.828
11	0.076	27.00	206.26	399.58	1.6831		0.828

Table 5.8 Thermodynamic parameters of 2-stage HPU at $T_c = 95°C$. Cooling capacity = 28 kWt; Thermal power = 34.3 kWt; CM1 = 3.06 kW; CM2 = 3.25 kW; EC = 0.084 kWt; R123 flow rate in EV, $\dot{m}_{EV} = 0.203$ kg/s, in C, $\dot{m}_C = 0.204$ kg/s; COP 4.44.

	Parameters						
Points	***P*, MPa**	***T*, °C**	***V*, dm³/kg**	***h*, kJ/kg**	***S*, kJ/(kg K)**	***x*[a]**	$\dot{m}_i/\dot{m}$
1	0.076	27.00	206.26	399.58	1.6831		0.997
2″	0.144	43.82	113.19	409.74	1.6831		0.997
2	0.144	50.43	116.10	414.64	1.6984		0.997
3	0.144	50.41	116.09	414.63	1.6984		1.000
4″	0.271	68.41	63.03	425.37	1.6984		1.000
4	0.271	74.97	64.66	430.54	1.7134		1.000
5″	0.271	58.00	60.37	417.17	1.6740		1.000
5	0.271	58.00	0.73	262.31	1.2064		0.997
6	0.271	58.00	0.73	262.31	1.2064		1.000
7	0.144	37.61	15.21	262.31	1.2089	0.132	0.003
8	0.144	44.61	113.54	410.33	1.6850		0.003
9	0.271	57.61	0.73	261.90	1.2052		0.997
10	0.076	20.00	47.22	261.90	1.2137	0.233	0.997
11″	0.076	20.00	200.68	394.67	1.6666		0.997
11	0.076	27.00	206.26	399.58	1.6831		0.997

[a] $x = \dot{m}_g/\dot{m}$, the mass of vapor divided by the total mass of a liquid-vapor mixture.

Exergy loss in the low-pressure level compressor is:

$$\dot{E}_{Dk}^{CM1} = \dot{m}_{EV} T_{env}(s_2 - s_1) \tag{5.25}$$

Exergy loss in the electric motor of the low-pressure compressor is:

$$\dot{E}_{Dk}^{CM1} = \dot{m}_{EV} \dot{W}_{CM1}(1 - \eta_{el})/\eta_{el} \tag{5.26}$$

where $\dot{W}_{CM1}$ is the isentropic work of the low-pressure compressor; η_{el} is electromechanical efficiency of the compressor.

Exergy loss in the high-pressure level compressor is:

$$\dot{E}_{Dk}^{CM2} = \dot{m}_C T_{env}(s_4 - s_3) \tag{5.27}$$

where $\dot{m}_C$ is the R123 mass flow rate at the high pressure level.

Table 5.9 Results of calculations for the two-stage HPU.

Parameters and units	Results				
Ambient temperature, t_{env}, °C	−15	−10	−5	0	8
Heat production, $\dot{Q}_{HS}$, kW	95	80.8	68	52	34
Cold production, $\dot{Q}_{EV}$, kW	66	57.4	49.5	40	28
Temperature of HTF after HPU, t_h, °C	90	83	75	64	49
Condensation temperature, t_c, °C	95	90	85	72	58
Temperature of the heat transfer in the return line, t_c, °C	66.2	63	58	51	40.5
Mass low rate through heating loop, $\dot{m}_{HS}$, kg/s			0.954		
Thermal water temperature at evaporator input, t_{input}, °C	30	30	30	30	30
Thermal water temperature at evaporator outlet, t_{output}, °C	22	22	22	22	22
Evaporation temperature, t_{ev}, °C	20	20	20	20	20
Mass flow rate of geofluid through evaporator, $\dot{m}_{geo}$, kg/s	1.96	1.71	1.47	1.193	0.835
Low-pressure compressor drive power, $\dot{W}_{LP}$, kW	12.8	10.4	8.41	5.62	3.06
High-pressure compressor drive power, $\dot{W}_{HP}$, kW	16.3	12.9	10.2	6.36	3.25
Economizer thermal power, $\dot{Q}_{EC}$, kW	16.0	11.4	7.87	2.96	0.084
Intermediate pressure, $P_{int.}$, MPa	0.231	0.218	0.206	0.175	0.144
R123 mass flow rate in LP stage, $\dot{m}_{EV}$, kg/s	0.542	0.463	0.393	0.304	0.203
R123 mass flow rate in HP stage, $\dot{m}_C$, kg/s	0.655	0.543	0.448	0.324	0.204
Isentropic efficiency of compressors	0.763	0.757	0.750	0.721	0.675
COP	3.27	3.46	3.67	4.34	5.44
Coefficient of energy use COP_{HPU}	3.07	3.26	3.43	4.0	4.75

Exergy loss in the electric motor of the high-pressure compressor is:

$$\dot{E}_{Dk}^{CM2} = \dot{m}_C \dot{W}_{CM2}(1 - \eta_{el})/\eta_{el} \tag{5.28}$$

where $\dot{W}_{CM2}$is the isentropic work of the low-pressure compressor.

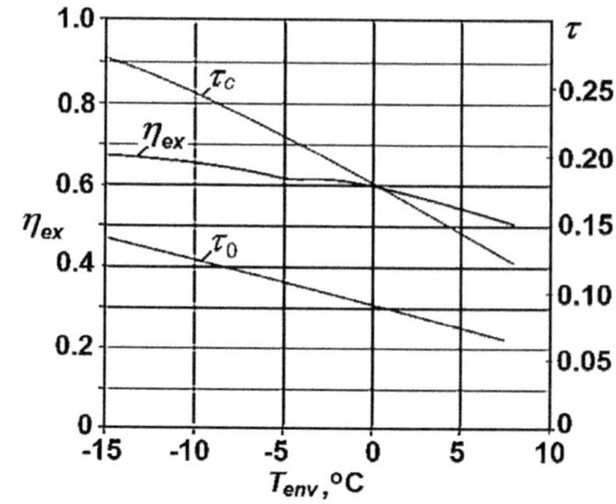

Fig. 5.37 The dependence $\eta_{ex} = f(T_{env})$ and $\tau = f(T_{env})$ on the ambient air temperature for the HPU system shown in Fig. 5.36.

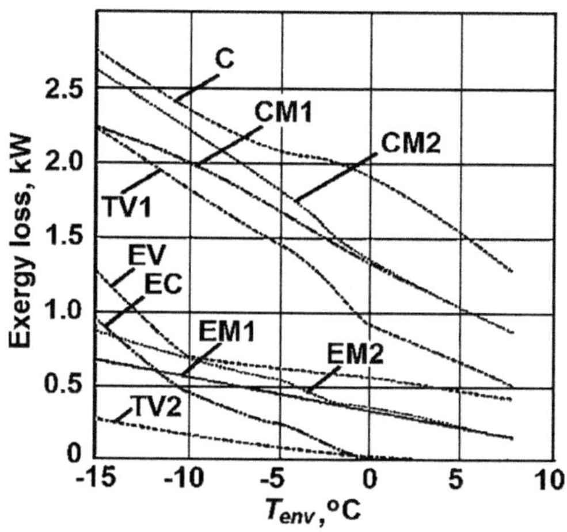

Fig. 5.38 Exergy losses in the HPU elements versus the ambient air temperature showing the effect of different seasons on the exergy losses. EM1 and EM2 are the electric motors that drive CM1 and CM2 compressors, respectively.

The formula for calculating the loss of exergy for the condenser has the following form:

$$\dot{E}_{Dk}^{COND} = \dot{m}_C(h_5 - h_6 - T_{env}(s_5 - s_6)) - \dot{Q}_{HPU}\tau_c \quad (5.29)$$

where $\dot{Q}_{HPU}$ is thermal capacity of the HPU; τ_c is the exergy temperature Carnot function of a heat receiver.

Loss of exergy in the economizer EC is:

$$\dot{E}_{Dk}^{EC} = T_{env}[\dot{m}_{EV}(s_9 - s_6) + (\dot{m}_C - \dot{m}_{EV})(s_8 - s_7)] \tag{5.30}$$

where $(\dot{m}_C - \dot{m}_{EV})$ is the R123 mass flow rate through the economizer.

Exergy loss in a throttle valve at the low-pressure side is:

$$\dot{E}_{Dk}^{TV1} = \dot{m}_{EV} T_{env}(s_{10} - s_9) \tag{5.31}$$

For the throttle valve at the high-pressure side, the loss of exergy is determined by:

$$\dot{E}_{Dk}^{TV2} = (\dot{m}_C - \dot{m}_{EV})\ T_{env}(s_7 - s_6) \tag{5.32}$$

The overall exergy efficiency for the system is:

$$\eta_{ex} = \frac{\dot{Q}_{HPU}\tau_{cond}}{\sum \dot{W}_{comp} + \dot{Q}_0\tau_0} \tag{5.33}$$

5.7 Summary

Geothermal energy, either alone or in conjunction with heat pumps, is providing heating for residents and businesses in cities all over the world. Historical accounts of very old systems in Europe and in the United States show that these systems have a long life, give reliable service, and are environmentally friendly as well as economic. Greenhouses warmed by geothermal fluids allow year-round crop growing even in severe climates. A wide variety of fish and other water creatures such as alligators are raised in geothermally-heated ponds. The outflow from geothermal electric power plants can be used as the heating medium. Such facilities are good businesses and can often be promoted as tourist attractions. The use of heat pumps with geothermal energy offers a multiplier effect to achieve high performance even under cold ambient temperatures.

Nomenclature

$\dot{E}$	exergy loss
C	specific heat capacity
k	isentropic index
$\dot{m}$	mass flow
p	pressure
q	specific heat transfer
q^*	specific cooling capacity
$\dot{Q}$	heat

h	enthalpy
s	entropy
T	temperature
V	volume
$\dot{W}$	power

Greek letters

η	efficiency
τ_c	carnot factor
ξ	exergetic weight
χ	exergy analysis

Subscripts

BWSWD	Boise Warm Springs Water District
C	condenser
CB	City of Boise
COP	coefficient of performance
CM	compressor
EM	electric motor
HPU	heat pump unit
SICH	State of Idaho Capital Mall
SC	subcooler
EC	economizer
TV	throttle valve
IC	intercooler
EV	evaporator
P	pump
HS	heating system
RH	regenerative heat exchanger
WPS	water purification system

Review questions

1. If the waste brine from a geothermal power plant is 120 °C and the lowest temperature allowable during cooling to prevent scaling is 40 °C, list and briefly describe the possible direct heat uses that might be made with the brine.
2. Describe the role of legends and myths of ancient peoples in exploration activities for geothermal prospects.
3. The hot springs at Chaudes-Aigues in France have been known for millennia. Their longevity and constancy make them scientific and technical curiosities. Write a short essay discussing their origins and what conditions have led to their very long life.

4. In view of the different climate conditions in Iceland and The Netherlands, write a short essay contrasting the basic design considerations for greenhouses in these two place with emphasis on the most influential factors in each case.
5. Describe the advantages and disadvantages of shell-and-tube versus plate-type heat exchangers, specifically for geothermal greenhouse duty.
6. Section 5.5.2 describes an aquaculture facility in Iceland, an island country with a large fishing industry. List and briefly describe as many other similar countries as you can where geothermal-assisted aquaculture is being done or have the potential to do so.
7. Explain why one would use two-stage refrigerators and heat pumps? Give the reasons for the transition from one-stage to a two-stage compression?
8. Show on the T-s diagram an ideal two-stage cycle and the areas that represent the reduction of work and increase of cooling capacity compared to a one-stage cycle.
9. Why are different working fluids used in the lower and upper cycles of a cascade unit?
10. Compare cascade and two-stage refrigerators. Show which machine has the higher theoretical cooling coefficient of performance COP.
11. How does the installation of the additional equipment listed below influence the cooling COP of the cycle: (i) subcooler, (ii) regenerative heat exchanger, and (iii) economizer?
12. What is the reason for using an economizer in the two-stage heat pump system in Fig. 5.36?
13. Compare the thermodynamic efficiency of a cycle with two-stage compression and incomplete intermediate cooling with a single-stage cycle.
14. How can the operating mode of a refrigeration system with partial intercooling be transformed to full intercooling operation mode (RHE circuit)?
15. Show the flow diagram and thermodynamic T-s diagram of a two-stage refrigeration machine with full intercooling and with parallel throttling.
16. Repeat Question 15 except for a cycle with full intercooling and successive throttling.

Exercises

1. With reference to the data in Table 5.3, estimate the thermal supply capacity (in MWth) of the City of Boise, Idaho, using a mass-weighted geofluid supply temperature for the three production wells and the stated reinjection temperature. You may assume a specific heat of 4.19 kJ/kg K for the geofluid.
2. Estimate the average thermal supply capacity (in MWth) of the U.S. Veteran's Administration geothermal heating system for the period 1997–2002 assuming a reinjection temperature of 45 °C. You may assume a specific heat of 4.19 kJ/kg K for the geofluid.

Acknowledgments

One of the authors (RDP) thanks the following individuals for answering questions and providing additional information, data, and materials regarding their geothermal projects: Del Eytchison, Manager, Boise Warm Springs Water District, Boise, Idaho; Jon Gunnerson, Geothermal Coordinator, Public Works, City of Boise, Idaho; Ric Johnston, Facilities Manager, Division of Public Works, State of Idaho, Boise, Idaho; and Radboud Vorage, Project Director, Greenhouse Geo Power, IJsselmuiden, Netherlands.

References

[1] DiPippo R. Geothermal power plants. In: Principles, applications, case studies and environmental impact. 4th ed. Oxford, England: Butterworth-Heinemann: Elsevier; 2016.
[2] Lund JW, Boyd TL. Direct utilization of geothermal energy 2015 worldwide review. In: Proc. World geothermal congress 2015, Melbourne, Australia, April 19–25, 2015.
[3] Cataldi R, Suarez MC. History of geothermal energy in the world. Aracne; 2018 [in English and Italian].
[4] Gibert JP, Jaudin F. Using geothermal waters in France: the district heating system of Chaudes-Aigues from the middle ages. Chap. 18. In: Cataldi R, Hodgson SF, Lund JW, editors. Stories from a heated earth – our geothermal heritage. Sacramento, CA, USA: Geothermal Resources Council and International Geothermal Association; 1999. p. 287–305.
[5] Chaudes-Aigues. French Wikipedia; 2018 [in French]. https://fr.wikipedia.org/wiki/Chaudes-Aigues.
[6] "Centre thermoludique," Caleden – centre Thermoludique de Chaudes-Aigues. 2018. https://www.caleden.com/index.php/centre-thermoludique.
[7] May ML. Right from Hades: water and politics in Boise, Idaho [Thesis, Master of Arts in History]. Boise State University; 2015.
[8] Davis A. Boise City Dept. of Arts & history, Natatorium; 2013.
[9] Kleibrink Thompson P. The natatorium: still making a splash after all these years. Greenbelt Magazine, Boise, Idaho 2013. https://greenbeltmagazine.com/2013/05/01/the-natatorium-still-making-a-splash-after-all-these-years/.
[10] Idaho geothermal history - a detailed history. Governor's Office of Energy Resources; 2008. https://oemr.idaho.gov/wp-content/uploads/idaho_geothermal_history.pdf.
[11] Nelson LB, Niemi WL, Stoker RC. Geothermal resource exploration in Boise, Idaho. U.S. DOE, Idaho National Engineering Lab., Rep. No. EGG-2011. 1980.
[12] Wood SH, Burnham WL. Geologic framework of the Boise warm springs geothermal area, Idaho. Geological Society of America Centennial Field Guide - rocky mountain section. 1987. p. 117–22.
[13] Boise Warm Springs Water District Map. 2018. http://bwswd.com/district-map/.
[14] Boise Warm Springs Water District – about the district. 2018. http://bwswd.com/about-the-district/.
[15] Waag CJ, Wood SH. Evaluation of the Boise geothermal system. Final report to Idaho Department of Water Resources, Boise, Idaho. 1987.
[16] Neely KW. Production history for the State of Idaho Capitol Mall geothermal system - 1983–1994. Boise, Idaho: Idaho Department of Water Resources; 1995.
[17] Gunnerson J. City of Boise direct use geothermal heating district. PowerPoint presentation. 2018.
[18] Higginson RK, Barnett JA. Investigation of the Boise, Idaho geothermal system. Rep. to the Idaho Dept. of Water Resources. 1987.
[19] Boise VA medical center map. 2018. https://www.boise.va.gov/docs/BVAMC_MAP2.pdf.
[20] District Heating Systems in Idaho. Idaho Geothermal Resources, Idaho Dept. of Water Resources. http://www.idahogeothermal.org.
[21] Boyd TL. Geothermal greenhouse information package. Oregon Inst. of Technology; 2008.
[22] Lund JW. Geothermal district heating. PowerPoint presentation. 2015.

[23] Rafferty K. Greenhouses. Chapter 15. In: Lienau PJ, Lunis BC, editors. Geothermal direct-use engineering and design guidebook. Geo-heat Center, Oregon Inst. of Technology; 1991. https://www.osti.gov/servlets/purl/5859999.

[24] Vorage R. Geothermal heat for greenhouses and project development in The Netherlands. Presentation at Der Geothermie Kongress, 3 November 2015, Essen, Germany.

[25] Vorage R. Fotoboek - Aardwarmteproject Koekoekspolder (Photobook – Cuckoo Polder Geothermal Project). 2012 [in Dutch].

[26] Hodgson SF. Warm beneath the wind, geothermal Holland. Geothermal Resources Council - Bulletin 2016;45(2):30–7.

[27] Smith KC. A Layman's guide to geothermal aquaculture. Oregon Inst. of Technology, Geo-Heat Center; 1981.

[28] Juillerat L. Hot water holds many opportunities for S. Oregon farms. Capital Press; 2017. http://www.capitalpress.com/Profit/20170912/hot-water-holds-many-opportunities-for-s-oregon-farms.

[29] Hayes A, Johnson WC. Geothermal aquaculture: a guide to freshwater prawn culture. Oregon Inst. of Technology; 1980.

[30] Ragnarsson A. Geothermal energy in aquaculture. In: Presented at "Short course VI on utilization of low- and medium-enthalpy geothermal resources and financial aspects of utilization". Santa Tecla, El Salvador: UNU-GTP and LaGeo; 2014.

[31] Lind L, O'Brien L, Bell J. Geothermal Energy Helps to Grow Prawns. GNS Science Case Study.http://www.geo-exchange.ca/en/UserFiles/File/GeoConneXion/Case%20-StudyFINAL%20FULL%20VIEW_revised.pdf.

[32] Huka Prawn Park History. http://www.hukaprawnpark.co.nz/history.

[33] Huka Prawn Park. https://en.wikipedia.org/wiki/Huka_Prawn_Park.

[34] Crocodiles of the World. Oxfordshire, UK. https://www.crocodilesoftheworld.co.uk/conservation/conservation-status/.

[35] Blevins J. Colorado Gators boasts Bruce, the largest alligator in the west. The Denver Post 2016. https://www.denverpost.com/2016/07/22/colorado-gators-largest-alligator-in-the-west/.

[36] Tosun DD. Crocodile farming and its present state in global aquaculture. Journal of Fisheries Sciences 2013;7(1):43–57. http://www.fisheriessciences.com/fisheries-aqua/crocodile-farming-and-its-present-state-in-global-aquaculture.php?aid=549.

[37] ASHRAE handbook – fundamentals. 2009.

[38] Kreith F, West RE. Energy efficiency. Handbook. Taylor and Francis Group; 1997.

[39] Haywood RW. Analysis of engineering cycles. Oxford: Pergamon press; 1975.

[40] Morosuk T. Cycle method in design of refrigeration machines and heat pumps. In: Proceedings of the V Minsk international seminar, heat pipes, heat pumps, refrigerators, Belarus; 2003. p. 391–8.

[41] Granryd E. A regenerative refrigeration cycle. In: Proc. 14th int. Congr. Refrig., Moscow; 1975. p. 251–61.

[42] Dincer I. Thermodynamic, exergy an environmental impact. Energy Sources 2000;22: 723–32.

[43] Hellman J, Kong H, Meurer C. In: Buchwald Dr H, editor. Refrigeration and air-conditioning technology: Solkane Pocket manual. Hannover: Heinel Welbeagentur GmbH; 2004. 285 p.

[44] Gordon JM, Ng-Cornwall KC. Cool thermodynamics. The engineering and physics of predictive, diagnostic and optimization methods for cooling systems. England: MPG Books Ltd.; 2001. 276 p.

[45] Hasegawa H, Saikawa M, Hashimoto K, Iwatsubo T. Development of two-stage compression and cascade heating heat pump system for hot water supply. ASHRAE Transactions 1996;102:248−56.

[46] Feidt M. Thermodynamic and optimization of reverse cycle machines, refrigeration, heat pumps, air condition, cryogenic. In: Bejan A, Mamut E, editors. Thermodynamics and the optimization of complex energy systems; 1998. p. 205−30. Neptune (Romania).

[47] Domanski PA, Didion DA, Doyle JP. Evaluation of suction-line/liquid-line heat exchange in the refrigeration cycle. International Journal of Refrigeration 1994;17(7):487−93.

[48] Domanski PA. Theoretical evaluation of the vapour compression cycle with a liquid-line/suction-line hear exchanger, economizer, and ejector, NISTIR-5606. US: National Institute of Standards and Technology; 1995. p. 1−31.

[49] Klein SA, Reindl DT, Brownell K. Refrigeration system performance using liquid-suction heat exchangers. International Journal of Refrigeration 2000;23:588−96.

[50] Taras MF. Is economiser cycle justified for AC applications. ASHRAE Journal 2005; 47(7):38−44.

[51] Kelly S, Tsatsaronis G, Morosuk T. Advanced exergetic analysis: approaches for splitting the exergy destruction into endogenous and exogenous parts. Energy 2009;34:384−91.

[52] Henderson PC, Hewitt NJ, Mongey B. An economic and technical case for a compressor/expander unit for heat pumps. International Journal of Energy Research 2000;24:831−42.

[53] Hewitt NJ, Huang MJ, Nguyen QM. The development of an air source heat pump, to be presented at the "renewable energy in maritime climates" conference, Dublin, Ireland, 26th−28th April, 2006.

[54] Khan JR, Zubair SM. Design and rating of a dedicated mechanical subcooling vapour compression refrigeration systems. Proceedings - Institution of Mechanical Engineers 2000;214(Part. A):455−71.

[55] Mastrullo R, Mazzei P. Exergetic analysis of compound mechanical refrigeration systems. Journal of Heat Recovery Systems 1987;7(5):401−12.

[56] Morosuk T, Nikulshin R, Morosuk L. Entropy-cycle method for analysis of refrigeration machine and heat pump. Thermal Science 2006;10:111−24.

[57] Minh NQ, Hewitt NJ, Eames PC. Improved vapour compression refrigeration cycles: literature review and their application to heat pumps. In: International refrigeration and air conditioning conference at Purdue, July 17−20; 2006.

Geothermal energy in combined heat and power systems[(1)]

6

6.1 Introduction to geothermal CHP systems

Geothermal combined heat and power (CHP) systems are a means to capture the available energy in a geothermal resource in a highly efficient manner. The Second Law of thermodynamics and considerations of exergy provide a road map for efficient energy usage. In processes involving heat transfer, high efficiency is attained by closely matching the geothermal fluid with the working fluid that provides the heating. In some cases, the geofluid may be used directly as the heating fluid, but often a secondary working fluid is the ultimate heat carrier in the heated space. In this section we consider the basic configurations and basic thermodynamics of CHP systems.

There are ample opportunities for combined geothermal power and heating applications, as can be seen from Fig. 6.1 which shows only a few of the many possibilities. Typically power generation requires geofluids at temperatures greater than 100 °C. Discharge geofluid temperatures can be matched to a variety of direct heat applications, as is evident from the figure. Under certain circumstances, the geofluid may be used alone, straight from the source, to supply heat to a consumer needing thermal energy.

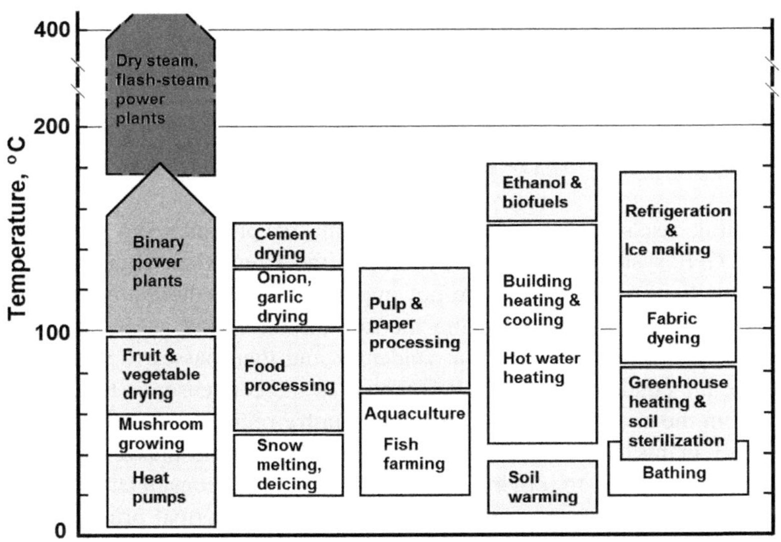

Fig. 6.1 Selected heating applications that may be combined with geothermal power plants.

(1) Written by Ronald DiPippo.

Low-Temperature Energy Systems with Applications of Renewable Energy
https://doi.org/10.1016/B978-0-12-816249-1.00006-6

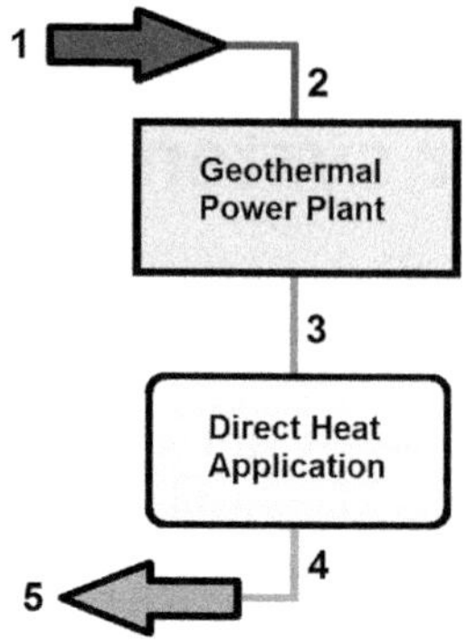

Fig. 6.2 Generalized flow diagram for a geothermal CHP system.

Whatever the heating application, a simplified general representation of a CHP system is shown in Fig. 6.2. The geothermal source (1) may be a drilled well or a hot spring; the geofluid is transmitted to the power plant (2) by a system of pipes (gathering system); within the power plant various processes are carried out depending on the type of plant, e.g., dry steam, flash steam or binary, and the geofluid is discharged (3) still carrying useful energy and exergy. The fluid then enters the heating application(s) where it supplies needed thermal energy for one or several applications, some of which are shown in Fig. 6.1; see Fig. 5.1 for a more complete listing. The geofluid is now in a low-energy state (4), close to the temperature of the surroundings and possessing very little exergy. Finally, the spent geofluid (5) is disposed of, typically via an injection well. The mass flow rates at the various state points in general will not be the same, as we will see in the case studies in Section 6.3.

6.2 Generic CHP system: thermo-economic analysis

Before tackling case studies of actual CHP systems, we present first a general analysis of a generic CHP system [1]. Figure 6.3 contains the essential elements of a CHP system. The details of each element are not important for the thermodynamic systems analysis shown below. State 3 leaving the turbine is a liquid, a vapor, or a liquid-vapor mixture that normally enters a condenser and then passes to a water cooling tower where the condensate is cooled for reuse in the condenser as the coolant. The overflow from the cooling tower passes to a drain for reinjection. For noncondensing plants, state 3, in this case a vapor, would be vented to the atmosphere, but this is rarely done nowadays owing to environmental or efficiency considerations. The heat exchanger HX transfers energy and exergy from the geothermal brine (state 7) to a fresh water supply. Various end user applications may take place including domestic hot water, space heating, or snow melting. Those details are not pertinent to this generic analysis.

The processes involved in the plant operation are shown in temperature-entropy (T-s) coordinates in Fig. 6.4.

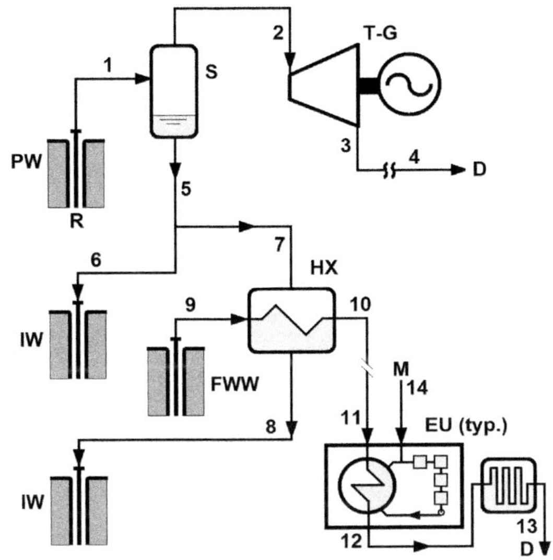

Fig. 6.3 Simplified schematic for a combined single-flash geothermal power plant combined with a hot water production facility, after [1]. *D*, drain or discharge to atmosphere; *EU*, end user; *FWWs*, fresh water wells; *HX*, heat exchanger; *IW*, injection wells; *M*, mains water; *PW*, production wells; *R*, geothermal reservoir; *S*, separator; *T-G*, turbine-generator.

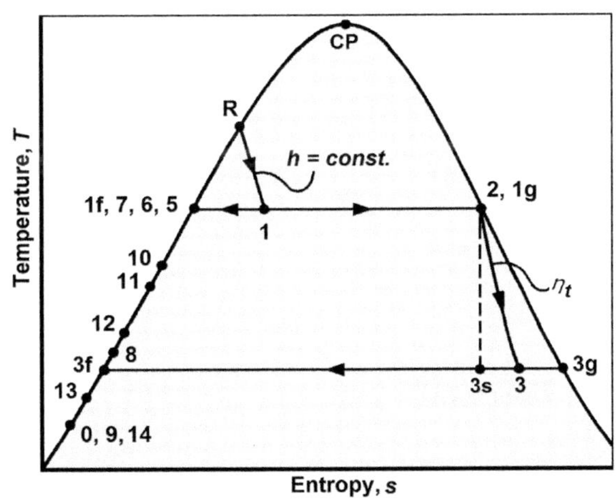

Fig. 6.4 Processes carried out for plant in Fig. 6.3 shown in *T-s* coordinates relative to the saturation line for water substance; shaded area is liquid-vapor region; not to scale.

The objective of this exercise is to optimize the amount of exergy that is produced by the CHP plant. This includes the power generated by the turbine-generator plus the exergy delivered to the hot water heat exchanger HX. The exergy losses incurred by the hot water production and distribution system will not be considered here, but will be examined in the case study in Section 6.3.1.

First we write out the equations for the rate of exergy at each of the state points shown in Figs. 6.3 and 6.4.

$$\dot{E}_R = \dot{m}_R(h_R - T_0 s_R - e_0) \tag{6.1}$$

$$\dot{E}_1 = \dot{m}_R(h_1 - T_0 s_1 - e_0) \tag{6.2}$$

$$\dot{E}_2 = \dot{m}_R x_1(h_{1g} - T_0 s_{1g} - e_0) \tag{6.3}$$

$$\dot{E}_3 = \dot{m}_R x_1(h_3 - T_0 s_3 - e_0) \tag{6.4}$$

$$\dot{E}_4 = 0 \tag{6.5}$$

$$\dot{E}_5 = \dot{m}_R(1 - x_1)(h_{1f} - T_0 s_{1f} - e_0) \tag{6.6}$$

$$\dot{E}_6 = \dot{m}_R(1 - x_1)(1 - y)(h_{1f} - T_0 s_{1f} - e_0) \tag{6.7}$$

$$\dot{E}_7 = \dot{m}_R(1 - x_1)y(h_{1f} - T_0 s_{1f} - e_0) \tag{6.8}$$

$$\dot{E}_8 = \dot{m}_R(1 - x_1)y(h_8 - T_0 s_8 - e_0) \tag{6.9}$$

$$\dot{E}_9 = 0 \tag{6.10}$$

$$\dot{E}_{10} = \dot{m}_9(h_{10} - T_0 s_{10} - e_0) \tag{6.11}$$

$$\dot{E}_{11} = \dot{m}_9(h_{11} - T_0 s_{11} - e_0) \tag{6.12}$$

$$\dot{E}_{12} = \dot{m}_9(h_{12} - T_0 s_{12} - e_0) \tag{6.13}$$

$$\dot{E}_{13} = 0 \tag{6.14}$$

$$\dot{E}_{14} = \dot{m}_{13}(h_{14} - T_0 s_{14} - e_0) \tag{6.15}$$

$$\dot{E}_{15} = \dot{m}_{13}(h_{15} - T_0 s_{15} - e_0) \tag{6.16}$$

where

$$e_0 = h_0 - T_0 s_0 \tag{6.17}$$

$$x_1 = \dot{m}_2/\dot{m}_1 = (h_R - h_{1f})/(h_{1g} - h_{1f}) \tag{6.18}$$

$$s_1 = s_{1f} + x_1(s_{1g} - s_{1f}) \tag{6.19}$$

$$h_3 = \frac{h_2 - A\left[1 - \dfrac{h_{3f}}{h_{3g} - h_{3f}}\right]}{1 + \dfrac{A}{h_{3g} - h_{3f}}} \tag{6.20}$$

$$A = 0.5\ \eta_{t,d}(h_2 - h_{3s}) \tag{6.21}$$

$$s_{3s} = s_2 \tag{6.22}$$

$$x_{3s} = (s_2 - s_{3f})/(s_{3g} - s_{3f}) \tag{6.23}$$

$$h_{3s} = h_{3f} + x_{3s}(h_{3g} - h_{3f}) \tag{6.24}$$

$$y = \dot{m}_7/\dot{m}_5 = \dot{m}_7/\dot{m}_R(1 - x_1) \tag{6.25}$$

The term $\eta_{t,d}$ represents the isentropic efficiency of a steam turbine operating entirely with superheated (i.e., dry) steam. Owing to the moisture that forms during the expansion of saturated vapor, the actual geothermal steam turbine efficiency will be less than $\eta_{t,d}$. The Baumann rule was adopted to show this degradation and leads to Eq. (6.20). For computational purposes, $\eta_{t,d}$ was taken as 0.88; thus the A-factor becomes

$$A = 0.44\ (h_2 - h_{3s}) \tag{6.26}$$

The factor y in Eqs. (6.7)–(6.9) determines what fraction of the separated geofluid is sent to the heating facility. This must be determined by a detailed assessment of the reservoir and the reinjection needed to maintain reservoir pressure without excess cooling of the formation, along with the needs of the heating applications. For the present analysis, a value of 2/3 has been arbitrarily assigned to y, i.e., 1/3 of the separated geofluid is reinjected; thus

$$\dot{m}_7 = (2/3)\ \dot{m}_5 = (2/3)\ \dot{m}_R(1 - x_1) \tag{6.27}$$

An optimization will be conducted to seek the value of the flash temperature T_1 that yields the maximum total exergy output of the plant, i.e., the greatest sum of electrical power and separated brine exergy (which is delivered to the HX):

$$\dot{E}_{out} = \dot{W}_e + \dot{E}_7 \tag{6.28}$$

where the electrical output is

$$\dot{W}_e = \eta_G \dot{W}_T = \eta_G \dot{m}_R x_1 (h_2 - h_3) \tag{6.29}$$

Several additional assumptions were made to facilitate the calculations:

1. Pressure losses in piping are neglected.
2. All plant components are perfectly insulated.
3. Pure water is used to model the geothermal fluid.
4. All liquid state points are assumed saturated.
5. Refprop [2] was used to determine the thermodynamic properties of water.
6. Generator efficiency is 0.975.
7. Reservoir temperature is 220 °C.
8. Condenser temperature is 40 °C.
9. Geofluid total mass flow at wellheads is 350 kg/s.
10. Dead state: 20 °C and 0.10 MPa.
11. Monetary value of electricity is 0.10 US$/kWh.
12. Monetary value of heat exergy sent to the HX is $0.10 \times H$ US$/kWh, where $0 \leq H \leq 1$.

An Excel spreadsheet was written to simulate the processes of the plant with Refprop as an "add-in" and calculations were performed for the range of flash temperatures from 120 °C to 180 °C to find the optimum thermodynamic operating point. If the exergy associated with the two energy outputs from the plant (the electricity and the brine heat sent to the HX) is optimized thermodynamically, then the optimization leads to the optimal separator temperature and the highest combination of electrical and heat output; see Fig. 6.5. From Fig. 6.6, it is evident that a stand-alone power plant has a utilization efficiency that peaks at 26.9% at 124 °C. If that plant is augmented by a direct heat system, the efficiency increases to 38.6%, a 43% gain. When the exergy of the separated brine is put to use, the utilization efficiency increases continuously as the separator temperature rises. However, to achieve the overall best performance, essentially all the geofluid should be sent to the HX and no electricity would be generated. However, this is not a feasible scenario since the monetary value of heat will always be less than the value of the electricity.

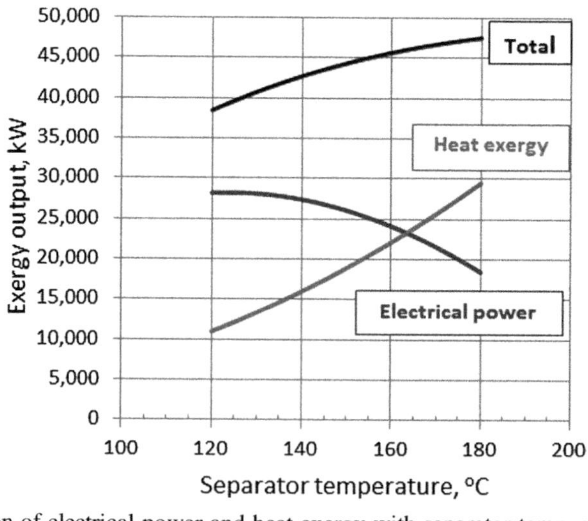

Fig. 6.5 Variation of electrical power and heat exergy with separator temperature.

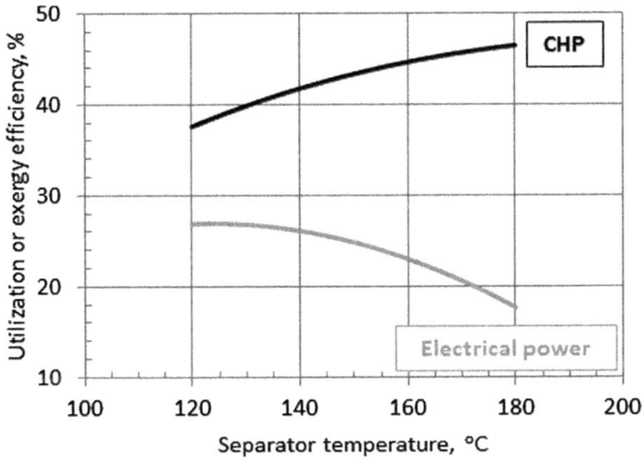

Fig. 6.6 Thermodynamic utilization efficiency: stand-alone power plant and CHP.

A parametric study in which the H-factor (see assumption 12) was included reveals an interesting set of results. Fig. 6.7 shows the revenues per hour of plant operation as a function of the H-factor. When H = 1 (equal value) the curve is monotonically increasing with the separator temperature. When H = 0 (no value for the heat), the result is simply the optimization of the stand-alone single-flash geothermal power plant. For intermediate values of H, there is an optimum point where the combined revenue from electricity and heat, based on exergy, is a maximum. Figs. 6.8 and 6.9 give the optimum results.

The conclusion may be drawn from Fig. 6.9 that about 22 million US$ can be earned per year by such a plant operating with an 85% capacity factor, if the electricity is worth 0.10 US$/kWh and the heat value is 20% of the electricity value. Depending on the capital cost of the whole facility, this may or may not be sufficient to warrant the investment.

6.3 Cycle choices for geothermal CHP plant

Combined heat and power plants can start with either geothermal steam plants or organic Rankine cycles at the high-temperature end. In the first case, the steam plant is designed primarily for power generation and the direct heat application is added after some years of proven operation. In the second case, the plant is usually designed as a CHP plant from the beginning, although a power unit may be added to a long-standing direct heat plant to make fuller use of the resource.

6.3.1 Dry- and flash-steam plants

Geothermal steam plants have a history of more than 115 years of operation dating back to 1904 when the first geothermal plant went into operation at Larderello, Italy. There is a vast supply of literature on steam plants of all types [3]. The main

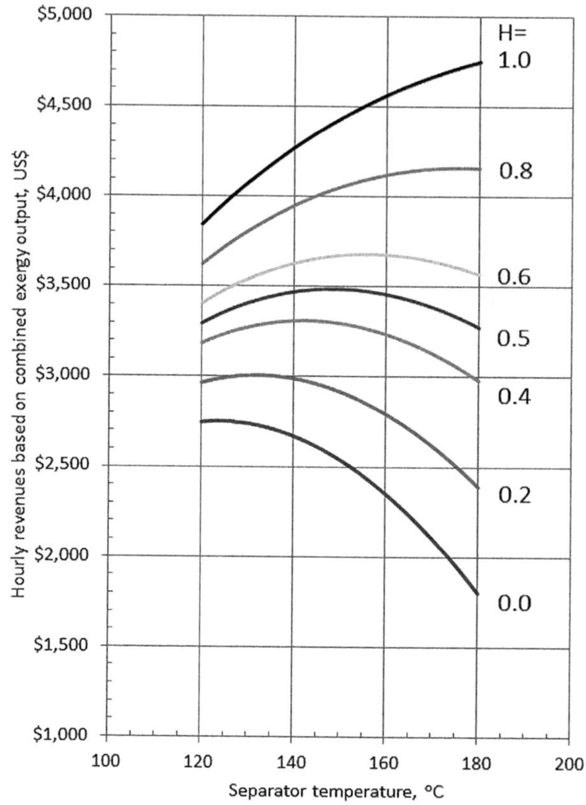

Fig. 6.7 Optimization results for CHP using 220°C geothermal resource.

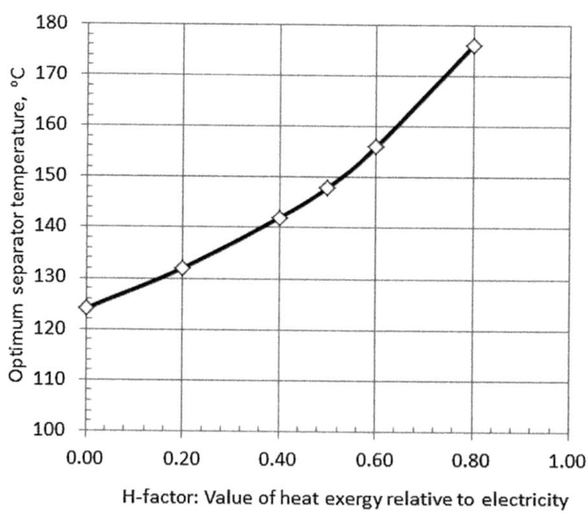

Fig. 6.8 Optimum separator temperature as a function of the H-factor.

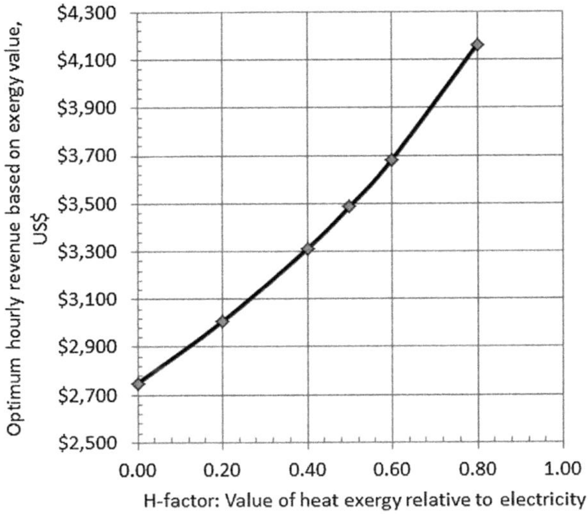

Fig. 6.9 Optimum hourly revenues for CHP plant as a function of the H-factor.

consideration for using such plants in conjunction with a direct heat application is the availability of liquid geofluid as a byproduct of the power operation. On the one hand, for the relatively rare dry steam plants, such as at Larderello and The Geysers, California, U.S., there is very little condensate and nowadays it is fully reinjected to help maintain the reservoir. On the other hand, the vast majority of geothermal fields produce low-quality liquid-vapor mixtures, where the steam is separated for use in the turbine and the liquid is either reinjected or flashed a second time to produce low-pressure steam that drives a low-pressure turbine. The liquid that remains after separation or flashing is usually reinjected. Thus, in the case of flash-steam plants there is an opportunity to add direct heat uses, subject to certain constraints related to the potential for chemical precipitation and scaling.

The process flow diagrams for single- and double-flash plants and their corresponding thermodynamic state diagrams are shown in Figs. 6.10–6.13 [4]. For both types of

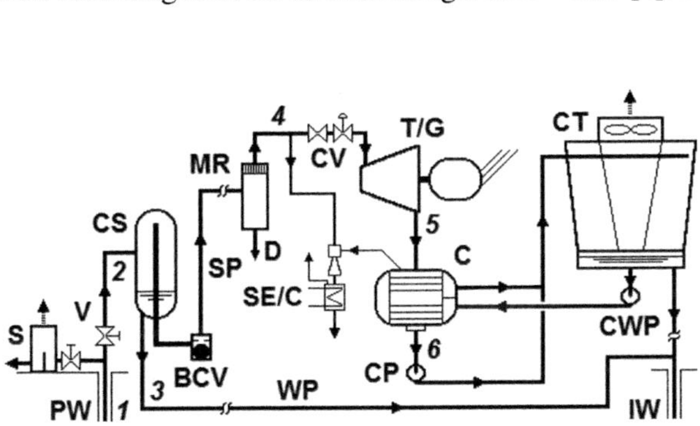

Fig. 6.10 Schematic flow diagram for single-flash geothermal power plant [4]. See Nomenclature list.

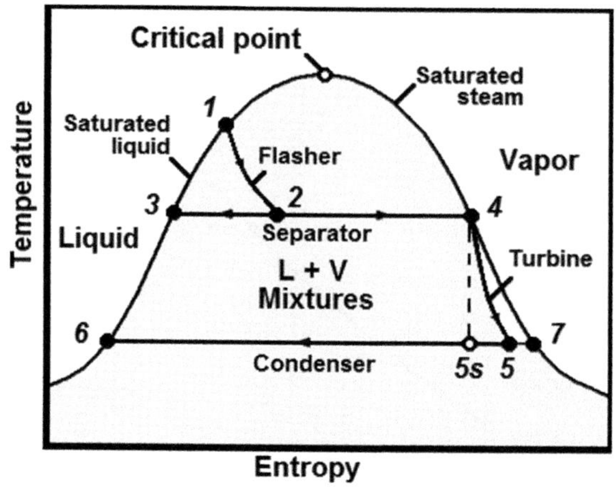

Fig. 6.11 Process diagram for single-flash plant in temperature-entropy coordinates; not to scale, after [4].

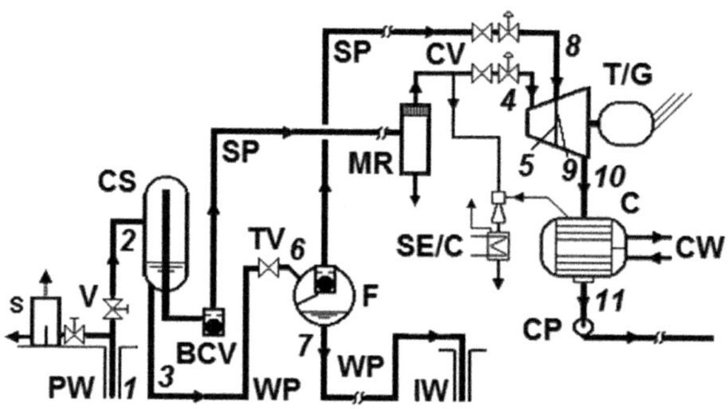

Fig. 6.12 Schematic flow diagram for double-flash geothermal power plant [4]. See Nomenclature list.

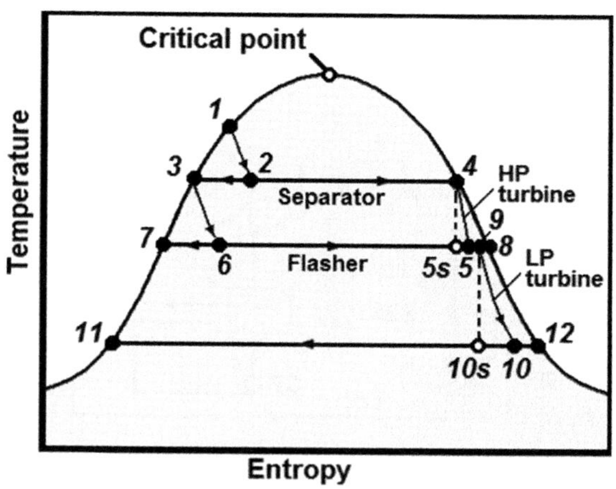

Fig. 6.13 Process diagram for double-flash plant in temperature-entropy coordinates; not to scale, after [4].

plant, fluid at state 3 is a possible candidate for direct use. However, for a double-flash plant the full flow rate at state 3 is best used to generate the low-pressure steam owing to the relatively high value of the electricity over that of hot water. State 7 is the best fluid available for direct use in that case, and the temperature is typically sufficiently high for many applications. For example, a high-temperature resource found in New Zealand, Kenya or Indonesia may be 320 °C, and with a condensing temperature of say 50 °C, the roughly optimum division of temperature by the "Equal-temperature-split rule" [4] leads to a separator at 230 °C and a flasher at 140 °C. From Fig. 6.1, having geofluid at 140 °C opens up many possibilities for direct heat usage. The main drawback to such usage, from the point of view of the power plant, is the possibility of loss of reinjection capacity. The lower the temperature of the separated geofluid, the greater the likelihood of chemical scaling in the surface piping and in the injection wells. Thus, for each resource the lower limit on geofluid temperature must be determined to prevent such scaling. Even with such a limit, there are typically many types of direct heat applications that can safely be entertained.

6.3.2 Organic Rankine cycles

Organic Rankine cycles (ORCs) tend to be deployed at low-to-moderate-temperature resources. Applications are not limited to geothermal but may include any number of waste heat recovery situations. Among all geothermal plants in the world, binary plants rank first in total number of units in operation, and constitute about 35% of all plants [4]. ORCs follow the basic Rankine cycle commonly used in fossil-fuel power plants, but without many of the technical features used there for improving the thermal efficiency of the cycle. Fig. 6.14 shows a simple flow diagram for an ORC and the process diagrams are given in Fig. 6.15 using two sets of thermodynamic coordinates.

The cycle working fluid, a low-boiling-point substance, follows a closed path through five main components: a preheater (PH), evaporator (E), turbine (T), condenser (C), and condensate pump (CP). The geofluid generally is pumped from

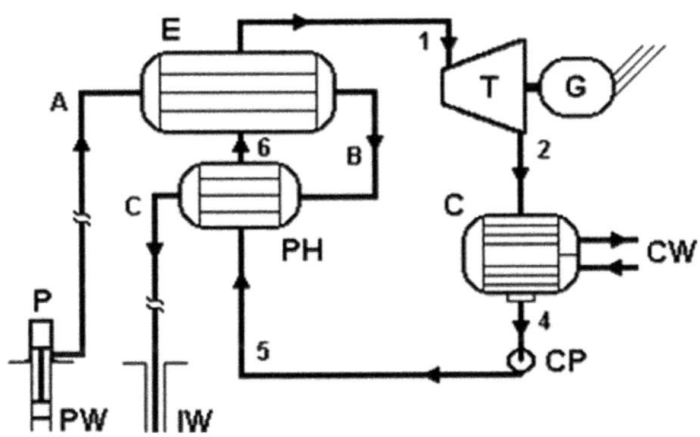

Fig. 6.14 Simplified flow diagram for a basic ORC plant [4]. *A-B-C,* geofluid states; *C,* condenser; *CP,* condensate pump; *CW,* cooling water; *E,* evaporator; *G,* generator; *PH,* preheater; *T,* turbine.

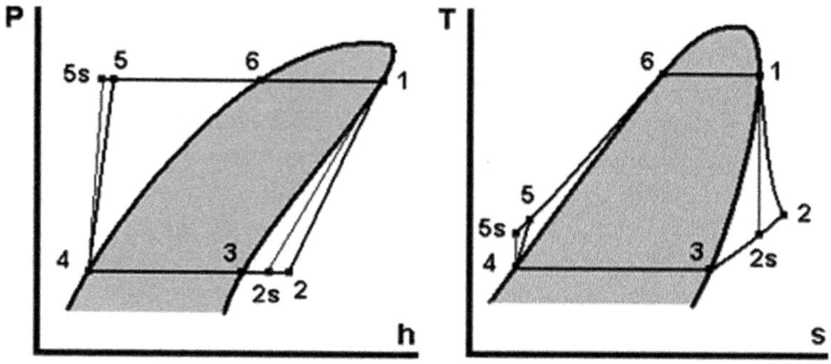

Fig. 6.15 ORC processes in pressure-enthalpy (P-h) and temperature-entropy (T-s) coordinates; not to scale [4].

the reservoir by means of a well pump, either mechanical, shaft driven or electrically powered submersible, reaching the surface at state A (Fig. 6.14). It passes through the evaporator where it brings the working fluid to a saturated or slightly superheated vapor condition, then through the preheater where it supplies the sensible heat to raise the working fluid to a saturated liquid state, called its bubble point. From that point, the geofluid is available for direct heat usage prior to being reinjected. The temperature of the geofluid here is typically significantly lower than in the case of steam plants, but can still be useful for space heating or spas, among other applications.

ORCs can be divided into two categories depending on the pressure in the heat exchangers: subcritical and supercritical. Figs. 6.14 and 6.15 are drawn for the subcritical case, i.e., processes 5-6-1 are at a pressure lower than the critical pressure; note the critical point lies at the top of the saturation curves. There is a fundamental difference between these two categories: a supercritical cycle allows for a much closer match between the temperature of the geofluid and the working fluid in the PH and E heat exchangers, and results in a more efficient transfer of heat, i.e., a smaller loss of exergy.

This is evident in Fig. 6.16 which shows the heat transfer processes for the two cases in a temperature-heat transfer diagram. It is schematic but illustrates that the mean effective temperature difference (METD) is much larger for the subcritical case because of the effect of the pinch-point, ΔT_{pp}, which occurs along a much smoother curve in the supercritical case owing to the continuous shape of the isobar. The shaded areas are proportional to the METD. The narrower gap between the heating and cooling curves means there is less irreversibility and less loss of exergy during the heat transfer. For this particular example, the working fluid is raised to a higher temperature and the geofluid is cooler at the discharge (state C) in the supercritical case, but the latter observation is not a general conclusion. It follows that a larger heat exchanger is need to transfer the same amount of heat when the METD is smaller, which implies higher cost. Thus, thermodynamic efficiency comes at a higher capital cost of equipment.

ORCs may be used instead of flash-steam units to form a CHP system, as shown in Fig. 6.17. Bypass pipelines should be used to allow the hot water plant to operate when the power plant is unavailable or when the demands for heat and electricity vary.

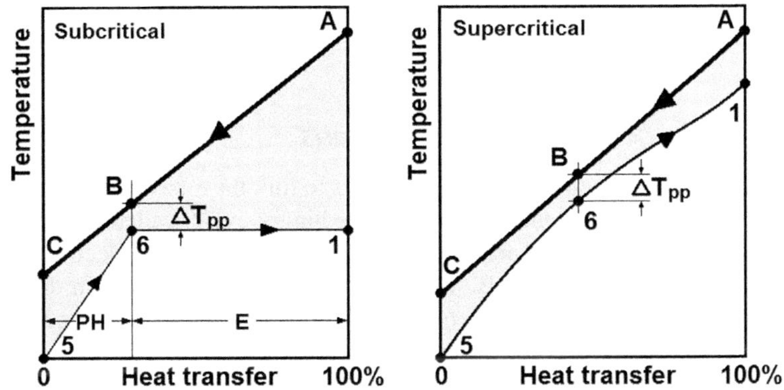

Fig. 6.16 Comparison between subcritical and supercritical ORC cycles regarding heat transfer. Note that the geofluid inlet (state A) and the working fluid inlet (state 5) have the same temperature, and the pinch-point temperature difference, ΔT_{pp}, is kept constant.

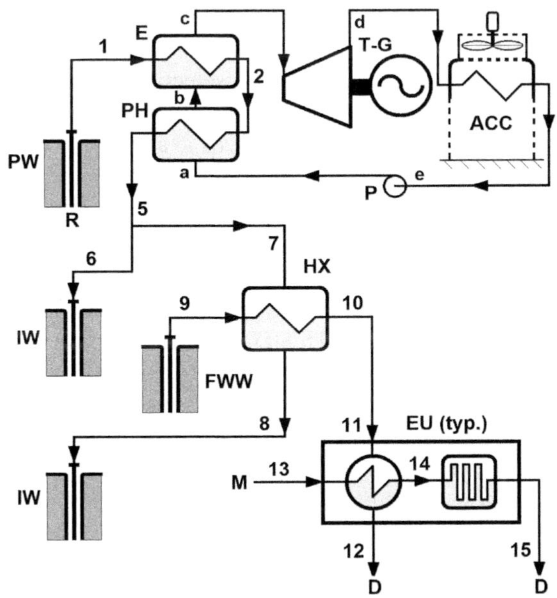

Fig. 6.17 Simplified schematic for a binary geothermal power plant combined with a hot water production facility to serve end users. See Fig. 6.3 for nomenclature; new items. *ACC*, air-cooled condenser; *E*, evaporator; *P*, pump; *PH*, preheater; *T-G*, turbine-generator.

The possibilities for direct heat applications in conjunction with a binary plant increase if a heat recuperator in used in the Rankine cycle. The recuperator captures heat from the turbine exhaust prior to the condenser and feeds it back into the working fluid after the feed pump and before the preheater. However, this is only feasible with retrograde working fluids (see Section 6.4). The recuperated heat lowers the thermal load in the heat exchangers, thereby allowing the geofluid to emerge at a higher temperature than without a recuperator, and thereby opens more direct heat applications for possible use.

6.3.3 Cascade geothermal CHP stations

It is common practice in geothermal power plants to link flash-steam plants and binary plants in a cascade arrangement. This allows the binary cycle, called a bottoming cycle, to capture energy in the separated brine to generate additional electricity without the need for more production wells. This is similar to the CHP system described in Section 6.2, except that a binary cycle (see Fig. 6.14) would receive the separated brine at state 5 in Fig. 6.3 first, followed by the direct use applications. Of course, the type of direct heat use is more restricted in this case owing to the lower temperature available after the binary plant.

6.4 Working fluid selection criteria in CHP stations

We now consider the important matter of the selection of the working fluid. This design decision has great implications for the performance of an organic Rankine cycle (ORC) binary plant. While there are many choices available for working fluids, there are also many constraints on that selection that relate to the thermodynamic properties of the fluids, as well as considerations of health, safety, and environmental impact.

6.4.1 Thermodynamic factors

Table 6.1 lists some candidate fluids and their relevant thermodynamic properties; pure water is included solely for comparison. Clearly all of the candidate fluids have critical temperatures and pressures far lower than water. Furthermore, since the critical pressures of the hydrocarbons are reasonably low, it is appropriate to consider the feasibility of supercritical cycles.

Binary mixtures of these fluids have also been studied for use in geothermal binary plants. In particular, the thermodynamic properties of 90% i-C_4H_{10} and 10% i-C_5H_{12} were determined by the U.S. National Bureau of Standards (predecessor of NIST) when it was chosen as the working fluid for the Heber Binary Demonstration plant in California in the 1980s [5]. Mixtures evaporate and condense at variable temperature, unlike pure fluids that change phase at constant temperature. This means that subcritical-pressure boilers for mixed fluids can be better matched to the brine cooling curves, in a manner similar to, but not exactly like, supercritical pure fluids. Water-ammonia mixtures, commonly used in absorption refrigeration cycles, are used in Kalina cycles which have been applied to low-temperature geothermal resources [6].

The liquid-vapor saturation pressure-temperature curve is another important property for binary cycle working fluids. The curve must be appropriate for the temperature of the geothermal fluid. Figure 6.18 shows the (P-T)-saturation lines for the substances in Table 6.1, except for water which is not a candidate for geothermal binary plants.

Table 6.1 Thermodynamic properties of some candidate working fluids for binary plants [4].

Fluid	Formula	T_c, °C	T_c, °F	P_c, MPa	P_c, lbf/in^2	P_s @ 300 K, MPa	P_s @ 400 K, MPa
Propane	C_3H_8	96.95	206.5	4.236	614.4	0.9935	n.a.
i-Butane	i-C_4H_{10}	135.92	276.7	3.685	534.4	0.3727	3.204
n-Butane	C_4H_{10}	150.8	303.4	3.718	539.2	0.2559	2.488
i-Pentane	i-C_5H_{12}	187.8	370.1	3.409	494.4	0.09759	1.238
n-Pentane	C_5H_{12}	193.9	380.9	3.240	469.9	0.07376	1.036
Ammonia	NH_3	133.65	272.57	11.627	1686.3	1.061	10.3
Water	H_2O	374.14	705.45	22.089	3203.6	0.003536	0.24559

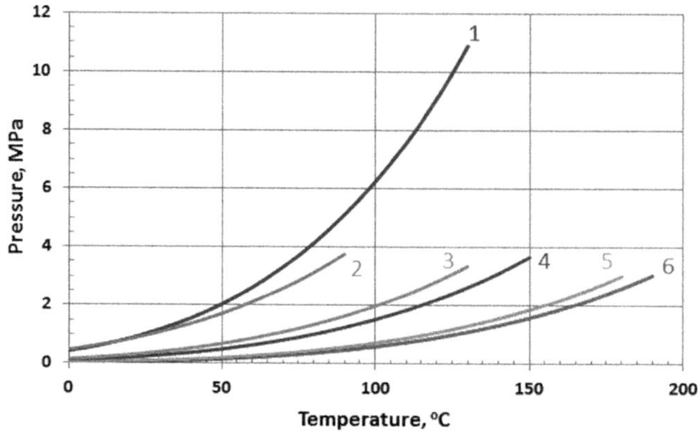

Fig. 6.18 Saturation (liquid-vapor) curves for candidate working fluids. *1*, ammonia; *2*, propane; *3*, isobutane; *4*, butane; *5*, pentane; *6*, isopentane; data from Refprop [2].

Ammonia (1) and propane (2) are suitable for low-temperature resources, isobutane (3) and butane (4) for low-to-moderate temperatures, and pentane (5) and isopentane (6) for moderate-to-high temperatures.

Another important characteristic of binary cycle candidate fluids is the shape of the vapor saturation curve as viewed in temperature-entropy coordinates; see Fig. 6.19. The saturated vapor curve for water (shown as the thin, normal line) has a negative slope everywhere, but certain hydrocarbons and refrigerants show a positive slope for portions of the saturation line (called retrograde). That is, there exists a local minimum in the entropy at some low temperature, T_m, and a local maximum in entropy at a higher temperature, T_M. Retrograde fluids include butane, isobutane, pentane and isopentane. These fluids exhibit retrograde behavior over the following temperature ranges, $T_m \rightarrow T_M$: C_4H_{10}, $-3\ °C \rightarrow 127\ °C$; i-$C_4H_{10}$, $-3\ °C \rightarrow 117\ °C$; C_5H_{12}, $-3\ °C \rightarrow 177\ °C$; i-$C_5H_{12}$, $-13\ °C \rightarrow 177\ °C$. Since T_m is lower than any temperature encountered in geothermal binary plants, for our purposes these fluids

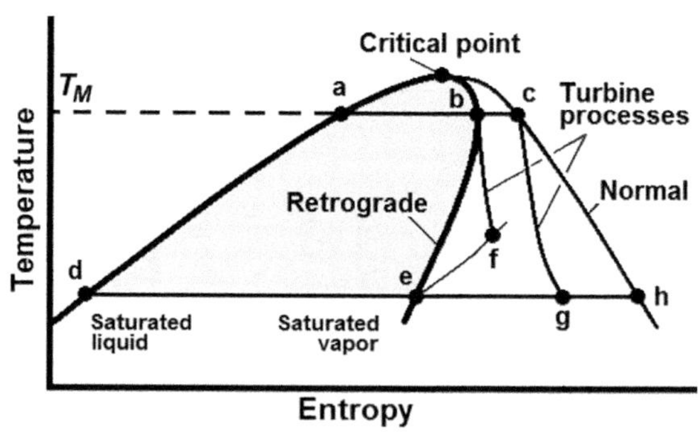

Fig. 6.19 Schematic temperature-entropy diagram contrasting normal and retrograde saturated vapor curves and turbine expansion processes, after [4].

can be taken as having saturated vapor lines similar to that shown in Fig. 6.19 (heavy line). This has major implications for ORC cycles.

Normal fluids, such as water, require considerable superheat, extending the isobar a-b-c upwards, to avoid excessive moisture at the turbine exhaust, state g. This is always designed into fossil-fuel steam power plants. However, retrograde fluids allow expansion from the saturated vapor line directly into the superheated region, process b-f, thereby avoiding any moisture during the turbine expansion process. Moisture during expansion in a turbine causes damage to the blades and leads to lower isentropic efficiency. It is possible to carry out an expansion process starting from the vapor saturation line between the critical point and point b, or from a supercritical initial state, such that the expansion line passes through the wet region (shaded) for a portion of the process, emerging into and continuing in the superheated region, without suffering any wetness penalty in efficiency. Apparently, the fluid remains in a metastable vapor state while passing through the wet region by staying close to saturation line.

6.4.2 Environmental, health and safety factors

The environmental, safety and health properties of potential working fluids must be considered in addition to the thermodynamic characteristics. These include flammability, toxicity, ozone depletion potential (ODP), and global warming potential (GWP). Table 6.2 displays these properties for the fluids in Table 6.1, plus two chlorofluorocarbons (R12, R114) that once were used as working fluids in binary plants, three hydrofluorocarbons (R32, R134a, R245fa), and carbon dioxide (CO_2).

Table 6.2 Environmental and health properties of substances [7].

Fluid	Formula	Toxicity	Flammability	ODP	GWP
R12	CCl_2F_2	Non-toxic	Non-flammable	1.0	4500
R114	$C_2Cl_2F_4$	Non-toxic	Non-flammable	0.7	5850
Propane	C_3H_8	Low	Very high	0	3
i-Butane	i-C_4H_{10}	Low	Very high	0	3
Butane	C_4H_{10}	Low	Very high	0	3
i-Pentane	i-C_5H_{12}	Low	Very high	0	3
Pentane	C_5H_{12}	Low	Very high	0	3
R32	CH_2F_2	Low	Low	0	675
R134a	$C_2H_2F_4$	Very low	Non-flammable	0	1300
R245fa	$C_3H_3F_5$	Very low	Non-flammable	0	1020
Carbon dioxide	CO_2	Non-toxic	Non-flammable	0	1.0
Ammonia	NH_3	Toxic	Lower	0	0
Water	H_2O	Non-toxic	Non-flammable	0	–

The ODP is normalized at 1.0 for refrigerants R-11 and R-12 which are the worst in this regard. The GWP is normalized at 1.0 for carbon dioxide. Owing to their high ODP and GWP, both R-12 and R-114 have been banned from use by the Copenhagen Amendment (effective as of 1994) to the Montreal Protocol (signed in 1987, effective as of 1989). The binary plant at Paratunka in Russia that was installed in 1967 used R-12 as its working fluid in conjunction with a very low temperature geofluid, ~81 °C [8].

All of the hydrocarbon candidate fluids obviously are flammable and necessitate appropriate fire protection equipment on site, over and above the usual requirements for any power plant.

6.5 Optimization of geothermal CHP cycles

In Section 6.2 we performed a system optimization for a CHP plant to determine which separator temperature yielded the highest revenues from combined output of heat and power when the two outputs had different monetary values. The individual components of the plant were not examined. The efficiency of exergy transfer is an important consideration in the design of each component of the plant. In this section, we examine this aspect of the optimization process.

6.5.1 Thermal and exergetic efficiencies

Thermal efficiency is a long-established measure of performance for power plants operating on a cycle in which the working fluid passes through a series of components undergoing various processes and periodically returns to some "initial" state. During the cycle of processes, heat and work may be exchanged with the surroundings. The thermal efficiency is defined as the ratio of the net work output to the heat input:

$$\eta_{th} = \frac{\dot{W}_{net}}{\dot{Q}_{in}} \tag{6.30}$$

Exergy efficiency is a more recent measure of plant performance that is defined as the ratio of the net exergy output to the exergy input associated with the heat input. Exergy is the maximum theoretical output allowed by the Second Law of thermodynamics. It applies to heat transfer, work transfer, and mass flows. The exergy associated with heat transfer is the work that could be delivered by using that heat in an ideal Carnot cycle operating between the temperature of the heat source and the ambient heat sink or the atmosphere. Thus the equation is:

$$\dot{E}_Q = \dot{Q}\left[1 - \frac{T_0}{T}\right] \tag{6.31}$$

where T_0 is the ambient temperature or dead-state temperature, in absolute degrees.

The exergy of a work transfer is the maximum amount of work that could be delivered in the absence of any dissipative phenomena, i.e., the exergy is equal to the work itself:

$$\dot{E}_W = \dot{W} \tag{6.32}$$

The exergy of any given stream of mass $\dot{m}$ at a temperature T and pressure P is the work that could be produced by changing the state of that mass by means of an ideal reversible process from its initial state to a final state where it is in thermodynamic equilibrium with its surroundings. Thus, if a fluid exists at (T, P) and is brought to (T_0, P_0) via a steady ideal process, the exergy of the fluid in its initial state is

$$\dot{E}_m = \dot{m}\{h(T,P) - h(T_0,P_0) - T_0[s(T,P) - s(T_0,P_0)]\} \tag{6.33}$$

These equations may be used to calculate the overall plant exergy efficiency, applied to the type of plant involved. For a flash-steam plant receiving a stream of geofluid from a well, the formula is

$$\eta_{ex} = \frac{\dot{W}_{net}}{\dot{E}_{in}} = \frac{\dot{W}_{el}}{\dot{E}_{gf}} \tag{6.34}$$

where the subscripts *el* refers to the electrical power delivered from the plant and *gf* refers to the geofluid at the wellhead condition. In some cases, the initial state of the geofluid in the reservoir may be an appropriate inlet state. For a binary plant operating on a cycle, one may use Eq. (6.34) for the plant performance, or the following formula for the cycle itself:

$$\eta_{ex} = \frac{\dot{W}_{net}}{\dot{E}_{in}} = \frac{\dot{W}_{el}}{\dot{E}_{Q}} \tag{6.35}$$

The derivations of these equations can be found in any thermodynamics textbook. It is important to stress that absolute temperature must be used in all calculations involving temperature.

6.5.2 Exergy loss in system components

Each component of a geothermal plant may be analyzed using the exergy concept to determine how efficiently it is performing. There is no accepted, standard definition for the exergy efficiency of a component. One approach is to perform an exergy accounting, keeping track of all input and output exergies, and then simply calculate the ratio of the output sum to the input sum. When this is done, it will always be observed that the output sum is less than the input sum. The difference is the exergy destroyed by the irreversibilities present in the component. This outcome is unlike the accounting of energy which always balances in accordance with the First Law of thermodynamics or conservation of energy. Exergy is not conserved and the size of the destruction is proportional to the amount of irreversibility present. Those losses can be addressed by design modifications to improve the performance, albeit at some monetary cost. Thus, for any component the exergy destruction $\dot{E}_D$, can written as

$$\dot{E}_D = \sum_{in} \dot{E}_m + \sum_{in} \dot{E}_Q + \sum_{in} \dot{E}_W - \sum_{out} \dot{E}_m - \sum_{out} \dot{E}_Q - \sum_{out} \dot{E}_W > 0 \tag{6.36}$$

where the terms are found from Eqs. (6.31–6.33).

On this basis, the exergy efficiency of a given component can be written as

$$\eta_{ex,i} = \frac{\sum \dot{E}_{out,i}}{\sum \dot{E}_{in,i}} = 1 - \frac{\dot{E}_D}{\sum \dot{E}_{in,i}} < 1 \tag{6.37}$$

where the exergy terms appropriate for the component are calculated. This approach is *pro forma* and can be applied to any component or subsystem of the plant.

An alternative approach may be more useful, namely, determining the exergy of the output which the component was designed to produce, and then comparing that to the input exergy that is needed to produce that output. This requires an examination and understanding of how the component operates, and may lead to better means of improving its performance. As an example, consider the turbine in Fig. 6.3. It is designed to generate power $\dot{W}_T$, and receives a mass flow of geothermal steam at state 2. The exhaust at state 3 carries a certain amount of exergy with it, but it is not used further; eventually it ends up in the surroundings. From a functional point of view, the turbine generates pure exergy (electrical power) while being driven by the exergy in the fluid at state 2. Thus, a functional exergy efficiency would be

$$\eta_{ex,T} = \frac{\dot{W}_T}{\dot{E}_2} = \frac{\dot{W}_T}{\dot{m}_2[h_2 - h_0 - T_0(s_2 - s_0)]} \tag{6.38}$$

Another functional definition might be the ratio of the output power to the change in the exergy of the steam as it passes through the turbine. This would be appropriate if some further use could be made of the geofluid at state 3. Thus, an alternative functional efficiency would be

$$\eta_{ex,T} = \frac{\dot{W}_T}{\dot{E}_2 - \dot{E}_3} = \frac{\dot{W}_T}{\dot{m}_2[h_2 - h_3 - T_0(s_2 - s_3)]} \tag{6.39}$$

Similar definitions can be developed for all the components in the CHP plant shown in Fig. 6.3.

6.5.3 Exergy losses in CHP plants

The system shown in Fig. 6.3 will be used to illustrate the exergy losses in a CHP plant. To facilitate this presentation, Table 6.3 shows all the thermodynamic properties needed to carry out the exergy efficiency calculations. The optimum point in Fig. 6.7 for the case of H = 0.5 is chosen for this exercise. Numbers shown in ***bold, italic face*** were assumed; all others were found using Refprop or calculated. A pressure loss of 10% was assumed for fluids passing through heat exchangers. The End User receives only enough hot water from the distribution system to meet its needs. The circulating hot water within the End User was taken as 1.2 kg/s; the calculated hot water distribution rate is nearly identical for this case.

With the exergy rates now determined at all state points and neglecting work requirements for water pumps, the exergy accounting can be carried out. Each component will be studied separately.

1. Separator, S

$$\dot{E}_{D,S} = \dot{m}_1 e_1 - \dot{m}_2 e_2 - \dot{m}_5 e_5 = 66.42 - 38.93 - 27.49 = 0.00 \text{ MW}$$

This result says that the separator has no exergy destruction because it merely divides the two-phase liquid-vapor mixture into its constituent parts, with each part retaining its original exergy. If we account for pressure losses in the separator, there would be some loss of exergy.

2. Turbine, T

$$\begin{aligned}\dot{E}_{D,T} = \dot{m}_2 e_2 - \dot{m}_3 e_3 - \dot{W}_T &= 38.93 - 7.15 - 52.835 \times (2743.5 - 2244.6)/1000 \\ &= 5.42 \text{ MW}\end{aligned}$$

This is about 14% of the exergy supplied to the turbine at state 2; in that sense, the turbine has a functional exergy efficiency of 86%. Viewed as a percentage of the change in exergy by the steam passing through the turbine, it would be 83% efficient. The turbine delivers 26.36 MW to the generator, of which 25.7 MW is sent out to the grid.

3. Heat exchanger, HX

$$\dot{E}_{D,HX} = \dot{m}_7 e_7 + \dot{m}_9 e_9 - \dot{m}_8 e_8 - \dot{m}_{10} e_{10} = 18.33 + 0.02 - 3.24 - 7.60 = 7.51 \text{ MW}$$

This is a significant loss, amounting to about 41% of the incoming exergy. That is, only 59% of the incoming exergy, essentially all of it from the separated brine, is transferred to the fresh water for use in the hot water distribution system. This component should be examined to find ways to improve the operation.

4. End user, EU

Various applications may be occurring at the end user such as radiant floor heating of residences, domestic hot water production, snow melting, greenhouse heating, etc. The transfer of exergy may happen in a hot water tank, under the surface of a driveway or sidewalk, in an air coil or in soil at a greenhouse, etc. In all cases, at the low temperatures

Table 6.3 State-point properties for optimum CHP using H = 0.5; see Section 6.2.

State point	Temperature, °C	Pressure, kPa	Quality	Entropy, kJ/kg.K	Enthalpy, kJ/kg	Exergy, kJ/kg	Mass flow rate, kg/s	Exergy rate, MW
R	***220***	2319.6	***0***	2.5177	943.58	208.42	***350***	72.95
1	148	451.18	0.1510	2.5813	943.58	189.77	350	66.42
2	148	451.18	***1***	6.8552	2743.5	736.80	52.84	38.93
3	***40***	7.38	0.8633	7.2053	2244.6	135.27	52.84	7.147
5	148	451.18	0	1.8214	623.56	92.519	297.17	27.49
6	148	451.18	0	1.8214	623.56	92.519	99.06	9.164
7	148	451.18	0	1.8214	623.56	92.519	198.11	18.33
8	***70***	406.06	Subcooled	0.9549	293.37	16.342	198.11	3.238
9	***20***	200.00	Subcooled	0.2964	84.100	0.100	195.19	0.020
10	***100***	180.00	Subcooled	1.3071	419.23	38.936	195.19	7.600
11	***85***	135.00	Subcooled	1.1345	356.08	26.384	1.197	0.032
12	***50***	121.50	Subcooled	0.7038	209.44	6.031	1.197	0.007
13	***20***	***100***	Subcooled	0.2965	84.006	***0***	1.197	0
14	***20***	***100***	Subcooled	0.2965	84.006	***0***	–	0
0	***20***	***100***	Subcooled	0.2965	84.006	***0***		0

involved, the amount of exergy exchanged is likewise very low. From Table 6.3, it may be seen that about 32 kW of heat exergy is released by the distributed hot water altogether. If the hot water tank raises the domestic water temperature from, say, 25 °C to 60 °C, it would gain about 13 kW and the transfer would be about 52% efficient, ignoring any other uses outside the building.

6.6 Case studies

There is widespread use of geothermal CHP systems across the world [9]; only two plants have been selected for discussion. The next two sections present case studies based on the experiences of the author of this chapter (RDP). The first one is a medium-size geothermal flash-steam plant that supplied separated brine to a hot water heating plant for distribution to a municipality, and the other is a small binary plant combined with a heating system for a college campus. The case studies draw upon the author's earlier works on these two systems [10,11].

6.6.1 Kakkonda-Shizukuishi, Honshu, Japan

The Kakkonda-Shizukuishi CHP facility was built in the late 1980s and at its peak generated 50 MW of electrical power and supplied 1050 t/h (292 kg/s) of hot water to industrial, commercial and residential end users. It was the largest project of its type in Japan at the time. Kakkonda Unit 1 began operating as a stand-alone, single-flash power plant in 1978. The plant is a fairly standard design with five production and reinjection well pads, with a set of three cyclone separators at Pad B, just below the power station. Steam was sent from the well pads to pad B and on to the power house via a network of pipes; the separated brine was reinjected back to the reservoir via wells on the pads. The available surface area for construction is limited owing to the rugged mountainous terrain on both sides of the Kakkonda River that flows past the plant site. This necessitated directional drilling from a few well pads, situated relatively close to the river, to minimize interference among production and injection wells.

Figure 6.20 shows the power house and well pad B, one of the closest to the plant. After several years of successful operation, it was decided to add a hot water production facility (HWPF) to complement the electrical power by capturing some of the energy from the separated brine that had been completely reinjected. The town of Shizukuishi, whose center lies some 17 km southeast of the power plant, would be the beneficiary of the heat. Given the rugged terrain at the plant site, the HWPF was sited about 1.5 km away from the power station in the direction of Shizukuishi; see map Fig. 6.21.

Owing to difficult topography between the power plant and the town, the hot water pipeline passes over bridges and through tunnels constructed for this purpose; Fig. 6.22 shows the path in schematic form. Figure 6.23 shows the distribution pipelines within Shizukuishi town. The pipelines take advantage of gravity as the town lies about 420 m lower than the HWPF.

Fig. 6.20 Kakkonda 50 MW Unit 1 as seen in 1979. Well pad B is in foreground with cyclone separators at right and silencers at left; photo from H. Nakamura, 1978.

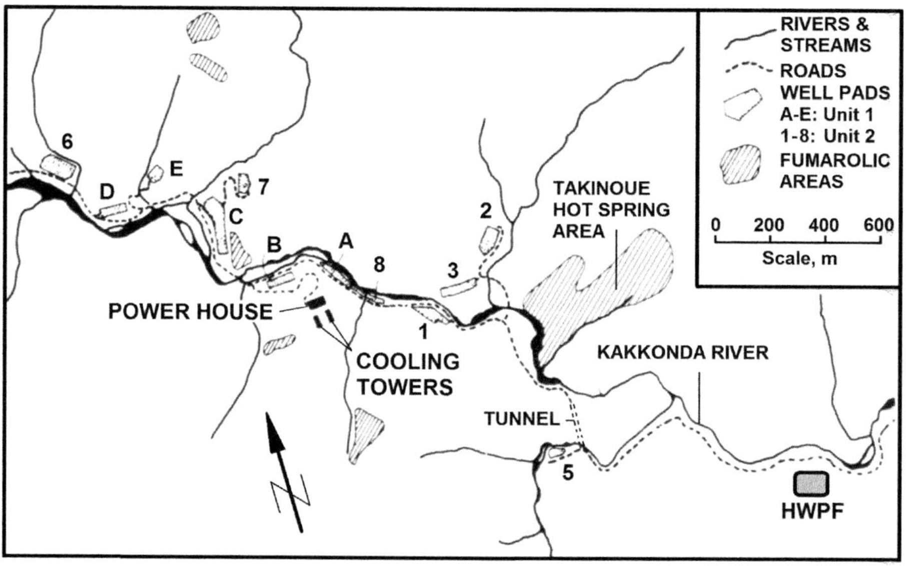

Fig. 6.21 Map of the Kakkonda-Shizukuishi power plant and HWPF area, c. 1987 [10].

Table 6.4 shows the temperatures and flow rates at key points along the flow path; see Fig. 6.22. The list of users of the hot water is extensive and includes: private homes (radiant heating), apartments, condominiums, resort hotels, tourist attractions, a hospital and school, town government buildings, day-care centers, an industrial park, greenhouses, botanical gardens, fish farming, swimming pools and snow melting. A total of 2670 private homes and 23 public facilities were served [10].

The HWPF, nearing completion in 1986, is shown in the photograph, Fig. 6.24. The operating principle is flash evaporation of the hot brine in five stages to obtain clean

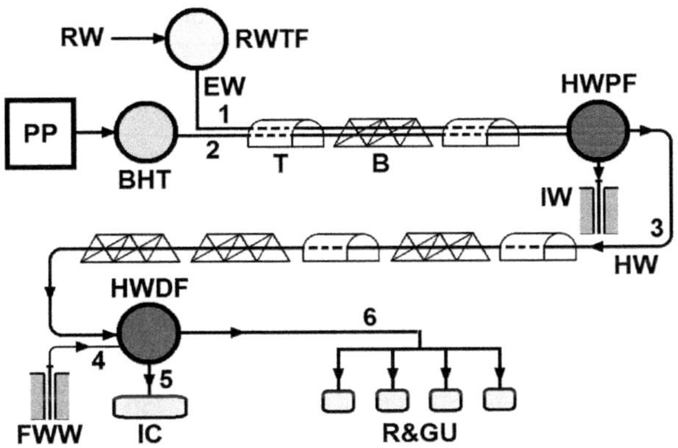

Fig. 6.22 Schematic of Kakkonda-Shizukuishi CHP pipeline and distribution plan, redrawn from [12]. *B*, bridge; *BHT*, brine holding tank; *EW*, exchange water; *FWW*, fresh water well; *HW*, hot water; *HWDF*, hot water distribution facility; *HWPF*, hot water production facility; *IC*, industrial consumers; *IW*, injection well; *PP*, Kakkonda geothermal power plant; *R&GU*, residential & general users; *RW*, Kakkonda river water; *RWTF*, river water treatment facility; *T*, tunnel. See Table 6.1 for data on numbered locations.

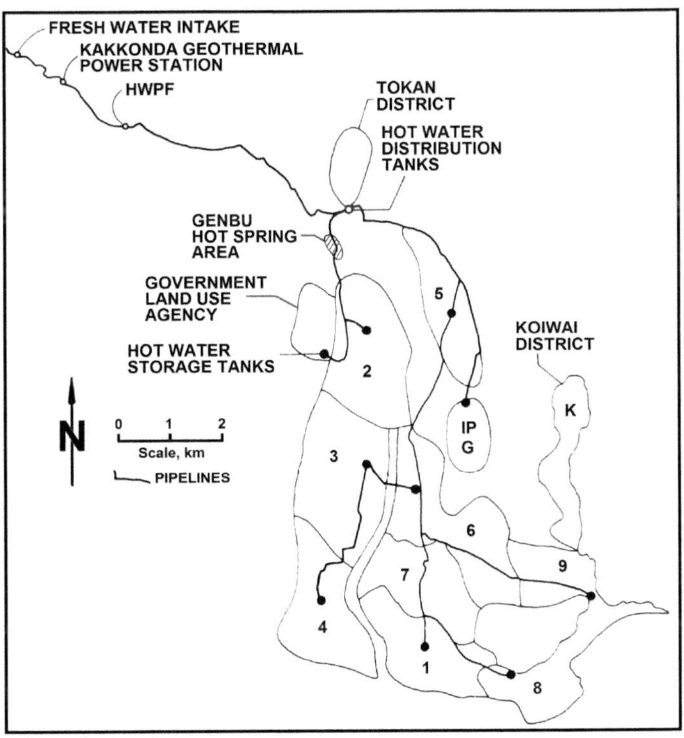

Fig. 6.23 Hot water pipelines in Shizukuishi town and the districts served by the system [10].

Table 6.4 Temperatures and mass flow rates at key points in the system shown in Fig. 6.22.

Location	Item	Temperature, °C	Mass flow, t/h	Mass flow, kg/s
1	Exchange water	15	710	197.2
2	Geothermal brine	145	1000	277.8
3	Produced hot water	115	800	222.2
4	Fresh water for dilution	~15	250	69.4
5	Industrial consumption	115	300	83.3
6	Residential & general usage	85	750	208.3

Fig. 6.24 Kakkonda-Shizukuishi HWPF in 1986, looking generally south; photo by R. DiPippo. Five upper steam pipes run from the flash evaporator (horizontal vessel at top rear) to the direct-contact heat exchanger (horizontal vessel middle). The vertical vessel at the right is a deaerator.

steam at ever lower temperatures which heats fresh water from the Kakkonda River via five heat exchangers, four of which are direct contact type (see Fig. 6.25). Thus, the treated river water mixes with the flash steam to form hot water of sufficient purity for municipal use.

The process flow diagram for the overall system is shown in Fig. 6.25. Multiple production, injection, and fresh water wells were used.

The utilization efficiency of the HWPF can be calculated based on the exergy of the separated brine coming from the power plant and the exergy of the hot water sent to the heating districts. From Table 6.4, the ratio of the exergy rate at location 3 to that at location 2 is a measure of the efficiency. Thus, using a 15 °C dead state, we find

$$\eta_U = \frac{\dot{E}_3}{\dot{E}_2} = \frac{13,234.1}{26,730.8} = 0.495 \tag{6.40}$$

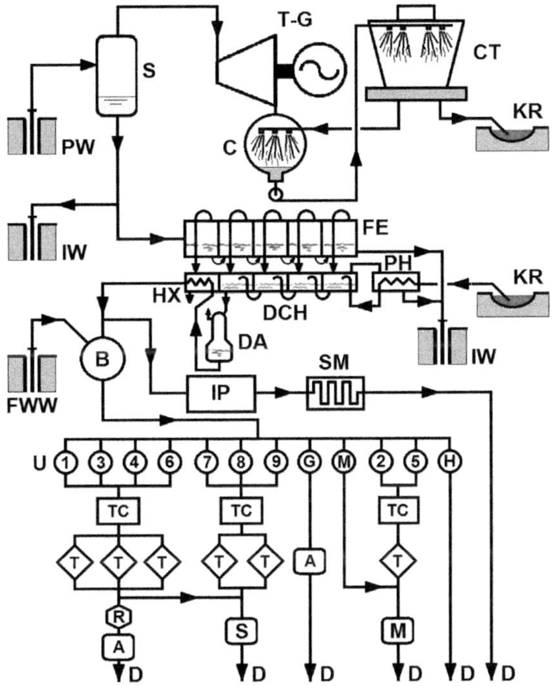

Fig. 6.25 Kakkonda-Shizukuishi CHP flow diagram, after [12]. *1–9*, heating districts; *A*, aquaculture; *B*, blending tank; *C*, condenser; *CT*, cooling tower; *D*, drain; *DA*, deaerator; *DCH*, direct contact heater; *FE*, flash evaporator; *FWWs*, fresh water wells; *G*, botanical gardens; *H*, hotels, motels, tourist facilities; *HX*, heat exchanger; *IP*, industrial park; *IW*, injection wells; *KR*, Kakkonda river; *M*, miscellaneous users; *PW*, production wells; *R*, radiant heating in private homes; *T*, total use in private homes; *TC*, temperature conditioning; *U*, direct heat users.

This system was sponsored as a national project by the Japan Agency for Natural Resources and Energy and by the Ministry of International Trade and Industry (MITI). It was carried out by the Iwate Prefectural Government. The project was in force from 1980 to 1988 [13]. The official name of the project was "The Verifiable Investigation Project for Geothermal Water Supply." The last stage that included the actual operation of the system was called the "Verifiable Investigation for Practical Use." The plant and equipment cost was 68 million US$. This expensive project was built and operated for about ten years, after which a feasibility study was carried out to determine if the plant should be continued. The decision to terminate the project was greatly influenced by the price of oil at the time; a steep decline in oil prices rendered the plant uneconomic to continue as a commercial venture. Ironically, it was the steep rise in oil prices during the two Oil Shocks of 1973 and 1979 that had motivated the study and construction of the project [Pers. comm., A. Suzuki, May 29, 2018]. Based on satellite imagery from Google Earth (Fig. 6.26), the facility was intact and possibly in operation in 2004, but by 2014 it had largely been dismantled.

Fig. 6.26 HWPF: (L) April 5, 2004; (R) April 16, 2014; *Google Earth* images.

The concept of CHP geothermal plants is used widely in Japan. The following power plants, among others, are all coupled to direct heat usage in nearby communities: Otake, Hatchobaru [14], Matsukawa, Suginoi [15] and Kirishima-Kokusai [16]. Given the huge number of hot spring spas (onsen) in Japan, it is now becoming popular to add small binary power units to the existing spas by capturing heat from the highest temperature geofluid, typically around 90−100 °C. This water is too hot for bathing and is available for power generation without disturbing the activity at the spa. After being cooled by say 30−40 °C, the geofluid is at a temperature is appropriate for balneology and no additional wells are needed.

6.6.2 Oregon Institute of Technology (OIT), Klamath Falls, U.S.

The Oregon Institute of Technology (OIT) is located in Klamath Falls, in south-central Oregon, U.S., about 28 km north of the border with California, at the far northwest edge of the Basin and Range Province; see Fig. 6.27. The school was established in 1947 to train veterans returning from WWII. The original location was in an idle Marine Corps hospital about three miles northeast of Klamath Falls, but was relocated to its present site in 1964. There are many direct uses of geothermal heat in the city and this was a principal reason for the campus move. It acquired its OIT name in 1973 and in 1975 the Geo-Heat Center was created to support the development of the natural geothermal resource in the area for practical purposes.

A campus geothermal district heating system has been in operation since 1964 and currently provides heat to 11 buildings, cools five buildings, and supplies domestic hot water and snow melting on sidewalks; see Fig. 6.28. The heating covers 55,700 m^2, cooling 25,800 m^2 and snow melting 214 m^2. Previously, the campus used an oil-fired boiler for heating; the geothermal system saves roughly $225,000 annually in heating cost [17].

In 2010, the first geothermal power plant was added to the district heating system to provide electrical power for the campus. The unit was a small 280 kW (nom.) binary plant designated as "PureCycle" by the manufacturer. The unit is housed in a separate

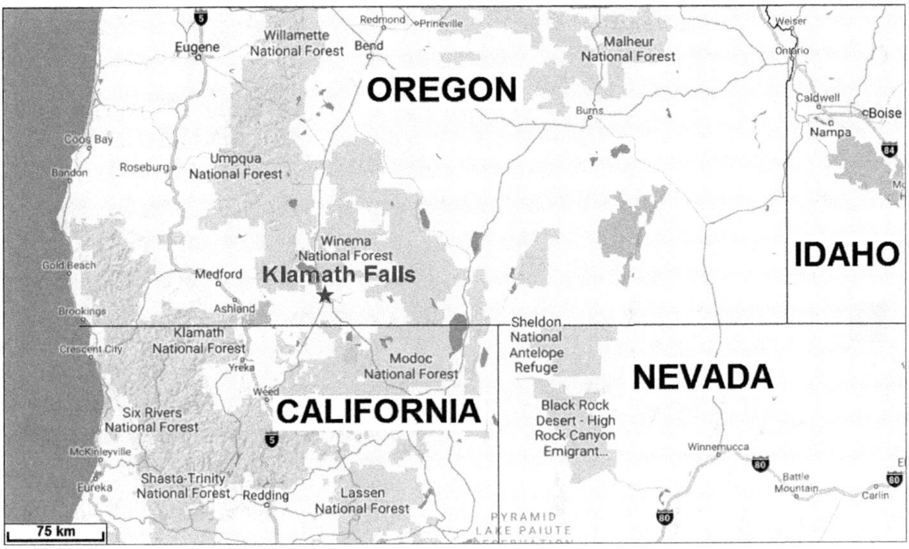

Fig. 6.27 Location map for Klamath Falls, Oregon, U.S., modified from *Google Maps*.

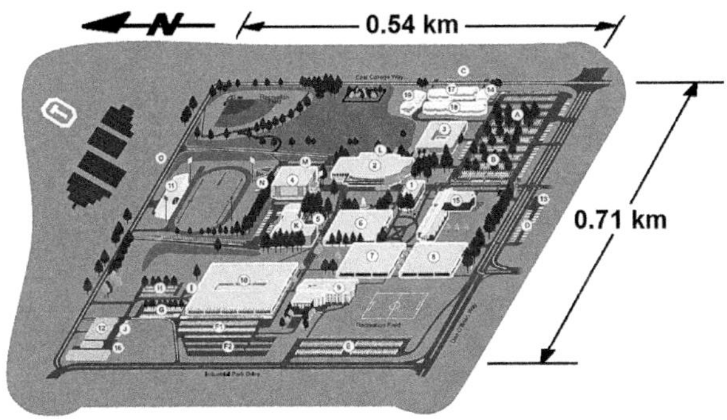

Fig. 6.28 Campus map of Oregon Institute of Technology, modified from http://www.oit.edu/visitors-info/about/campus-map.

building just outside the southeast corner of the campus; see Fig. 6.29. A case study of Plant 1 was presented in Ref. [11]. The temperature of the geofluid that is pumped from an average depth of 536 m was only 91 °C. The actual net cycle power was about 226 kW of which 148 kW was needed to run the well pumps, leaving 78 kW for campus use. The cycle itself used R245fa as the working fluid and showed a thermal efficiency of about 8%. The geofluid temperature leaving the power plant was 73 °C and 57 °C leaving the heat exchanger of the district heating plant, going to two reinjection wells. Fig. 6.30 shows the compact power unit. The capital cost was $890,000 or $3180/kW; it saved $23,000 per year and offset 669,000 kWh of electricity purchases.

A second geothermal power plant was installed in 2014, comprised of two binary units: a 1000 kW and a 750 kW unit, mounted side-by-side in an adjacent new

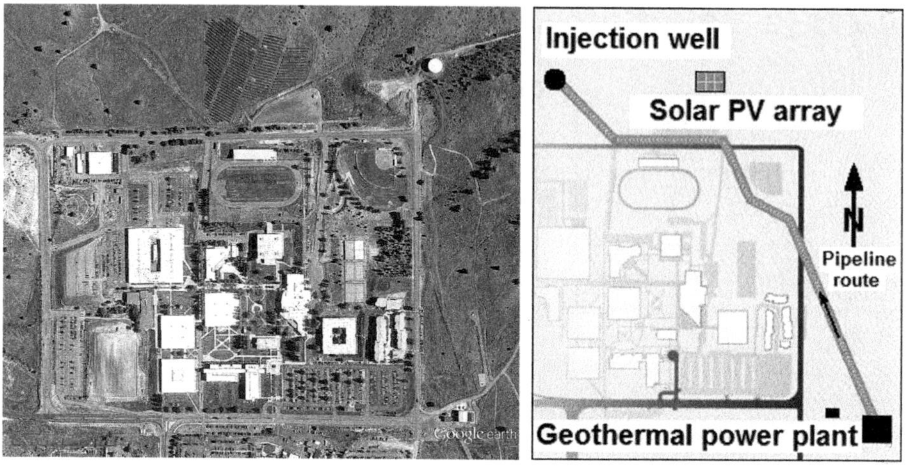

Fig. 6.29 *Google Earth* aerial view of OIT campus, June 2, 2016 (left); Schematic map of geothermal power plant, reinjection pipeline, and solar PV array (right), modified from [18].

Fig. 6.30 Plant 1 280 kW (nom.) "PureCycle" binary unit [18].

building; see Fig. 6.31. The capital cost was $14,700,000 or $8400/kW; it saves $400,000 per year and offsets 7,646,000 kWh of electricity purchases.

The simple payback periods for both plants may seem long by commercial standards but the campus will exist for decades to come, thereby amortizing the costs over many years. Furthermore, there were several grants from various entities that reduced the actual OIT outlay for the units.

Figure 6.32 shows the power plant complex and the sites for the three production wells that supply the hot geofluid. The wells are fairly closely spaced with separations as follows: wells 2 to 5, 156 m; wells 2 to 6, 143 m; and wells 5 to 6, 121 m. They are vertical wells that intersect a steeply dipping fault system lying east of the plant. The wells were deepened in preparation for the new Plant 2 from an average depth of

Fig. 6.31 Plant 2 with 1000 kW and 750 kW binary units. Large pipes at the top carry cooling water to the condensers; the lower vessels are the evaporators (large diameter) and preheaters (small diameters) for the units [18].

Fig. 6.32 Plants 1 and 2 with an installed capacity of 2030 kWe, modified from *Google Earth*.

488 m to 1618 m. Each well enclosure houses the wellhead and the mechanical pump and motor.

The OIT CHP system is now complete. Besides geothermal, the campus also has a large 2000 kW photovoltaic (PV) solar system to augment the electricity produced by the geothermal units; see Fig. 6.29. The intermittent nature of the PV means it is used as a peaking system.

The OIT campus is fully heated by geothermal fluid and about 35% of its electricity needs are supplied by its geothermal binary units. When the PV system is producing, about two-thirds of the electrical requirement of the campus is being met by renewable energy.

6.7 Summary

Geothermal combined heat and power (CHP) systems are a very effective way to utilize the available energy in a geothermal resource. CHP systems may be designed with any of several geothermal power generation schemes including dry steam, flash steam, binary or hybrid type plants. The fundamental principles of thermodynamics are applied to CHP systems and working equations are derived that allow analysis to be carried out to determine optimum performance for various conditions that may be encountered in a geothermal resource. There is a wide assortment of possible working fluids for use in geothermal CHP that may be chosen based on technical and environmental factors. Examples are given showing optimum performance results that achieve the best thermodynamic and economic performance for a range of price ratios for electricity and heat. Exergy analysis reveals where thermodynamic losses occur and suggests how to improve efficiency by eliminating irreversibilities. Two historically important case studies are included of CHP systems: a large municipal district heating system in Japan where a power plant supplied waste hot brine to heat fresh water for many heating needs for a nearby town, and a unique university campus system in the United States that provides both electricity and heat for all of the buildings including some snow-melting for walkways.

Nomenclature

General Note: See Nomenclature for Chapter 1; new symbols for this chapter are shown below.

A	Baumann rule factor, Eq. (6.22)
H	exergy value factor
y	flow split ratio, Eq. (6.26)

Subscripts

ex	exergy
th	thermal
U	utilization

Review questions

1. List the kinds of geothermal energy conversion systems that may be used in conjunction with a direct-heat application.
2. What kinds of direct-heat applications are appropriate for geofluid supply temperatures: (a) less than 40 °C; (b) between 80 and 120 °C; and (c) between 130 and 150 °C?
3. What technical factors would limit the lowest temperature for which a geofluid might be used for a particular direct-heat application.

4. With reference to Table 6.1, which of the fluids listed there would be technically feasible candidates for a bottoming *supercritical* binary cycle power plant, if waste brine from a flash plant was available at a temperature of 140 °C? Explain your rationale.

Exercises

1. A small geothermal CHP system is based on two deep wells, one producer and one injector. The geofluid from the producer is divided into two streams, one goes to the power plant and one goes to the heating plant; see Fig. 6.E1. The fluid ratio depends on the electrical and heating demands. The geofluid leaving the power plant mixes with the other stream before entering the heating plant. An auxiliary gas-fired heated is available as needed. With reference to the figure, typical conditions are shown below:

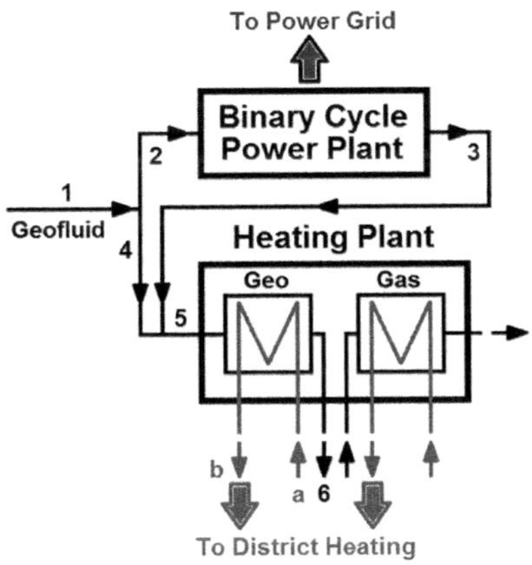

Fig. 6.E1 CHP system.

$T_1 = 98\ °C;\ \dot{m}_1 = 32\ kg/s;\ P_1 = 250\ kPa;\ \dot{m}_2 = 27\ kg/s; T_6 = 40\ °C$

The binary power cycle uses perfluoropentane, C_5Fl_{12}, as the working fluid. The evaporator pressure is 0.5 MPa and the condensing temperature is 50 °C. The preheater-evaporator pinch-point temperature difference is 5 °C and occurs at the bubble point of the C_5Fl_{12}. The dead state is at 20 °C and 0.1 MPa.

Calculate the following:

Part A. Geofluid temperature leaving the power plant, T_3.

Part B. Mass flow rate of the C_5Fl_{12}.

Part C. Power generated by the turbine if it has an isentropic efficiency of 0.88 and the generator efficiency is 0.975.

Part D. Power needed to run the feed pump if it has an isentropic efficiency of 0.80 and the drive motor efficiency is 0.95.

Part E. Net power delivered by the power plant to the grid.

Part F. Thermal power delivered to the district heating plant from the geofluid.

Part G. Thermal efficiency of the binary cycle.

Part H. Exergetic efficiency of the preheater-evaporator.
Part I. Overall exergetic efficiency of the geothermal CHP system.

2. A university campus has a small geothermal CHP system. The hot geofluid is received (state A) at a temperature of 125 °C, a pressure of 0.30 MPa, and a mass flow rate of 40 kg/s. All of it passes first through a binary power plant to generate electricity for the buildings and to run the geothermal well pumps. Cooling water is available at a temperature of 5 °C (state C). The binary cycle uses R245fa as the working fluid. The pinch-point temperature differences in the evaporator and condenser are 4 °C and 8 °C, respectively. The geofluid leaving the power plant goes directly to the heating plant. The plant specifications are as follows:

$$Geofluid: \ T_A = 125\ ^\circ C;\ P_A = 0.3\ MPa;\ \dot{m}_A = 40\ kg/s.$$

Binary cycle: working fluid = R245fa; evaporator pressure = 1.25 MPa; condenser pressure = 0.25 MPa; turbine inlet (state 1) is saturated vapor; condenser outlet (state 4) is saturated liquid; $\Delta T_{pp,e} = 4$ °C; $\Delta T_{pp,c} = 8$ °C; turbine isentropic efficiency = 0.85; feed pump isentropic efficiency = 0.75.

Cooling water: inlet temperature $T_C = 5$ °C.
Calculate the following:
Part A. Mass flow rate of the R245fa.
Part B. Power generated by the turbine if the generator efficiency is 0.97.
Part C. Power needed to run the feed pump if the drive motor efficiency is 0.95.
Part D. Net power delivered if the well pumps require 125 kW from the power plant.
Part E. Thermal efficiency of the binary cycle.
Part F. Geofluid temperature leaving the power plant, T_B.
Part G. Cooling water mass flow rate.
Part H. Cooling water discharge temperature.
Part I. Maximum thermal power delivered to the campus heating system if the geofluid reinjection temperature must be above 65 °C.

3. With reference to Figs. 6.8 and 6.9 for the 220 °C geothermal resource, consider a case where the H-factor is 0.24.
Part A. What separator temperature yields the optimum system design point based on combined exergy-value output?
Part B. Assuming the plant is set up to operate at the optimum condition, determine the total combined exergy output for electricity and heat, in kW.
Part C. If the plant costs $5500/kW, where the kW amount includes the combined electrical and heat exergy output, and the plant is assumed to operate at its design capacity for 8,000 h/y (i.e., 91% capacity factor), how many years would be needed to recover the capital cost (simple pay-back)?
Part D. Assess and comment on the result from Part C, in light of the fact that annual operating & maintenance costs have not been accounted for.

Acknowledgments

The author of this chapter thanks Anna Suzuki, Assistant Professor, Tohoku University, Sendai, Japan, for providing information on the Shizukuishi project.

References

[1] DiPippo R. Exergy analysis of combined electricity and direct-heat geothermal flash-steam plants. Geothermal Resources Council Transactions 1987;11:411–6.

[2] Thermodynamic and Transport Properties Database (REFPROP): version 9.1. National Institute of Standards and Technology (NIST), US Department of Commerce, https://www.nist.gov/srd/refprop.

[3] Lund JW. 100 years of geothermal power production. In: Proc. thirtieth workshop on geothermal reservoir engineering, SGP-TR-176. Stanford, California: Stanford University; January 31–February 2, 2005.

[4] DiPippo R. Geothermal power plants: principles, applications, case studies and environmental impact. 4th ed. Oxford, England: Butterworth-Heinemann: Elsevier; 2016.

[5] Gallagher JS, Linsky D, Morrison G, Levelt Sengers JMH. Thermodynamic properties of a geothermal working fluid; 90% isobutane-10% isopentane. NBS Tech. Note 1234. Washington, DC: National Bureau of Standards, U.S. Gov. Printing Office; 1987.

[6] Liebowitz HM, Mlcak HA. Design of a 2 MW Kalina cycle binary module for installation in Húsavík, Iceland. Geothermal Resources Council Transactions 1999;23:75–80.

[7] ASHRAE, Handbook fundamentals. Atlanta, GA: American Society of Heating, Refrigeration and Air-Conditioning Engineers, Inc.; 1997 [chapter 18].

[8] DiPippo R. Geothermal energy as a source of electricity: a worldwide survey of the design and operation of geothermal power plants. Washington, DC: U.S. Dept. of Energy, DOE/RA/28320-1, U.S. Gov. Printing Office; 1980.

[9] Lund JW, Chiasson A. Examples of combined heat and power plants using geothermal energy. In: Proc. 29th New Zealand geothermal workshop; 2007.

[10] DiPippo R. Kakkonda-Shizukuishi, a combined geothermal power- and heating-plant. Geothermal Hot Line 1987;17:24–9.

[11] Boyd T, DiPippo R. Technical assessment of the combined heat and power plant at the Oregon Institute of Technology, Klamath Falls, Oregon. Geothermal Resources Council Transactions 2012;36:1143–9.

[12] Shizukuishi District geothermal hot water supply study: conceptual plan. Brochure, Iwate Prefecture Geothermal Hot Water Study Office, Morioka, Iwate, Japan; 1985 [in Japanese].

[13] Horii S. Direct heat update of Japan. Geothermal Resources Council Transactions 1985;9:113–7.

[14] MHI. Hatchobaru geothermal power plant. Brochure H480-48GP02E1-B-0. Tokyo, Japan: Mitsubishi Heavy Industries, Ltd.; 1992.

[15] Kudo K. 3,000 kW Suginoi Hotel geothermal power plant. GeoHeat Center Bulletin 1996;17-2:7–8.

[16] Sugino H, Akeno T. 2010 country update for Japan. In: Proc. World geothermal congress 2010, Bali, Indonesia, April 25–29, 2010.

[17] Geothermal heating and cooling: how does geothermal heating and cooling work? Green Passive Solar Magazine. https://greenpassivesolar.com/sustainable-renewable-energy/geothermal-energy/.

[18] Geothermal Energy-Oregon Institute of Technology. Urban Ecology-CMU, Carnegie Mellon University, Pittsburgh, PA, USA. https://urbanecologycmu.wordpress.com/2016/11/01/geothermal-energy-oregon-institute-of-technology/.

Biofuels conversion: energy-saving processes and use of biogas[(1)]

7

7.1 World usage and basic processes of bioconversion technology

Biofuels hold a special place in the structure of renewable use of energy resources. Biofuel is an alternative fuel in the transport sector and a source of electricity from thermal power plants. Biofuel is seen as an important resource for ensuring energy security, developing agriculture, and reducing greenhouse gas emissions. Biofuel refers to fuel obtained from vegetable and animal raw materials, from products of life of organisms or organic industrial waste. Biogas is a product obtained in the process of anaerobic methane fermentation of biomass. The production and use of biogas refers to innovative energy, energy- and resource-saving, and environmentally-friendly technologies.

Today many countries have developed a biofuel industry that plays an important role in meeting energy needs while preserving and protecting the environment. Germany [1] and China [2] are the largest developers of biogas plants. More complete information on statistics on the worldwide development of biogas technologies is contained in the reports of the International Energy Agency, IEA, cited below.

In general, biomass is organic material related to waste agricultural and industrial production, household activities, as well as wastewater. Strictly speaking, the term biomass refers to living matter, typically per unit area or volume of habitat. The quantity may be expressed either as the weight of living matter (organisms) per unit area or as their volume per unit volume of the habitat. For the purposes of this chapter, biomass specifically refers to plant materials, crop waste products, or animal waste that can serve as feedstocks for processes that produce fuels which may in turn be converted into useful forms of energy. Biomass can provide significant savings in primary energy [3].

There are three basic categories of biofuels: liquid (e.g., bioethanol, biodiesel); solid (e.g., briquettes, fuel granules, chips, straw); gaseous (e.g., biogas, biohydrogen). The most promising direction in the production of biofuel is the production of biogas which will lead to the development of energy-saving technologies. The global biogas market is currently showing strong growth (Fig. 7.1). Figures 7.2 and 7.3 show the prospects for the development of the biogas industry in the Netherlands.

(1) Written by Andriy and Oleksandr Redko; edited by Ronald DiPippo.

Low-Temperature Energy Systems with Applications of Renewable Energy
https://doi.org/10.1016/B978-0-12-816249-1.00007-8

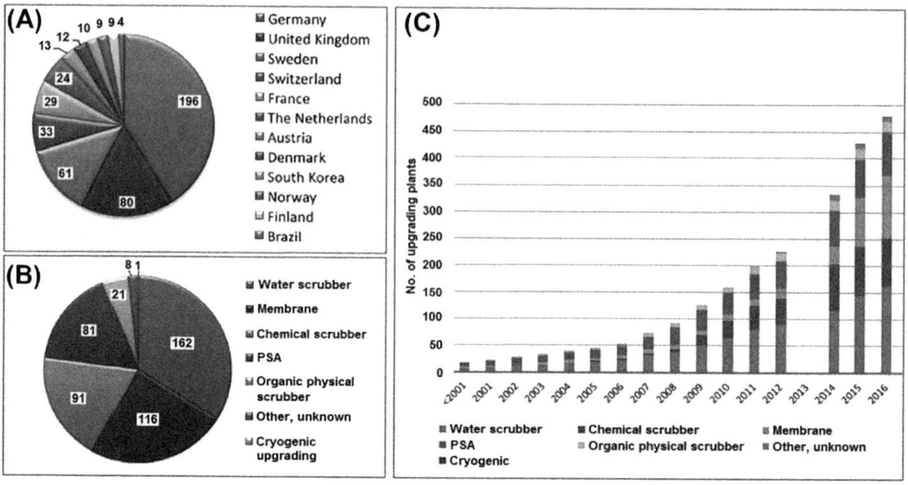

Fig. 7.1 (A) Number of biogas upgrading plants for IEA Task 37 member countries. (B) Breakdown of biogas upgrading technologies used at biomethane plants. (C) Biogas upgrading technologies growth up to 2016. Note: Upgrading refers to the process that converts the raw biogas into biomethane [4].

One of the world's largest biogas plants with direct infeed into the natural gas network is the Biogas-Park Könnern in Germany; see Fig. 7.4. The biogas park is located near the city of Könnern and covers an area of 17 ha. Vegetable raw materials are used as the feedstock: corn, cereals, grass, straw and silage; animal manure is also part of the input. When it was put into operation, 30 nearby farmers supplied the plant with about 120,000 tonnes of raw materials per year. The plant consists of four modules, each having four fermenters, and four digestate holding tanks. Additional site facilities include a technology building with separators, a central liquid pump unit, and a

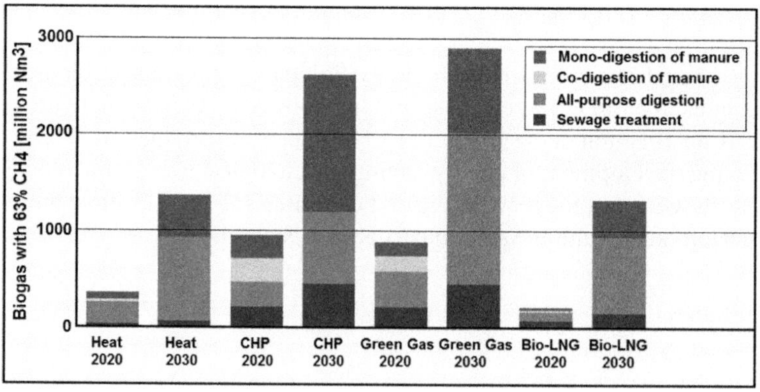

Fig. 7.2 Maximum potential per marketing route for biogas in the Netherlands [4].

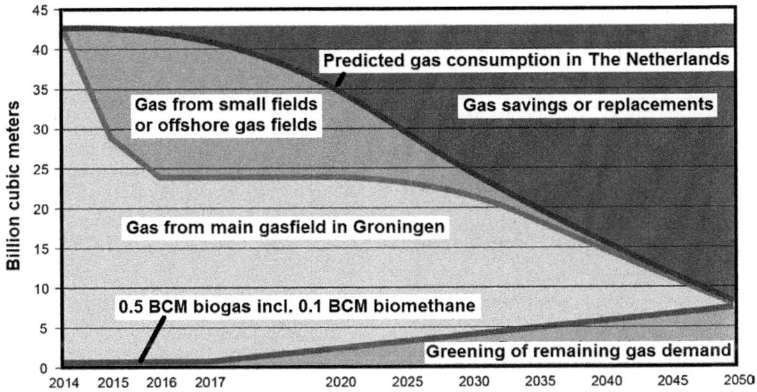

Fig. 7.3 Gas development route to 2050 for the Netherlands, modified from [4].

Fig. 7.4 Biogas-Park near Könnern, Germany [5].

biogas compressor. The annual production amounts to about 15 million standard cubic meters of biomethane which is enough to support roughly 10,000 households with heat and electricity or for 9000 automobiles each having an annual mileage of 30,000 km. The upgrading process is amine gas treatment. The capital cost of the plant, built in 2008, was approximately 31.5 million euros. The farmers collect the digestate waste and return it to their farms as fertilizer [6].

The following processes and activities take place in the biogas park [7]:

1. The raw biomass is kept in storage on site.
2. Biomass is extracted from storage and added to the fermentation tanks.
3. Various pumping operations between the insertion systems, fermenters, final proofer and product storage tanks.
4. Intermediate storage of the fermented digestate.
5. Removal of fermented digestate for transport to end-users for fertilizer on farms.
6. Generation of biogas in the gas-tight fermentation system.
7. Caching of the produced biogas in the gas storage system.
8. Applying heat to maintain temperature in the fermenters.

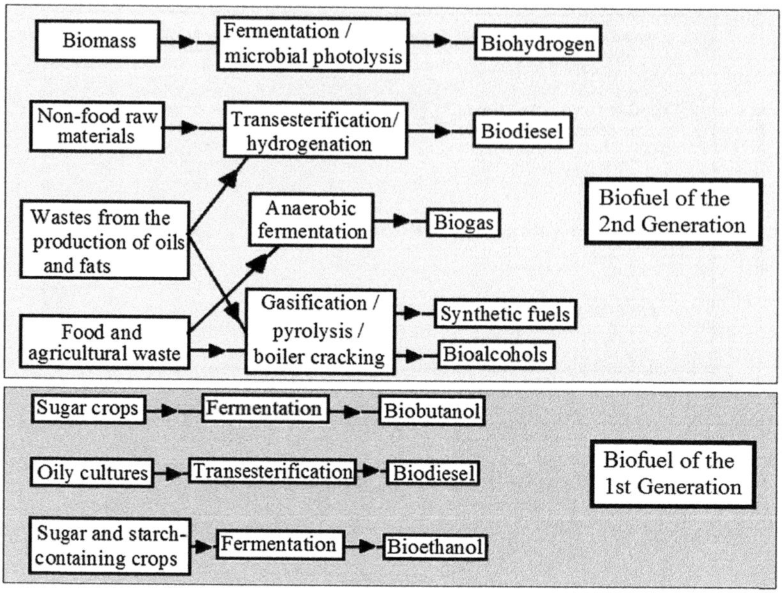

Fig. 7.5 Classification of biofuels by generation.

9. Upgrading of the biogas to biomethane via amine gas treatment.
10. Delivery of the biomethane to the public gas supply pipeline grid.

Classification of biofuels by first and second generations is given in Fig. 7.5.

The raw materials of the First Generation are agricultural crops with a high content of fats, starch and sugars. Vegetable fats are processed into biodiesel, while starches and sugars are converted into ethanol. Non-food remains of cultivated plants, grass and wood are called the Second Generation of raw materials. Such raw materials contain cellulose and lignin; these can be directly burned (as traditionally done with firewood), gasified (producing flammable gases), and processed by means of pyrolysis. Biofuel of the Third Generation refers to specially grown species of biomass.

The annual requirement of world energy is about 14 billion tonnes of oil equivalent. In Africa, the use of biomass for energy purposes is about 60% of total energy consumption; in Asia it is up to 40%; in Latin America it is about 30%. In many cases in these places, the biomass is likely firewood for heating and cooking. Owing to health and environmental impacts, this use of biomass is not a desired approach and should be distinguished from the more environmentally friendly applications described in this section.

7.2 Technical description of bioreactor processes

Bioconversion is a technological process that involves the treatment of organic mass for the purpose of obtaining high quality heat or fuel, as well as environmentally friendly organic fertilizers. Bioconversion is one of the most important renewable energy sources with the greatest potential. The use of indigenous unconventional energy sources will significantly reduce the dependence of countries on imported energy and negative environmental impact.

The increase in the cost of traditional energy sources and the further man-made impact on the natural environment both point to increasing the use of renewable alternative sources of energy. Up to 65% of renewable energy sources can be obtained from organic matter of agricultural origin.

An average of 20–25 m^3 volume of biogas is estimated to be to obtained nowadays from 1 m^3 of animal waste, but an amount of 30–35 m^3 is currently needed for technical and economic feasibility. This amount of gas can be obtained by combining animal excretions and household wastes with other raw materials that are distinguished by high content of dry organic mass [8–10].

Taking into account the type of finished product (solid, liquid or gaseous fuel), there are various methods of processing biomass, namely, thermal, chemical, thermochemical, biological and biochemical, in which thermochemical, biochemical and agrochemical processes are involved [9–12]. The choice of how to process organic waste from agricultural production, animal waste, and food production depends on the type of biomass, with the ultimate goal of obtaining solid, liquid or gaseous fuels in an economic and environmentally acceptable manner. Figure 7.6 shows the classification of main types of energy processes depending on the methods of processing organic mass.

The cyclic processes of biomass transformation in the technological system of biogas production are shown in Fig. 7.7.

Thermochemical processes include combustion, pyrolysis, gasification, and hydrogenation [4,8]. Burning is used to obtain heat from dry, homogeneous fuels. Biomass is heated to a temperature of 450–550 °C in the absence of air during pyrolysis, resulting in coke (charcoal), a solid residue, gases (CO, CO_2, H_2, H_2O, etc.), and liquid mass (alcohols, oils, etc.). During gasification there is heating (partial combustion) of fuel with a limited amount of air in the range of temperatures of 800–1600°C and pressures from 1.0 to 10.0 MPa. Synthesis-gas or syngas (CO, H_2, CH_4, in a corresponding

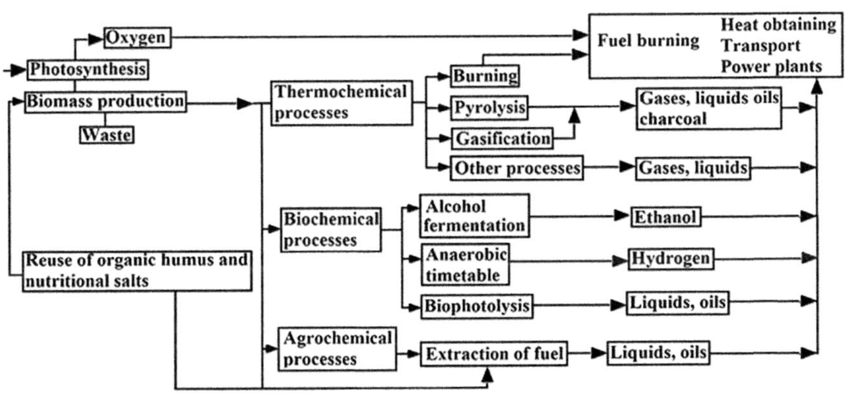

Fig. 7.6 Energy processes of biomass treatment.

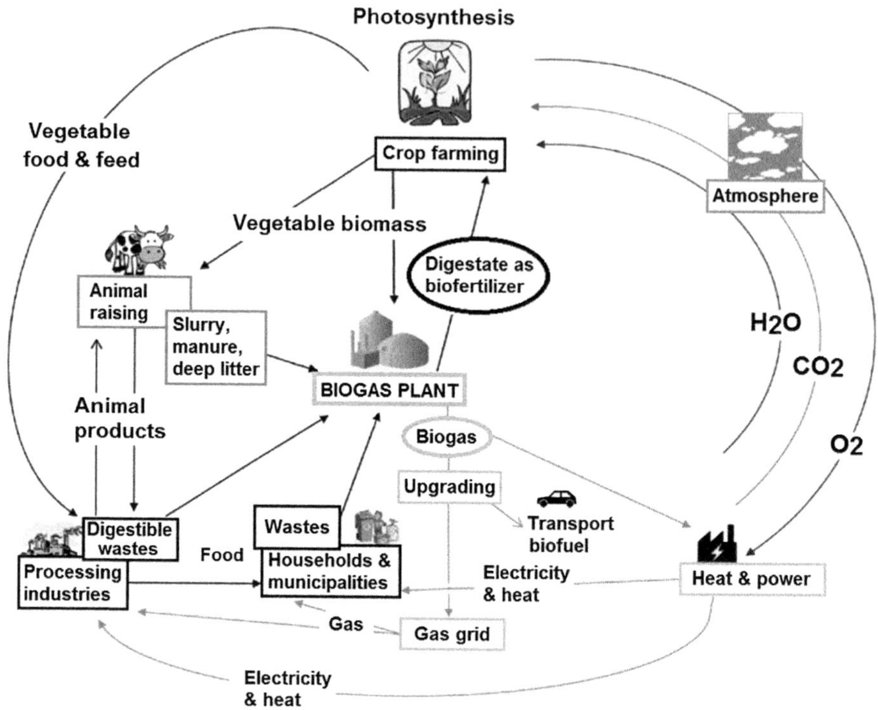

Fig. 7.7 An example of how a modern co-digestion biogas plant fits into the circular economy, modified from [13].

proportion) is formed during gasification. Hydrogenation occurs with different variants of preliminary preparation of raw materials and carrying out the process itself.

Biochemical processes include alcohol fermentation, anaerobic treatment, and biophotolysis. As a result of alcohol fermentation, ethanol is formed, which is an alternative or additive to gasoline. During anaerobic treatment, microorganisms in an oxygen-free medium absorb energy, directly process carbonaceous components and, as a result, produce carbon dioxide (CO_2) and methane (CH_4). The resulting mixture of CO_2 and CH_4, as well as associated gases, is called biogas. Biophotolysis is the decomposition of water into hydrogen and oxygen under the influence of light. The agrochemical process is the extraction of fuel.

The production of biogas from organic waste is based on its properties to extract combustible gas as a result of methane fermentation under anaerobic conditions. Biogas formed in methane fermentation is a mixture of 50−80% methane, 50−20% carbon dioxide, about 1% hydrogen sulfide, as well as a small amount of some other gases such as nitrogen, oxygen, hydrogen, ammonia, carbon dioxide, etc. Figure 7.8 shows the technological scheme of production of biogas and biomethane.

As can be seen from Fig. 7.8, biogas is the intermediate product of the conversion of organic wastes. The volumetric heat of combustion biogas is 21−25 MJ/m^3. Biogas is liquefied at a temperature of (−161.5 °C). To heat an average residential building

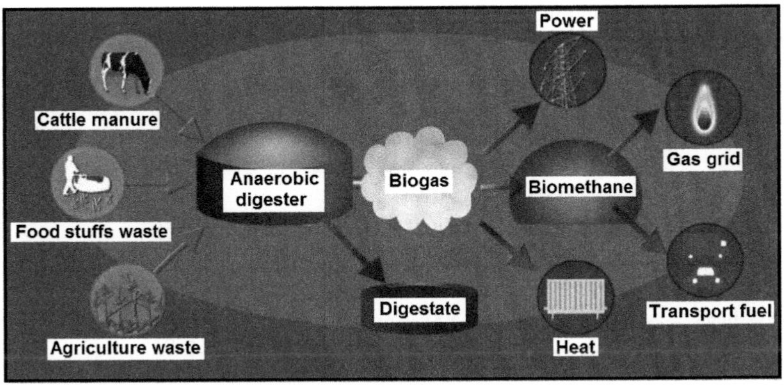

Fig. 7.8 Overview of anaerobic digestion and biomethane upgrading process, modified from [14].

requires about 60 m^3/y of biogas per 1 m^2 of living area, and to obtain 1 kWh of electricity requires 0.7–0.8 m^3 of biogas. It is recommended to use a mixture of 80–85% biogas and 15–20% liquid fuel for internal combustion engines [9,15].

Biomethane is the final product of the processes shown in Fig. 7.8. Biogas from anaerobic digestion can be upgraded to biomethane by a variety of methods (absorption, adsorption, membrane filtration, cryogenic separation). The process removes most of the CO_2 and impurities, resulting in a high-grade fuel that is typically over 90% methane and which meets the quality standards for injection into a natural gas pipeline grid.

The output and composition of biogas is determined by many factors, but depends mainly on the composition of the initial raw material. The output of biogas from various organic wastes of agricultural production and households as well as industrial wastes is given in Table 7.1.

The quality and quantity of the biomethane produced also depends on the content of inert substances; the more inert substances, the lower the methane output per unit mass of waste. Anaerobic bioreactors with immobilized microflora are more prevalent than aerobic bioregenerators. The main trends in the development of bioconversion systems are determined by the requirements of environmental protection and can be achieved through the use of both aerobic and anaerobic fermentation. Decisive factors in choosing a particular fermentation method are:

- primary and operational costs
- reliability of biogas equipment in operation
- requirements for maintenance and personnel
- efficiency of using the obtained products.

High efficiency of biomass treatment into energy products is achieved only with optimized parameters for the technical processes and the machines used for the production. The formalization of links between equipment parameters and process parameters adapted to existing raw materials will allow high efficiency of bioenergy production and increased biofuel quality.

Table 7.1 Biogas output and methane content from various kinds of waste.

Raw material	Biogas output, dm^3/kg dry matter	Methane content, %
Cattle humus[a]	200–300	50
Pig humus[a]	340–480	60–75
Horse humus[a] with straw	250	56–60
Potato tops	420	60
Corn stalks	420	53
Wheat straw	342	58
Sunflower husk	300	60
Silage	250	84
Fresh grass	360	52
Beet	430	84
Carrot Waste	250	60
Wood sawdust	220	51
Solid sewage sludge	570	70
Fecal sediment	250–310	60
Household waste and garbage	600	50

[a]Black, broken-down, stable organic matter in soil.

The hierarchy of ways to intensify the process of anaerobic fermentation of a substrate in bioreactors is shown in Fig. 7.9. Mixing with a rotational speed of 50–80 rpm provides the optimum mode [16,17].

To produce biogas, bioreactors with various design features are used. Indicators of the effective operation of such an installation are the productivity (the biogas output per unit volume of biomass), the duration of the working cycle, and energy consumption to ensure the intensification and thermostabilization of bioconversion. These indicators mainly determine the cost of biogas production.

The process of bioconversion of organic waste requires significant energy consumption, which can render the bioreactor uneconomic. Thermostabilization and intensification of anaerobic fermentation in bioreactors is an effective solution to the

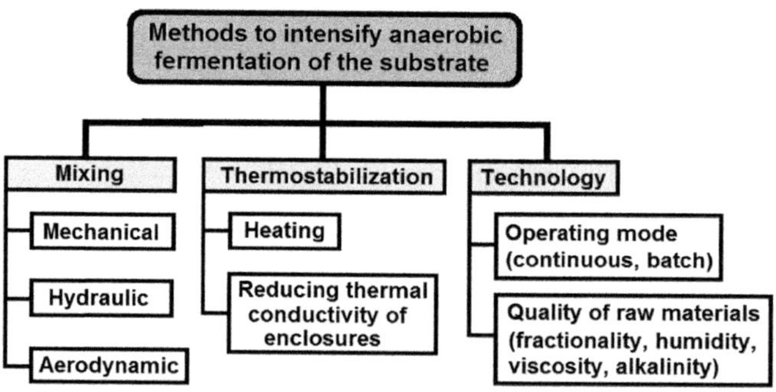

Fig. 7.9 Classification of factors for improving the energy efficiency of biogas production processes.

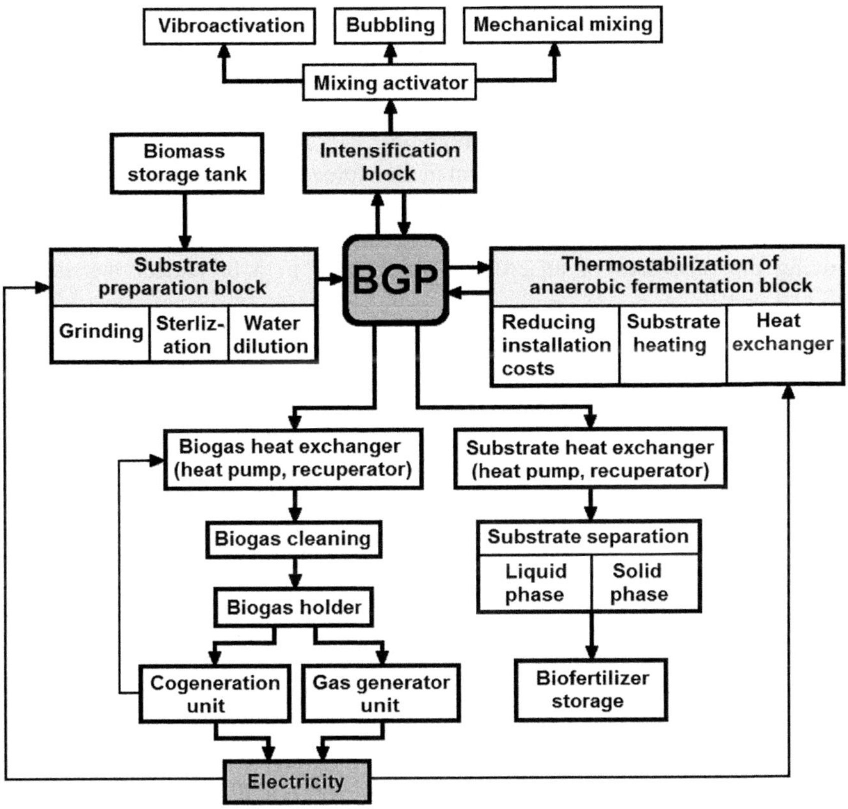

Fig. 7.10 Schematic energy-saving diagram of the bioconversion process.

problems of heat re-use and lower costs for the technological process of biogas production. The schematic energy-efficiency technological diagram of the bioconversion process in a biogas reactor plant (BGP) is shown in Fig. 7.10. The substrate from the storage tank after grinding and sterilization in the substrate preparation unit enters the bioreactor. The process of anaerobic fermentation is more effective when it is intensified and thermostabilized. Intensifying the process of biogas release is possible by mixing the organic mass by vibrating activation, bubbling or mechanical means. An important aspect of the stability of the thermal regime in the bioreactor is the heating of the substrate and the simultaneous thermal insulation of the walls of the reactor from fluctuations in ambient temperatures.

The heat of the output gas mixture and spent biomass is an effective source of energy that is reused for the processes inside the bioreactor. Such thermostabilization systems are provided by using a heat pump or heat exchanger. Due to the re-use of heat, high energy efficiency is achieved in the anaerobic fermentation process, and the bioreactor productivity is significantly increased.

To achieve high efficiency of bioreactors and obtain the maximum amount of biogas per unit volume of biomass, it is necessary to create optimal technical parameters in the bioreactor. The intensity of the fermentation process and, consequently, the formation of biogas is influenced by four groups of factors, namely, biological, physical, chemical, and organizational-technological.

The process of anaerobic fermentation in the bioreactor is influenced by several factors, the improvement of which leads to an increase in efficiency of use of raw materials for the generation of biogas and a decrease in cost of the entire process. Factors for improving the processes of anaerobic fermentation in a bioreactor are shown in Fig. 7.11 [8–10].

- Biological factors include: composition of fermented biomass (amount of proteins, fats, carbohydrates, lignins); composition of microflora (number and groups of microorganisms of the corresponding stage of decomposition); living conditions of microorganisms (content of harmful impurities).
- Physical factors include: temperature of digestion; bioreactor pressure; hydraulic mode.
- Chemical factors include: acidity of the medium (pH value); the content of volatile fatty acids in the digestible mass; the volume and composition of the resulting biogas.
- Organizational and technological factors include: dose of the daily loading of new portions of the digestible mass; ashless substance loading; the content of biomass substances which do not yield to treatment.

Depending on the temperature interval maintained in the bioreactor during the process, the following modes of fermentation are distinguished: cryophilic ($t < 20$ °C; $t_{opt} = 5–15$ °C), mesophilic ($t = 25–45$ °C; $t_{opt} = 32–42$ °C) and thermophilic ($t = 45–55$ °C; $t_{opt} = 48–51$ °C).

Cryophilic fermentation does not require special heating of the substrate, procedes at ambient temperatures, and is used at small individual installations in countries with warm climates. As a rule, such installations have a discrete nature of loading and unloading. The advantage of the cryophilic regime is the lack of heat energy consumption to maintain a stable temperature. Disadvantages include the slow reaction time of biological processes and low biogas output.

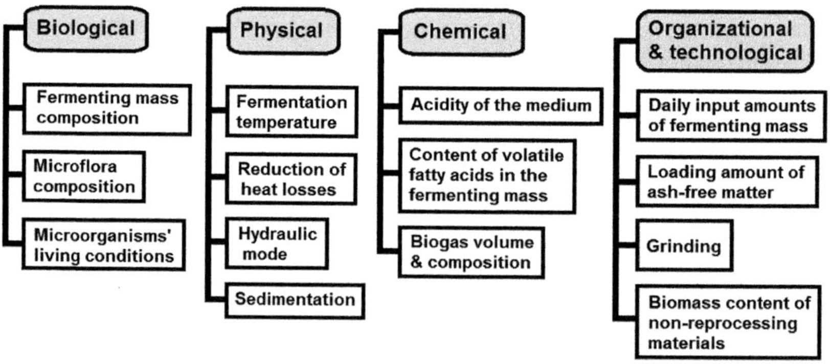

Fig. 7.11 Classification of methods to improve energy efficiency of biogas production.

Mesophilic fermentation, as confirmed by research and operational practice of installations, procedes in the most intensive way in the temperature range of 32–42°C. At the same time, the most active "working" bacteria are methanogenic ones with the maximum formation of biogas. However, the deviation from the optimal temperature interval by ±5 °C leads to a decrease in the biogas output by 2–2.5 times, which indicates the sensitivity of bacteria to the temperature of the environment and requires the implementation of precise controllers to maintain it. Heating and maintaining a stable temperature of fermentation is usually carried out by pumping heated water through special heat exchangers mounted in a bioreactor. Hot water is produced in a boiler unit, using part of the biogas for its operation.

Thermophilic fermentation offers the opportunity to obtain the maximum amount of biogas in a short period of time. The fermentation intensity is two times higher, and the residence time in the bioreactor is 50% shorter compared with mesophilic fermentation. However maintaining a relatively high temperature requires significant energy consumption.

From Fig. 7.11 it is clear that in addition to the properties and characteristics of the organic mass used for the biogas production process, physical factors in the bioreactor influence the operation, in particular, the temperature of fermentation and the hydraulic regime. These factors are important in improving the equipment for biogas production. By using the equipment with optimal parameters of heating and mixing the substrate in a bioreactor, it is possible to achieve an energy-efficient biogas production system.

The duration of the bioreactor working cycle to ensure the maximum biogas output from a unit volume of biomass is determined on the condition that:

$$\tau_{oc} = f(\tau_l, \tau_{pr}, \tau_f, \tau_d) \rightarrow \tau_{\min} \tag{7.1}$$

where τ_{oc} is the duration of the bioreactor operating cycle; τ_l is the time for loading biomass into a bioreactor; τ_{pr} is the time to prepare the biomass for the active phase of fermentation; τ_f is the time of active fermentation of biomass; τ_d is the time it takes to discharge waste biomass from the reactor.

The essential portion – about 80–90% – of the duration of the bioreactor operating cycle is the period of active fermentation of biomass, determined by the regime of anaerobic fermentation, quantitative and qualitative content of organic matter in the substrate, and energy costs to ensure the thermostabilization of the process of anaerobic fermentation. The main ways to reduce the period of active fermentation of biomass are the intensification and thermostabilization of the technological process of biogas production.

The anaerobic process that takes place in the bioreactor consumes a certain amount of energy, i.e., thermal energy to maintain thermal stability in the reactor and to preheat the substrate to the fermentation temperature, and the mechanical energy to mix the fluid in the containers and move material flows of the substrate. The operation of bioreactors requires a stable temperature regime under different environmental conditions. Fluctuations in the temperature of the substrate should not exceed the constrained limits for favorable fermentation of bacteria.

Fig. 7.12 HOST microferm biogas installation, Angers, France [18].

The technology of methane fermentation in a bioreactor requires compliance with the limits of temperature regimes, and the intensification of heat exchange between the heater and the substrate through mixing. A facility in Angers, France is shown in Fig. 7.12, and a diagram for the simulation of energy-saving heat and mass exchange processes in a bioreactor with a mixing device is shown in Fig. 7.13.

The mass exchange process (conservation of mass) in the bioreactor is described by the formula:

$$m_{sub} = m'_{sub} + m_g \tag{7.2}$$

where m_{sub} is specific mass of substrate [M_{sub} – substrate mass in the biogas plant] loaded into the reactor, kg; m'_{sub} is mass of the substrate after processing in the

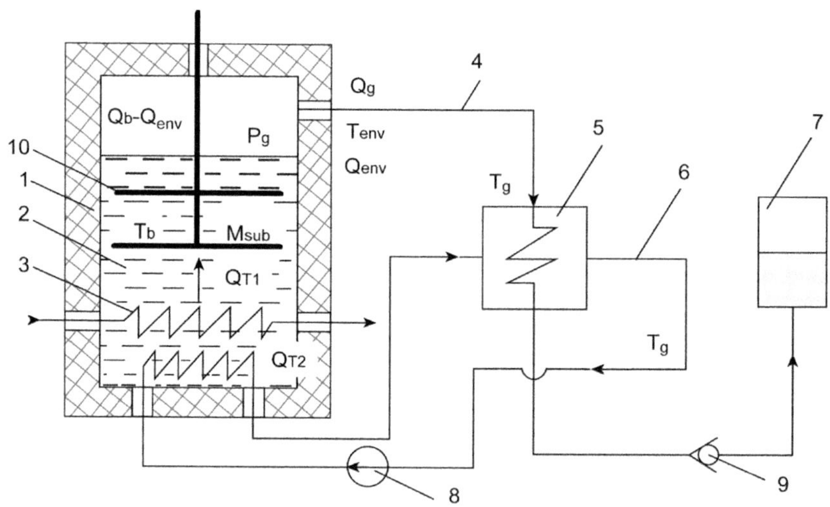

Fig. 7.13 System for the simulation of energy-saving heat and mass exchange processes. *1*, thermal insulation case; *2*, substrate; *3*, heater; *4*, gas pipeline; *5*, waste heat exchanger; *6*, main circuit of secondary heat transfer recirculation; *7*, biogas storage tank; *8*, pump; *9*, safety valve; *10*, mixing device; T_b, bioreactor temperature; T_{out}, temperature at the outlet of the bioreactor.

bioreactor, kg; m_g is mass of gas formed from the substrate due to anaerobic fermentation, kg.

Taking into account the specific weight of the components in the bioreactor whose density is uniform throughout the volume due to the mixing of the substrate, Eq. (7.2) becomes:

$$V_S\rho_S + V_L\rho_L = \rho_S\cdot\frac{dV_S}{d\tau}\cdot\tau + \rho_L\cdot\frac{dV_L}{d\tau}\cdot\tau + V_G\rho_G \tag{7.3}$$

where V_S and V_L are the volumes of the solid and liquid phases of the substrate at the bioreactor inlet, m^3; V_G is volume of gas formed as a result of anaerobic fermentation of the substrate, m^3; ρ_S and ρ_L are substrate solid and liquid phase densities, kg/m^3; and ρ_G is biogas mixture density, kg/m^3. The terms $\rho_S\cdot\frac{dV_S}{d\tau}\cdot\tau$ and $\rho_L\cdot\frac{dV_L}{d\tau}\cdot\tau$ represent the change in the mass of solid and liquid matter of the substrate after its treatment in a bioreactor.

The thermal regime in a bioreactor depends on the thermal insulation properties of its external walls, the additional energy supplied to the bioreactor to provide thermal stabilization, the regime of anaerobic fermentation, and any heat added to offset any heat losses. The heat balance equation of a bioreactor, assuming the temperature fields are uniformly distributed within its volume as provided by mixing of the substrate, is given by the relationship:

$$Q_b = \alpha_t A_b(T_b - T_{env}) + m_b c_m(T_b - T_{env}) + Q_h - Q_g \tag{7.4}$$

where Q_b is the heat entering the internal environment of the bioreactor, J; the first term on the right, $Q_{env} = \alpha_t A_b(T_b - T_{env})$, is the heat loss to the environment; α_t is the total heat transfer coefficient through the vessel and insulation of the reactor, W/(m^2 K); A_b is outer surface area of the bioreactor shell, m^2; T_b is the bioreactor internal temperature, °C; T_a is ambient temperature around the bioreactor, °C; m_b is mass of substrate, biogas mixture and air in the installation, kg; c_m is the given specific heat capacity of the substrate, biogas mixture and air in the bioreactor, J/(kg K); the term $m_b\ c_m$ $(T_b - T_{env}) = m_b\ c_m\Delta T$ represents the change in the thermal energy accumulated in the bioreactor; ΔT is current temperature difference during anaerobic fermentation; $Q_h = Q_{T1} + Q_{T2}$ is additional total heat entering the substrate thermostabilization unit (where Q_{T1} and Q_{T2} are the heat coming from the main heater and from the waste heat exchanger 5, respectively), J; Q_g is the heat of the biogas mixture removed from a bioreactor, J.

The bioreactor pressure influences the intensity of the fermentation process and, consequently, the conditions and the amount of biogas formation. An increase in pressure in the bioreactor has conflicting effects. On the one hand, increasing the total pressure and the partial pressure of carbon dioxide is a favorable condition for biogas production. On the other hand, increasing the pressure leads to an increase in the solubility of methane in water which inhibits its release into the gaseous phase. As a compromise, most bioreactors operate at a pressure that is slightly higher than atmospheric.

7.3 Energy-saving systems for biogas-producing plants

A modern biogas plant (BGP) is a complex of devices connected to each other by hydraulic, mechanical, electrical, and information communications. With high technology, biogas production management should be as computer-assisted and automated as possible in order to maximize efficiency. It is possible to achieve high productivity of the BGP with an integrated combination of all innovative solutions. Schematic flow diagrams for energy-saving systems for the production of biogas are shown in Fig. 7.14 (with a recuperative heat exchanger 13) and in Fig. 7.15 (with a heat pump 13), which ensure a reduction in energy costs for the production of biogas.

In Fig. 7.14, item 13 recovers heat from the waste sludge to preheat the feed organic material before entering the bioreactor 1. In Fig. 7.15, the preheating takes place in the condenser 18 of the heat pump 13. The substrate proceeds at a temperature of 10–15 °C into the bioreactor and heats the anaerobic process to 25–55 °C. Due to the heat released from the spent substrate in the evaporator 19, the spent substrate is cooled and proceeds to the separator 12.

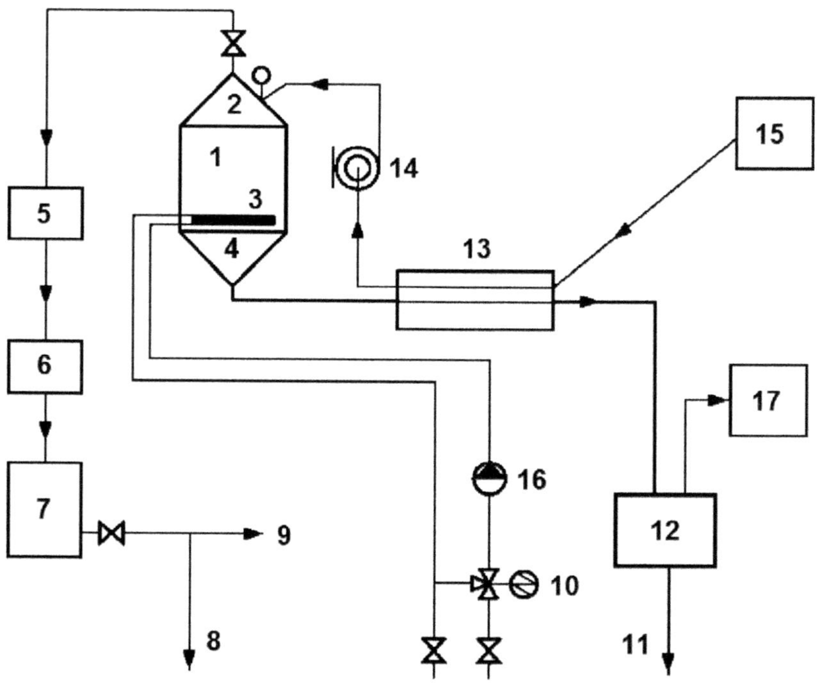

Fig. 7.14 Flow diagram of biogas production with a recuperative heat exchanger. *1*, reactor with top and bottom cones; *2*, top cone for biogas collection; *3*, tubular heating element; *4*, bottom cone for collecting sludge; *5*, sulfur removal device; *6*, carbon dioxide removal device; *7*, gas holder; *8*, pipeline sending gas for the needs of the economy; *9*, pipeline sending gas for the needs of the BGP; *10*, servodrive with a three-way valve; *11*, pipeline transporting liquid sludge; *12*, phase separator of the spent substrate; *13*, pipe-in-pipe heat recuperator; *14*, fecal pump; *15*, reservoir; collector of organic waste; *16*, circulation pump of the heat supply system; *17*, field for drying sludge.

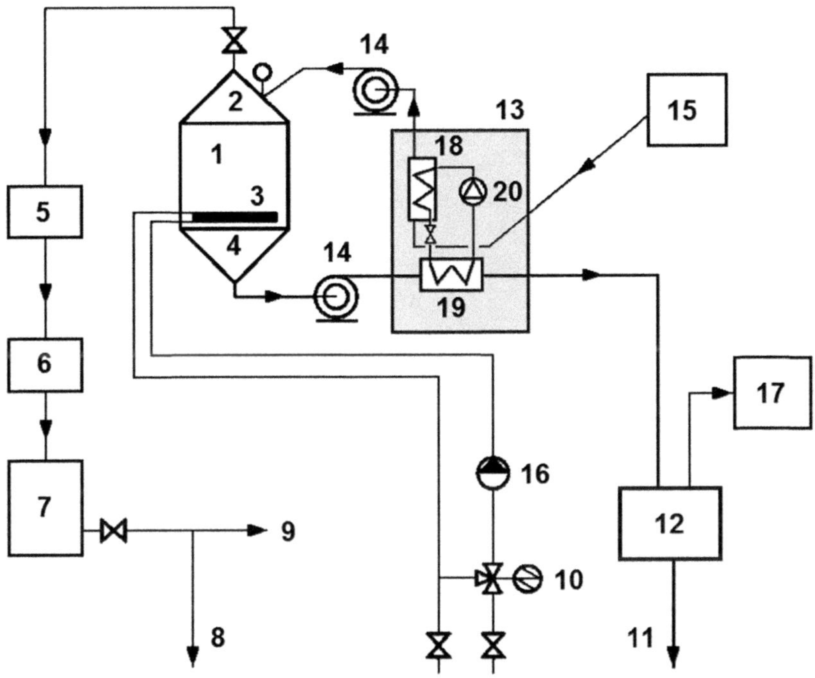

Fig. 7.15 Flow diagram of biogas production with a heat pump: See Fig. 7.14 for nomenclature; new items are: *13*, vapor-compression heat pump for heat recovery of waste sludge; *18-HP*, condenser; *19-HP*, evaporator; *20*, compressor with a thermo-regulation valve for the heat pump.

Figure 7.16 shows the biogas plant flow diagram with an absorption heat pump (ABHP) and an internal combustion engine (ICE) driving an electric generator; biogas is used as fuel for the ICE. The substrate proceeds to the bioreactor 4 by means of the pump 2 from the receiving tank 1. At the same time, 95 °C hot water from the ICE cooling jacket enters the heating system of the bioreactor, providing the required temperature setting, while the exhaust gases of the ICE provide the heat supply to the ABHP. The refrigerant of the ABHP cooling circuit is used to heat the inflow to the operating temperature of the thermophilic bioreactor (45–55 °C). The working fluid of the ABHP enters the sump with a temperature of 35–37 °C, where it cools the spent substrate. The cooled water produced in the absorption machine is used to cool the effluent in the chamber 5.

7.4 Bioreactor calculations

The amount of heat that can be obtained from a biogas plant is determined by the formula:

$$Q = \eta q_b V_b \frac{M_{dry}}{\rho_{dry}} \tau \tag{7.5}$$

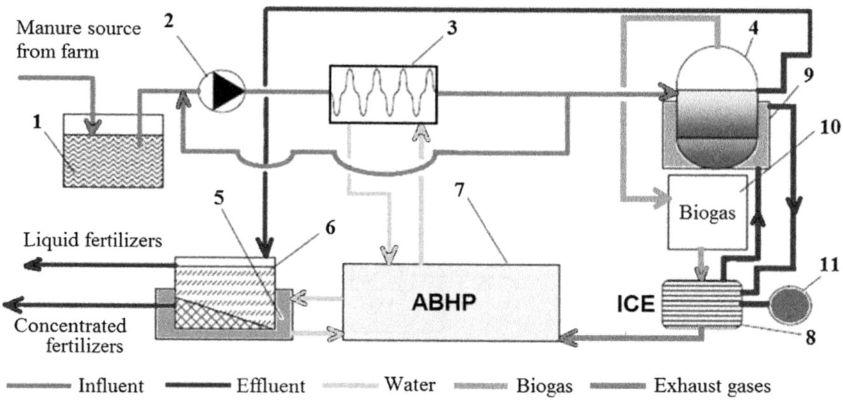

Fig. 7.16 Technical flow diagram showing the use of the waste heat of the effluent aided by an absorption heat pump. *1*, storage tank; *2*, substrate loading pump; *3*, spiral heat exchanger of the influent/water type; *4*, bioreactor; *5*, effluent cooler; *6*, effluent settling vessel; *7*, absorption heat pump; *8*, internal combustion engine; *9*, thermal jacket of the bioreactor; *10*, gas holder; *11*, electric generator.

where q_b = the volumetric heat of biogas combustion (16–22 MJ/m^3); η = the boiler COP, ρ_{dry} = the density of dry organic matter, kg/m^3;V_b = the specific volume of biogas released from dry organic matter per day, m^3/kg; M_{dry} = mass of dry organic matter, kg; and τ = the time of complete exchange of liquid in the reactor, days.

The volume of the substrate that is fermented in the biogas reactor is:

$$V_{sub} = 0.8\, V_{reac} \tag{7.6}$$

Depending on the amount of dry matter in the organic mass, the volume of the substrate is determined by:

$$V_{sub} = \frac{M_{dry}}{\rho_{dry}} + \frac{M_w}{\rho_w} \tag{7.7}$$

where ρ_w = the density of water that is necessary for the effective fermentation of the substrate, since the amount of dry matter should be 6–10% of the substrate volume, kg/m^3; M_w = the mass of water, kg.

To calculate the daily output of biogas, the following equation may be used which takes into account the type of biomass and the technological characteristics of the biogas reactor [9,11,19]:

$$V_\delta = \frac{\vartheta_{\lim} S_{org.sub.}}{\tau}\left(1 - \frac{K}{(\tau\omega_{\max} - 1 + K)}\right) \tag{7.8}$$

where V_δ = daily output of methane CH_4 (m^3/day with a 1 m^3 reactor); τ = duration of fermentation, days; $\vartheta_{\lim}$ = limited emission of methane per day (m^3/d from 1 kg of organic substances), (see Table 7.1); $S_{org.sub.}$ = concentration of organic substances at the outlet, kg/m^3; K = a kinematic coefficient of the substrate that takes into account its parameters, e.g., for cattle humus $K = 0.8 + 0.001e^{0.06S}$; for pig humus $K = 0.5 + 0.0043e^{0.09S}$, where S is the concentration of the substrate; ω_{max} = maximum growth rate of microorganisms, 1/day, which depends on the temperature of fermentation T_f:

$$\omega_{max} = 0.013\ T_f - 0.129 \tag{7.9}$$

The volume of the biogas plant is given by:

$$V_{BGP} = \tau V_{sub} \tag{7.10}$$

where V_{sub} is the volume of liquid mass of substrate supplied to the plant during one day.

The biogas output can be determined from Table 7.2. Depending on the mode of fermentation, the load on the bioreactor is 2.5–5.0 kg/(m^3·day) in mesophilic mode and 4.5–6.5 kg/(m^3·day) in thermophilic mode.

7.5 Efficiency of biofuel usage in energy supply systems

A bioconversion technology project requires an ecological and economic performance evaluation at the conceptual phase of the project. Such an assessment is carried out using the results of modeling studies of the dynamic environment of the proposed project including an economic feasibility study to minimize the cost of the bioconversion products.

The payback period for bioconversion investments and the cost of its products are determined by the dynamic environment of the project. The main factors of the dynamic environment are the capital costs to build the bioreactor and the operating costs for providing the technical process to produce the gas in an efficient manner. Profitability of bioreactors is dependent on the energy consumption for the intensification and thermostabilization of the biogas production process from organic waste. One of the indicators that determines the ecological and economic efficiency is the cost given by:

$$\Delta B_b = \frac{\sum B_b}{P_b} \rightarrow \text{minimum} \tag{7.11}$$

where $\sum B_b$ represents the total cost to implement an innovative bioconversion project and for operating the technical process of biogas production; P_b is the bioreactor productivity, determined by the quality of organic raw materials and methods of intensification and thermostabilization of the anaerobic biogas production process.

Table 7.2 Specific value of biogas output from a dry organic matter depending on temperature and duration of fermentation.

Process temperature, T, °C	Duration of fermentation, τ, day	Output of CH_4, m^3/kg	Process temperature, T, °C	Duration of fermentation, τ, day	Output of CH_4, m^3/kg
25	5	0.02244	44	5	0.11859
	10	0.05440		10	0.26995
	20	0.11626		20	0.42893
28	5	0.02168	48	5	0.20318
	10	0.09173		10	0.30812
	20	0.17480		20	0.47844
32	5	0.06323	50	5	0.17917
	10	0.16112		10	0.28513
	20	0.32998		20	0.35795
34	5	0.10007	54	5	0.11993
	10	0.20217		10	0.30234
	20	0.34773		20	0.38571
36	5	0.06667	56	5	0.18468
	10	0.16490		10	0.37588
	20	0.29610		20	0.46938
40	5	0.11561	60	5	0.06183
	10	0.26510		10	0.07472
	20	0.41756		20	0.09384

Any bioconversion project will be the most ecologically and economically feasible provided that:

$$\sum B_b = \sum B_{cap} + \sum B_{oper} - \sum B_{env} - \sum B_{pr} \rightarrow \text{minimum} \tag{7.12}$$

where $\sum B_{cap}$ is capital and current costs to implement a bioconversion project variant; $\sum B_{oper}$ is operational costs when implementing variants; $\sum B_{env}$ is the value of the savings from compensation of environmental damages during the project; $\sum B_{pr}$ is revenue from the sale of bioconversion products.

Operating costs are the most variable parameters and should be minimized by optimizing the biogas production process:

$$\sum B_{oper} = \sum B_L + \sum B_{RM} + \sum B_{UT} + \sum B_T + \sum B_i \rightarrow \text{minimum} \tag{7.13}$$

where $\sum B_L$ is labor cost; $\sum B_{RM}$ is cost for raw materials; $\sum B_{UT}$ is cost of used raw materials; $\sum B_T$ is cost for thermostabilization; $\sum B_i$ is cost for intensification.

Thus, the effectiveness of bioconversion projects may be estimated by the productivity of the bioreactor, taking into account the costs for thermostabilization and intensification of the fermentation process. The economic efficiency of investments in innovative, energy-efficient bioconversion projects is determined by taking into account capital and operating costs for the proposed project. The ratio of the total costs for project implementation to the bioreactor productivity can be used as an indicator of the ecological and economic efficiency; the lower this ratio the better the outlook for the project. The results of the economic assessment of using bioreactors will indicate their profitability or unprofitability.

7.6 Systems for heat supply using thermochemical organic wastes

An alternative to using natural gas is low-grade solid fuels, such as peat, straw, woodworking waste, food processing residues, solid household waste of organic origin, etc. Biofuel is a fuel of biological origin, the burning of which does not contribute to an increase of carbon in the earth's atmosphere. Using local raw materials for fuel provides a high guarantee of independence and reliability of energy supply for thermal systems. Implementation of this strategy would require reconstruction of existing natural gas boiler plants to accommodate the new biofuels.

The heating value of organic waste used as fuel depends on the type of raw material and its humidity (Table 7.3) [9,10,20].

Let us consider the schematic diagram of a boiler operating on straw (Fig. 7.17). The boiler operates as follows: a bale of straw is loaded into the furnace and burned. The amount of air for straw combustion is controlled by a blowing fan 4, which has variable frequency drive and is controlled by the automation system. Uniformly supplying air through air line 3 creates uniform combustion. The water in the accumulator

Table 7.3 Ash content and heating value of agricultural crop residues used as fuels.

Type of residues	Ash content of dry matter, %	Lower heating value, MJ/kg	
		Dry matter	**20% moisture content**
Straw of grain crops	5.0	17.4	13.5
Sunflower husk	5.0	19.3	14.6
Straw of rape	5.9	17.0	16.0
Corn stalk	4.6	18.0	16.5
Straw briquettes	4.0	18.8	16.5

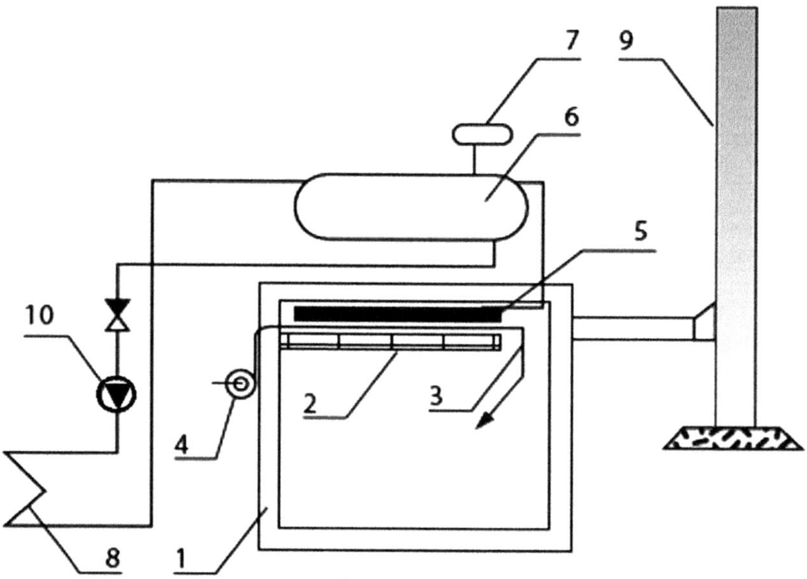

Fig. 7.17 Boiler operating with straw. *1*, boiler body; *2*, layer of refractory bricks; *3*, air line; *4*, blowing fan; *5*, circulating pipelines of a small circuit; *6*, tank/accumulator; *7*, expansion tank; *8*, heat exchanger of thermal energy consumers; *9*, stack; *10*, circulation pump in the heat supply circuit.

tank 6 is heated via a heat exchanger 5 which is protected from the open flame by a layer of refractory bricks 2. Water circulates naturally or is forced.

There are four basic ways to design the boiler connection:

1. Direct connection of the boiler to the system with a tank/accumulator (Fig. 7.17)
2. Direct connection to the heat supply system without a tank/accumulator
3. An independent connection through the heat exchanger (Fig. 7.18)
4. A cascade connection of a boiler operating with straw and a biogas boiler.

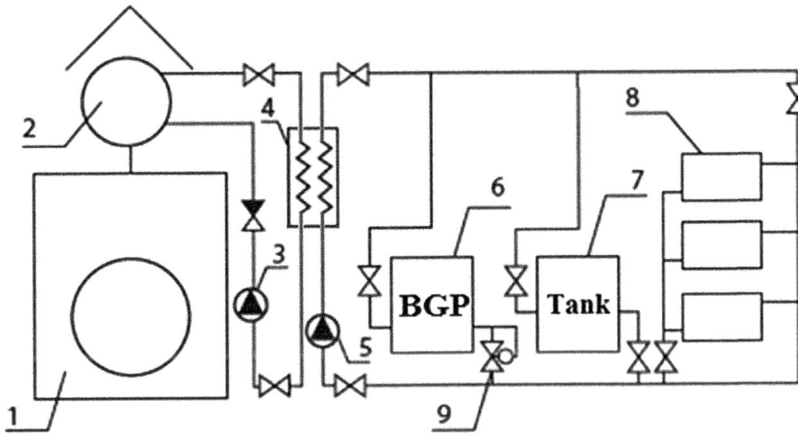

Fig. 7.18 Independent connection of a straw-fired boiler to consumers. *1*, boiler operating with straw; *2*, tank/accumulator; *3*, circulating pump of a boiler circuit; *4*, heat exchanger; *5*, circulating pump of the heating system; *6*, biogas reactor; *7*, tank/accumulator; *8*, consumers of thermal energy; *9*, regulating valve.

However, the use of a boiler without a tank-accumulator (Option 2) is not recommended, since it is difficult to adjust the boiler thermal power to follow a variable heat load demand.

The operation of the heat supply system is as follows: The boiler 1 generates a high-temperature heat transfer fluid (HTF), which is transported through piping systems with a pump 3. The boiler HTF transfers heat to a secondary heat transfer fluid in a counterflow heat exchanger 4. The temperature in the secondary circulating circuit depends on the selection of the heat exchanger. The circulating pump 5 is selected to compensation for the pressure losses in the system. The separation of hydraulic circuits allows the use of different heat transfer fluids, eliminates freezing of liquid, and prevents leakage from the heat transfer circuit. The operation of a biogas reactor can be arranged by installing a temperature flow limiter at the return pipeline to maintain the working processes in the bioreactor.

In the straw-fired heating systems in Fig. 7.17 the adjustment of hydraulic modes is carried out by valves with impulse lines, which do not require electrical energy and controllers, making these systems relatively inexpensive [21]. Limitations on the use of valves with impulse lines are set only by the manufacturer's maximum standard size. Figure 7.19 shows the biomass power plant. The characteristics of the boilers are given in Table 7.4.

7.7 Pyrolysis boiler systems

An effective way to solve complex problems of power engineering while maintaining a sustainable ecology is using the highly effective methods of chemical and thermal conversion of biomass into a combustible gas that can be used in cogeneration units. The

Fig. 7.19 Biomass power plant for burning wood pellets, wood chips and grated straw [23].

Table 7.4 Technical characteristics of Passat Energy A/S boilers for burning straw [22].

Passat type HO	**120**	**140**	**180**	**250**	**300**	**420**	**510**	**680**	**850**
Nominal output, kWt	140	160	210	249	345	490	595	790	990
Fuel consumption, kg/h	36	42	54	75	90	126	153	204	255
Min. return temp., °C	70								
Gas flow, m^3/h	362	421	337	460	548	761	920	1221	1522
Gas flow, Nm^3/h	230	266	534	730	870	1207	1460	1936	2414
Min. Efficiency, %	90								

See Ref. [22] for performance using of wood pellets and chips.

operation of gas generator pyrolysis boilers is based on the principle of biomass pyrolysis-combustion (or dry distillation) [12,20,24].

Pyrolysis occurs with the release of heat, which contributes to and improves the heating and drying of biomass in a boiler. When mixing the produced pyrolysis gas with the oxygen in air, a combustion process takes place in which flue gases at the outlet of the boiler contain practically no harmful impurities, and the amount of soot and ash formed is minimal. Emissions of carbon dioxide into the atmosphere are one-third those from burning organic substances in conventional boilers. The stability of the gasification process from biomass is provided by matching the geometric parameters of the pyrolysis boiler modes with the gas blowing mode, taking into account the chemical composition and such fuel parameters as volatile matter content, gas permeability, excess moisture, etc.

In relation to advanced solid fuel boilers, pyrolysis boilers have several advantages including:

- high efficiency (over 85%), achieved due to combustion of biomass without residue
- higher temperatures are achieved with pyrolysis burning of moist biomass
- less secondary air is needed to burn pyrolysis gas, improving boiler efficiency
- biomass is loaded into a combustion chamber only once or twice a day
- pyrolysis gas combustion is easily controlled from 30 to 100%
- easy maintenance and cleaning
- payback period is from one to four heating seasons.

The classification of pyrolysis boilers depending on the method of production and the designation of pyrolysis gas, as well as their design features is shown in Fig. 7.20.

A design feature of pyrolysis boilers is the presence of several combustion chambers (Fig. 7.21) that allows the most efficient and high quality use of the pyrolysis process. The simplest boiler has two combustion chambers. The first one, called the loading chamber, is designed directly for pyrolysis of fuel. In this chamber, the process of pyrolysis occurs with only a limited amount of oxygen, and the biomass decomposes to gases and solid residue. Once the biomass is decomposed, the gases enter the second chamber, where secondary air is fed to facilitate combustion. The chambers are separated by grates on which biomass briquettes are placed. Depending on the location of an additional heating chamber, boilers are distinguished as ones with an upper chamber and those with a lower chamber.

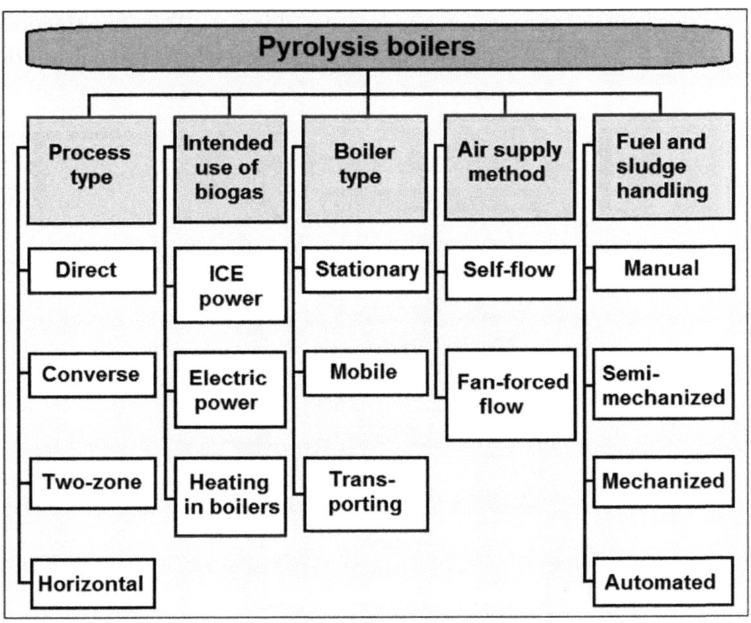

Fig. 7.20 Classification of pyrolysis boilers according to various factors.

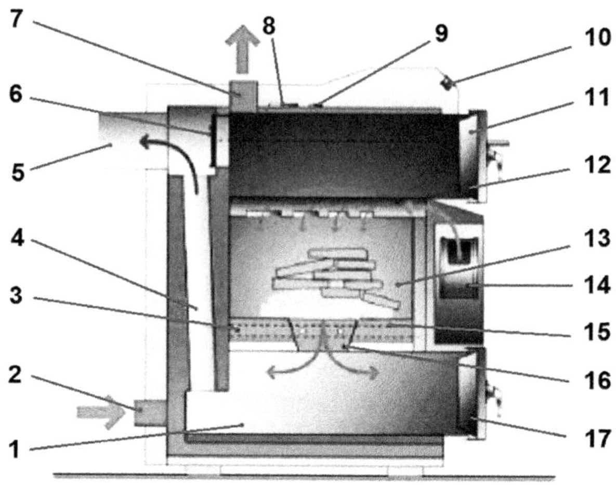

Fig. 7.21 Cross-section of a double-chamber pyrolysis boiler. *1*, combustion chamber; *2*, water feedback manifold; *3*, channel of secondary air supply; *4*, chimney conduit; *5*, flue; *6*, chimney choke; *7*, water supply manifold; *8*, temperature sensor; *9*, pressure sensor; *10*, regulator; *11*, upper door (boot); *12*, air supply channel; *13*, gasifier; *14*, fan; *15*, water heat exchanger; *16*, nozzle; *17*, lower door.

A boiler with a lower combustion chamber is quite popular (Fig. 7.22). Such a design makes it easier to put firewood into the combustion chamber, while the additional heating chamber exits directly into the flue tract. Smoke, succumbing to draft according to the laws of physics, moves into the chimney. This construction has its advantages, which consists in the fact that flue gases through special nozzles enter

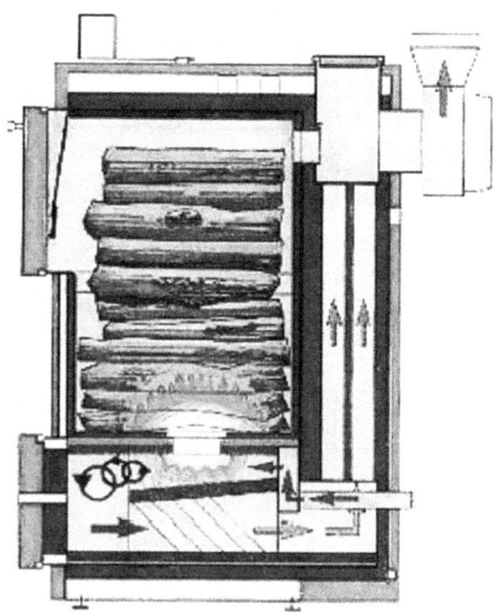

Fig. 7.22 Cross-section of a pyrolysis boiler with a lower combustion chamber.

the combustion chamber, where they are completely burned. Then they enter the chimney, where they are cooled down. Although this system sounds quite logical, there are some disadvantages. From the primary combustion chamber, ash enters the additional heating chamber, which requires more frequent cleaning. In addition, a strong draft must be created to pull the smoke down. A more costly designed chimney tract, which requires more material, is a disadvantage of such a system.

Pyrolysis boilers can be considered as a perfect alternative to direct combustion boilers. Compared to the latter, the pyrolysis version has a large combustion chamber, but this type of boiler requires the availability of an accumulator tank for storage of thermal energy. This is required in order to ensure the uniformity of the boiler operation, which is subject to fluctuations due to performance differences associated with different combustion efficiencies of the phases of fuel. All this makes the pyrolysis boiler an alternative in a system with a gas or electric boiler.

Thanks to the optimal internal system of the pyrolysis boiler, fuel consumption is lower than traditional boilers. However, the amount of raw materials consumed is influenced by external factors as well. Fuel consumption in a pyrolysis boiler depends on the following indicators:

- Humidity of firewood should not exceed 20%. At this humidity, the efficiency will be the highest. The system operates well even when loading firewood which has 50% moisture content, but a significant amount of the generated heat will be spent on drying the wood. This will significantly reduce the efficiency of a pyrolysis boiler.
- It is most economical to use wood as the fuel. Coal and peat do not allow as much pyrolysis gas to be obtained to provide a high level of heat transfer.

7.8 Summary

One important means of supplying the energy needs of the peoples of the earth is to make use of waste materials that can be chemically, biologically, and thermodynamically converted into fuels that can then be burned to provide heat or power. Specifically, organic waste products from agriculture or livestock farming, including vegetative scraps and manure, can be processed with a variety of techniques to generate biofuels such as biomethane or biodiesel. Thus, instead of contributing to the growing volume of landfill waste, these materials can produce useful fuels and thereby replace other conventional sources of these fuels. This chapter provides details on the processes and the reactors needed to carry out the energy conversions from waste to fuel. This is a rapidly growing field across all countries of the world since the raw materials are available essentially anywhere people have settled and cultivated the land and raised animals. The technology described in the chapter may be put to use wherever the raw materials are present and the economics are favorable.

Nomenclature

A area
B cost

c	specific heat capacity
COP	coefficient of performance
P	pressure
Q	heat loss
K	kinematic coefficient
m	specific mass
M	mass
S	concentration
T	temperature
V	volume

Greek letters

α	heat transfer coefficient
η	efficiency
ρ	density
θ	emission
τ	duration
w	growth rate

Subscripts

ABHP	absorption heat pump
BGP	biogas plant
ICE	internal combustion engine
HTF	heat transfer fluid

Review questions

1. What is the meaning of the term "biogas"?
2. What raw materials can be used to produce biogas?
3. How is biomethane produced?
4. What does the heating value of fuel from organic waste depend on?
5. Explain the operating principle of a boiler using straw.
6. How do boilers using straw and a biogas plant operate?
7. Explain the principle of the gas generator pyrolysis boiler operation.
8. Give the classification of pyrolysis boilers.
9. Describe the ways of intensifying the process of anaerobic fermentation of the substrate in a bioreactor.
10. Name the modes of substrate fermentation depending on the temperature.
11. Determine the factors for improving the energy-saving directions of the biogas production process.
12. Write the equation that describes the mass-exchange processes in a bioreactor.

13. Write the heat balance equation of a bioreactor and characterize its components.
14. Give examples of energy-saving constructive and technological systems for biogas production.
15. Explain the principle of operation of a biogas energy-saving plant with a heat exchanger and a heat pump.
16. How are energy losses determined in a biogas plant with a heat pump?
17. How do energy consumption for the operation of a biogas plant and the cost of using energy resources depend on the temperature regime of bioconversion?

Exercises

1. A farmer has 10 cows and wants to use their manure to produce biogas for heating a house with an area of 200 m^2. Determine the amount of biogas and the possibility of using it for heating.
2. Determine the volume of the bioreactor for Exercise 1, assuming that the fermentation time is 20 days.
3. The reaction rate in a biogenerator can be increased by a factor of two by heating the supplied substrate from 15−20 °C to 35−37 °C. How much does one need to increase the specific power of the heat generator to the supply heat?
4. Determine the amount of heat released during the following reaction which represents the fermentation of biomass: $C_{12}H_{22}O_{11} + H_2O \rightarrow 6\ CH_4 + 6\ CO_2$.

References

[1] IEA Task 37, energy from biogas. 2016.
[2] 13th FYP on agricultural biogas development in China.https://www.oav.de/fileadmin/user_upload/2_Termine/Allgemein/biogas_development_in_China.pdf.
[3] Statutory report in accordance with Norwegian authority requirements. Statoil; 2014. https://www.equinor.com/content/dam/statoil/documents/annual-reports/2014/Statoil-Statutory-report-2014.pdf.
[4] Wall DM, Dumont M, Murphy JD. Green gas − facilitating a future green gas grid through the production of renewable gas. IEA Bioenergy; 2018. p. 2 [Task 37].
[5] BIOGASPARK KÖNNERN-SÜD. https://www.fr-vechta.de/taetigkeitsbereiche/biogasanlagen/.
[6] Biogas plant in Könnern [in German]. https://www.biogas-kanns.de/links/biogas-atlas/koennern/447d14/.
[7] Nordmethan Die Biomethanraffinerie Könnern. Information to the public in accordance with annex V. 12. BImSchV for biogas plants the operating areas of the lower class. https://www.nordmethan.de/eigenanlagen/koennern.html.
[8] Baader B, Don E, Brennderfer M. Biogas: theory and practice. Moscow: Kolos; 1982. 570 p.
[9] Twidell J, Weir T. Renewable energy resources. 3rd ed. Oxfordshire, England; New York, NY: Routledge; 2015.
[10] Seadi T, Rutz D, Prassl H. Biogas. Handbook. Denmark: University of Southern Denmark Esbjerg; 2008. 125p.
[11] Gunter LI, Goldfarb LL, Metantenki M. Stroiizdat; 1991. 128 p.

[12] Abbasi T. Biogas and biogas energy. New York: Springer; 2012. 169 p.
[13] Al Seadi T, Stupak I, Smith CT. In: Murphy JD, editor. Governance of environmental sustainability of manure-based centralised biogas production in Denmark. IEA Bioenergy; 2018. p. 7 [Task 37].
[14] Lambert M. Biogas: a significant contribution to decarbonising gas markets? The Oxford Institute for Energy Studies; 2017. https://www.oxfordenergy.org/wpcms/wp-content/uploads/2017/06/Biogas-A-significant-contribution-to-decarbonising-gas-markets.pdf.
[15] Lindmark J, Thorin E, Bel Fdhila R, Dahlquist E. Effects of mixing on the result of anaerobic digestion: review. Renewable and Sustainable Energy Reviews 2014;401: 1030–47.
[16] Lemmer A, Naegele H-J, Sondermann J. How efficient are agitators in biogas digesters. Determination of the efficiency of submersible motor mixers and incline agitators by measuring nutrient distribution in full-scale agricultural biogas digesters. Energies 2013;6: 6255–73.
[17] Hashimoto AG. Effect of mixing duration and vacuum on methane production rate from beef cattle waste. Biotechol. Bioeng 1982;24:9–23.
[18] Microferm biogas plant. Angers, France. https://www.host.nl/en/case/angers-france/20.
[19] Chen T-H, Hashimoto AG. Effects of pH and substrate:inoculum ratio on batch methane fermentation. Bioresource Technology May-June;56(2–3):179–86.
[20] Saufer S. Biomass as a source of energy. Moscow: Mir; 1985. 368 p.
[21] Shen F, Tian L, Yuan H, Pang Y, Chen S, Zou D, Zhu B, Liu Y, Li X. Improving the mixing performances of rice straw anaerobic digestion for higher biogas production by computational fluid dynamics (CFD) simulation. Applied Biochemistry and Biotechnology 2013;171:626–42.
[22] Passat Energy A/S, Industrivej 16, Ørum, DK-8830 Tjele, Denmark. https://passatenergy.com/pdf/ho-product-information.pdf.
[23] Passat Energy A/S, Industrivej 16, Ørum, DK-8830 Tjele, Denmark. https://passatenergy.com/en/information-passatenergy/business-solutions/.
[24] Task 34 Pyrolysis of Biomass - IEA Bioenergy Task 34. http://task34.ieabioenergy.com.

Hybrid systems with renewable energy sources

8

8.1 Solar electric power stations: A brief overview with examples

In this chapter we will discuss the use of solar energy as one energy input to a hybrid system involving at least one other primary energy source. Although the focus of this chapter in low-temperature applications in keeping with the overall theme of the book, we will look at some solar applications that involve high-temperature systems. In this section, we will consider a small sampling of such systems to provide context for the low-temperature applications.

8.1.1 Solar insolation

The source of solar energy on earth is the flux of solar radiation entering the atmosphere, namely, 1367 W/m^2 (solar constant). After passing through the atmosphere and incurring losses, the maximum flux of solar radiation occurring at the equator and at sea level is 1020 W/m^2 [1]. Figure 8.1 gives the global direct normal solar irradiation (DNI), i.e., the long-term average solar energy in kWh/m^2 on a daily or yearly basis. By definition, the DNI is the amount of solar radiation received per unit area on a surface that is always held perpendicular to the sun's rays coming in a straight line from the sun at its current position in the sky. This value is most useful for designing solar systems that track the sun such as 2-axis tracking photovoltaic (PV) or concentrating solar parabolic trough (CSP) systems. For fixed, flat PV collectors such as might be found on large warehouse rooftops, a better design parameter is the global horizontal irradiance (GHI); see Fig. 8.2.

8.1.2 Large-scale photovoltaic power stations

Photovoltaic (PV) electric generating stations are direct energy conversion facilities that do not involve thermal energy conversion. Thus they operate at near-ambient temperature, and qualify as low-temperature power plants. These systems are routinely deployed at residences, typically on rooftops. Over the last few years, owing to the great increase in the number of units built and deployed, the unit price has decreased markedly – a consequence of an effect called the "learning curve."

Low-Temperature Energy Systems with Applications of Renewable Energy
https://doi.org/10.1016/B978-0-12-816249-1.00008-X

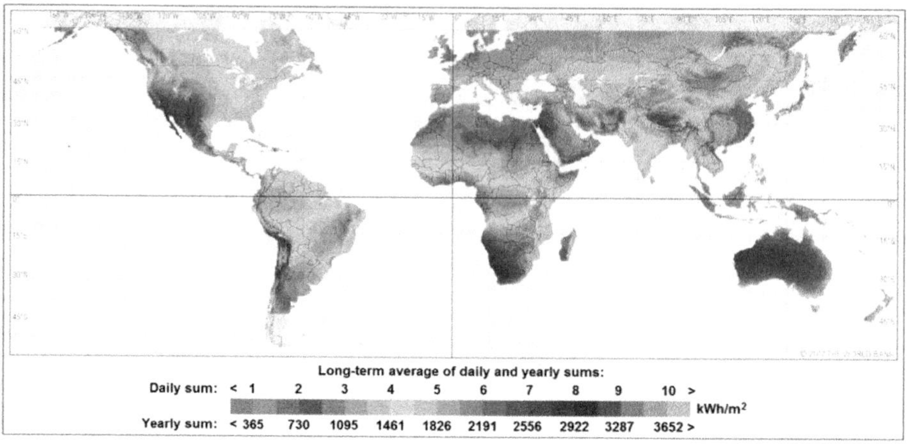

Fig. 8.1 Direct normal solar irradiation (DNI): ©2017 The World Bank, Solar resource data: Solargis [1], used with permission.

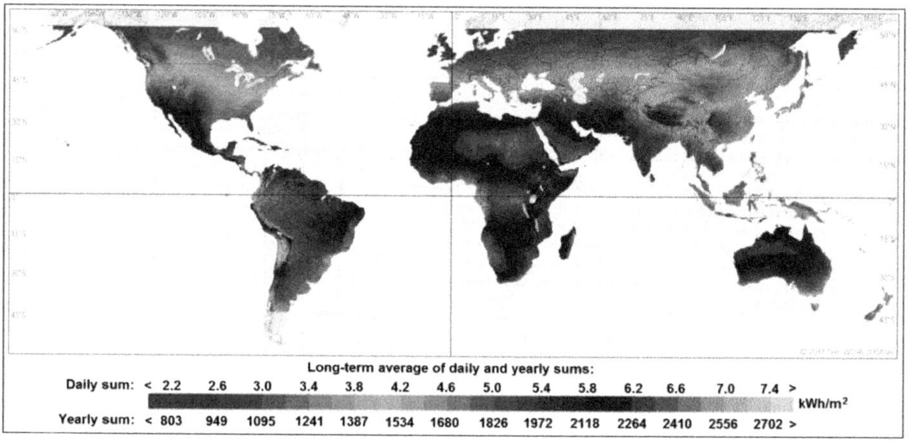

Fig. 8.2 Global horizontal irradiance (GHI): ©2017 The World Bank, Solar resource data: Solargis [1], used with permission.

Moreover, there has been a significant increase in utility-scale PV systems with installed peak capacities of 100 MW or greater. One example from many that could be selected is the Perovo solar photovoltaic (PV) power plant in Crimea (see Fig. 8.3). It has an installed capacity of 105.56 MW (100 MW, rated) and covers about 200 ha. There are 440,000 individual modules in the plant, and it is designed to generate 132.5 GWh of electricity per year. This is equivalent to a capacity factor of 15%. When it was completed in 2011, it was the largest PV plant by capacity in the world, but given the technological advances and improved economics in recent years, it now ranks only sixth in size. The financial viability of the project was greatly enhanced by a favorable feed-in tariff of 0.46 €/kWh, among the highest anywhere, which will be in place until the year 2030.

Fig. 8.3 Aerial view of the 105 MW Perovo solar PV power station in Crimea [2].

8.1.3 Floating photovoltaic power stations

Given that about 71% of the earth is covered by water and much of that is in areas with abundant sunlight (see Figs. 8.1 and 8.2), coupled with the absence of obstructions such as trees, buildings, etc., there is a growing movement to install PV systems on the surface of bodies of water. Since the PV panels cover the water surface, floating systems reduce evaporation and tend to preserve the aquatic environment. A further advantage of this form of deployment is the natural cooling effect afforded by the water and ambient air that keeps the performance of the PV cells operating at or very close to their design points by avoiding high temperatures that degrade their efficiency when installed on land.

The world's largest floating PV plant is located in China near the city of Huainan; it has a peak capacity of 40 MW; see Figs. 8.4 and 8.5. The water in this case is a pond associated with an abandoned coal mine and ranges in depth from about 4 to 10 m. A smaller, 20 MW, floating PV plant in the same area was previously the world's largest plant of this type [3].

The second largest PV power station in Japan is located at Kagoshima and has an installed peak capacity of 70 MW [4]; see Fig. 8.6. The official name of the plant is Kagoshima Nanatsujima Mega-Solar Power Plant; it came on-line November 1, 2013 following a 13-month construction period. It was the largest PV plant in Japan when it started operating, but was surpassed in 2014 by the 82 MW Oita plant. The

Fig. 8.4 Floating PV power plant near Huainan, China [3]. Photo by SunGrow.

Fig. 8.5 Google Earth image of the Huainan floating PV power plant deployed on waters of an abandoned coal mine in China, May 21, 2017.

Fig. 8.6 The Kagoshima Nanatsujima Mega-Solar Power Plant on Kyushu, Japan; Grand Sakurajima volcano can be seen in the background. Photo courtesy of Kyocera Corporation [6].

capital cost was \$275.5 M or about \$3900/kW. The plant produces 78.8 GWh of electricity per year for a capacity factor of 13%. While not exactly a floating power plant, it was built on land reclaimed from Kagoshima Bay. The plant covers an area of 11.8 ha and is comprised of 290,000 individual PV cell modules [5].

8.1.4 Co-located solar hybrid plants

Hybrid generation is based on using the advantages of renewable energy sources combined with traditional sources of generation. In many cases, plants using renewable energy sources are integrated with traditional generation plants which are located nearby, allowing the use of existing infrastructure, such as power transmission lines, substations, etc., thus reducing the cost of construction and operation. This mode of hybrid operation differs from thermodynamically integrated systems in which energy synergy is achieved and higher efficiency obtained; examples of such systems are presented in Sections 8.3 and 8.4.

There are several solar PV plants that have been built near hydroelectric stations, which are also renewable systems. Either adjacent flat land or the grassy slopes of the retaining dam (assuming proper orientation for good insolation) can serve as the site for the PV panels.

One example of a solar PV system co-located with a hydroelectric station is the 5 MW PV system at the Kotani dam near Toyotomicho, Himeji City, Japan; see Fig. 8.7 [7]. The construction began in December 2014 and the plant was commissioned in February 2016. The roughly 26 degrees-sloping surface of the dam faces south and has a length of about 150 m. The elevation at the top of the dam is about 158 masl, whereas at the base of the PV array it ranges from 93 to 107 masl. The 19,580 panels, each rated at 255 kWp, cover an area of about 3.2 ha. To facilitate construction, the mounting framework is made of light-weight aluminum that allowed an individual frame to be carried by a single worker. The construction cost was roughly $20 M or $4000/kW. The plant is expected to generate 5.5 GWh of electricity per year, for a capacity factor of 12.5% [7]. It should be noted that the Kotani dam does not generate electricity but serves as a holding reservoir for fresh water for residents and industry. Water is pumped from a nearby river when levels are high and stored behind the dam for later use.

Another project co-located with a hydro station is the plant at the Longyang Gorge on the Yellow River in Gonghe County, Qinghai Province, China. At that site one finds the largest co-located PV-hydro facility in the world: an 850 MW PV and a 1,280 MW hydroelectric generating station. The PV array is about 30 km west of the dam, across a portion of the Longyangxia reservoir; see Fig. 8.8 [7].

Fig. 8.7 Kotani Dam 5 MW solar PV installation near Himeji City, Japan [7].

Fig. 8.8 Co-located PV and hydro power plants at Longyang Gorge, China; modified Google Earth image, December 30, 2015.

8.2 Solar thermal heating systems

Solar energy may be used in a straightforward manner to provide thermal energy (heat) for a variety of applications. Collectors may be flat plates, either glazed or unglazed, or curved surfaces, typically parabolic troughs, although solar dishes that employ an array of small flat-plate mirrors are sometimes used. The troughs concentrate the sunlight on the focal line whereas the dish creates a hot spot at the focal point of the mirror system.

The application temperatures range from low (swimming pools), to medium (residences or commercial buildings) and to high (industrial processes and electrical power generation). To achieve high-temperatures, concentration of the sun's rays is necessary. When this is done for strictly thermal uses, the system process is called concentrated solar thermal (CST); when it is done to generate electricity, it is called concentrated solar power (CSP). Most CST systems are used to heat water, but a tiny number involve air heating. The collector areas devoted to water and air systems for 2015 are: 65,120.7 ha —water (99.7%); 174.3 ha – air (0.3%), for a grand total of 65,295.0 ha [8]. Table 8.1 gives the top five countries ranked by total solar thermal collector area, excluding concentrators, for both water and air systems. Thus, the five largest solar thermal countries constitute 83% of the total collector area in the world.

The growth of solar thermal systems over a recent 18-year span is shown in Fig. 8.9 [8]. The trend is beginning to slow down owing to advances in other thermal technologies such as heat pumps.

The largest solar thermal plant in the world, with a peak capacity of 110 MWth, supplies the Danish city of Silkeborg with 20% of its heating needs; see Fig. 8.10 [9]. As can be seen from the image, the land is not perfectly flat, so the collector panels follow the contour of the ground. The 12,436 panels have a collector area of 156,694 m^2, and there are four storage systems covering 16,000 m^2. Over the course of a year, the system produces 80 GWh of thermal energy and eliminates 15,700 tonnes of carbon dioxide emissions. The construction of megawatt-scale solar thermal plants significantly reduces the unit cost of heat supply. The economics are more

Table 8.1 Top five countries in solar thermal collector area, ha [8].

Rank	Country	Water	Air	Total
1	China	46,358.0	0.5	46,358.5
2	United States	2509.3	17.3	2526.6
3	Turkey	2133.5	0.9	2134.4
4	Germany	1933.6	2.6	1936.2
5	Brazil	1365.0	0	1365.0
	Totals	54,299.4	21.3	54,320.7

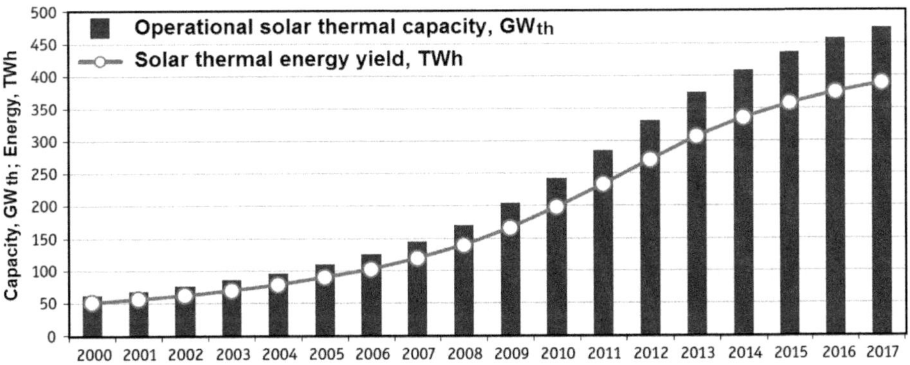

Fig. 8.9 Worldwide growth of installed solar thermal capacity and heat energy yield from 2000 to 2017 [8].

Fig. 8.10 Silkeborg, Denmark solar collector field, the largest in the world [9].

favorable when the system is installed at the time a building is constructed as opposed to being retrofitted on existing structures.

For additional material on solar thermal systems, the reader may consult Ref. [10].

8.3 Solar assisted heat pumps in heating systems

Today, there are arrangements that combine heat pumps with solar collectors of various designs (flat plate and parabolic troughs/vacuum tubes) for providing hot water and low-temperature heating. In this section we describe several ways heat pumps can be augmented using solar energy.

8.3.1 *Solar energy as a bottoming source of heat for heat pumps*

Figure 8.11 is a schematic diagram of a heat pump system with using solar radiation as a lower energy source. This avoids having to use ambient air or water at ambient temperature or a ground-coupled heat exchanger, both lower-temperature energy sources compared to the solar collector.

The heated room has heat losses to the environment Q_{HR}. The heat flux from the HP condenser Q_{cond} at the temperature of the coolant t_{cool} entering the heating system compensates for the loss. The conversion of incoming solar radiation into thermal energy is carried out in solar collectors in which an aqueous solution of propylene glycol is used as a heat transfer fluid, while water in a low-temperature heating system acts as

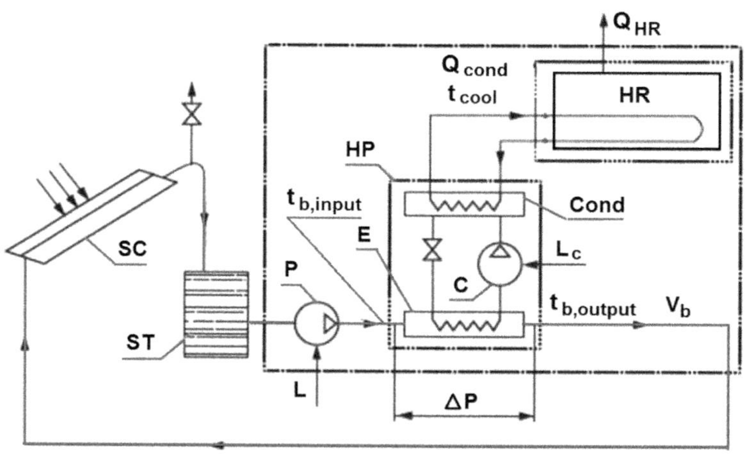

Fig. 8.11 Schematic diagram of heat pump unit using solar radiation as a heat source. *C,* compressor; *Cond,* condenser; *E,* evaporator; *HP,* heat pump; *HR,* heated room; *L,* work input to the pump drive to overcome the total pressure loss ΔP in the lower circuit; L_c, work input to the compressor drive; *P,* pump; *SC,* solar collector; *ST,* hot fluid storage tank [11].

the accumulating medium. A brine with a temperature $t_{b,inlet}$ and volumetric flow rate V_b is pumped from the solar collector to the evaporator of the heat pump. In the evaporator, the brine is cooled and its outlet temperature is $t_{b,outlet}$. The circuit feeds a storage tank, which is used to smooth out the temperature drops of the coolant during the day and provides more stable operation of the heat pump.

A question arises about the optimization of conditions of operation for a heat pump heating system. At a given heat load on the heat pump and temperature of the heat carrier in the heating system, which are both set by the objective of the heat supply itself, the temperature of the coolant at the evaporator outlet is ambiguous. This is because that the amount of heat extracted from the brine depends on: (a) the difference in temperature at the inlet and outlet of the evaporator, (b) coolant costs, and (c) the density of the incident flux of solar radiation.

When choosing the temperature of the brine at the entrance to the evaporator, it is necessary take into account that the amount of heat drawn from the solar collector depends on the type of collector, the volume flow rate of the brine, and the intensity of the incoming solar radiation, which in turn determine the efficiency of the collector. A change in the temperature of the brine leaving the evaporator leads to a change in the volumetric flow rate V_{brine} at a given value of the heat flux Q_{HP}, which, in turn, causes a change in the energy consumption of the pump compressor drive. However, the change in the brine temperature t_b at a constant value of temperature at the outlet of the condenser leads to changes in working conditions and energy consumption of the compressor drive. Because that the work consumption of the compressor drive and of the brine pump change in opposite directions, the problem arises to determine the optimal degree of cooling of the brine in the evaporator, which corresponds to the minimum specific total energy consumption of the heat pump system as a whole.

To meet the needs of the consumer, flat plate and parabolic trough/vacuum tube collectors are used in heating. The collector is protected by a reinforced housing covered with thermal insulation, to minimize heat losses. In the vacuum tube collectors, the absorber is built into the vacuum glass tube, allowing for the higher temperatures needed for heating buildings, i.e., the brine can be maintained at high temperatures in the collector, even with low ambient temperatures.

An analysis of the heat pump system described above using solar energy allows us to draw the following conclusions [12]:

- When using solar radiation as the lower energy source in the heat pump system, there is an optimal degree of cooling of the medium in the evaporator of the heat pump, which corresponds to the minimum total specific energy consumption of the system as a whole.
- The optimum cooling rate of the coolant in the evaporator depends on its temperature at the outlet of the solar collector (entering the evaporator) and in the coldest months of the year is in the range from 2.0 to 2.5 °C;
- Heat pump systems from an energy point of view are more appropriate using vacuum tube collectors because they allow raising the temperature of the brine at the evaporator inlet, and thus reduce the cost of external energy for the heating system. In this way, the coefficient of performance of the heat pump system can achieve values from 4 to 7.

8.3.2 Solar energy as a topping source of heat for ground-source heat pumps

Figure 8.12 shows a schematic diagram of a combined heat pump unit using the heat of soil and solar energy for a hot water heating system HWS.

The operating principle of this system is as follows. The conversion of incident solar radiation into heat energy is carried out in flat-plate solar collectors in which an aqueous solution of ethylene glycol is used as a heat-carrier, while the accumulating medium is water. The solution from a solar collector is pumped to the HP condenser where it is heated up to the required temperature in conjunction with the soil heat which is absorbed by an anti-freeze heat-carrier flowing through the pipes of the ground collector. The solution with temperature $t_{brine\ input}$ and volumetric flow rate $V_{GHE.}$ is fed by the pump from the ground collector to the evaporator of the heat pump. In the evaporator, the solution is cooled to an outlet temperature $t_{brine\ output}$. A heat exchanger is installed between the solar collector and the tank-accumulator, in which the salt solution is additionally cooled with cold water that allows the efficiency of the solar collector to be increased. The thermal energy obtained from these sources is accumulated in the tank-accumulator and is used when it is required to provide heat to the HWS.

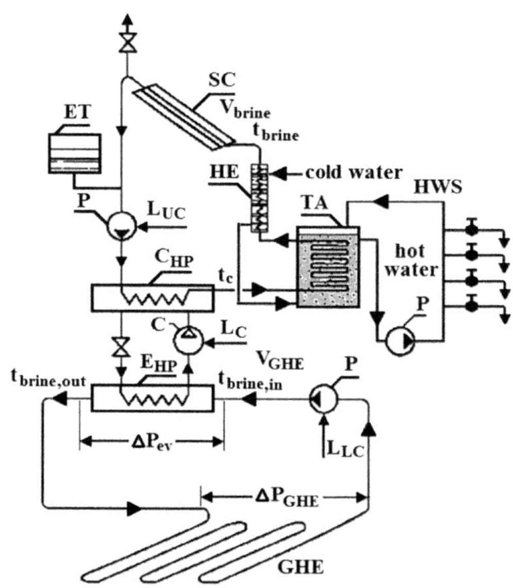

Fig. 8.12 Schematic diagram of a combined heat pump unit using soil heat and solar energy. *C,* compressor; C_{HP}, condenser; E_{HP}, evaporator; *ET,* expansion tank; *GHE,* ground heat exchanger; *HE*, heat exchanger; *HWS*, hot water supply; L_C, work input for HP compressor drive; L_{LC}, work input for pump drive in the lower HP circuit; L_{UC}, work input for pump drive in the upper HP circuit; *P,* pump; *SC*, solar collector; *TA,* tank/accumulator [11].

When using solar-assisted heat pumps in heating systems, there is the matter of optimizing the conditions of the combined heat pump system, namely, determining the optimal degree of cooling of the medium in the HP evaporator. The fact is, at a given thermal load on the heat pump and a temperature of the refrigerant in the HWS system, the temperature of the refrigerant at the evaporator outlet is ambiguous.

On the basis of the numerical analysis in Ref. [12], it is shown that the optimum degree of cooling of the brine in the evaporator is in a narrow range (1–2 °C) and increases with an increase in the factor *A* which depends mainly on the hydraulic resistance of the lower circuit, and decreases with an increase in the proportion of solar energy used to meet the hot water needs of the consumer. [Note: The factor *A* was defined and discussed in Chapter 2, Eq. (2.16).]

It is also shown that when the optimal cooling rate of the brine in the HP evaporator is provided, the minimum total energy consumption for the HWS system decreases sharply with an increase in the ratio, *K*, of heat fluxes of solar energy and the soil heat in a solar-assisted heat pump. In this case, the coefficient of performance for the heat supply system increases from 3.0 to 12.0 when the ratio *K* changes from 0.2 to 5.0. It should also be noted that in providing the optimal degree of cooling of the brine in the evaporator, the effect on the minimum external energy consumption of the combined pressure loss in the HP evaporator and in the GHE is insignificant.

8.3.3 Direct use of solar energy as a bottom source of heat for a heat pump

The use of solar collectors in combined heat pump systems substantially increases the capital investment in the heat supply system. However, there is an interesting alternative method to use solar energy in heat pumps, namely, by placing the heat pump evaporator into the solar collector. Thus, the evaporator and the collector are combined in one element; the concept was patented and is called the Bernier system [13]. In this way the extra cost of the hybrid system is minimized.

To simplify the system and to allow the possibility of heat pump operation in periods of cloudy weather, the solar collector does not have the usual glass cover and thermal insulation. Furthermore, lowering the temperature of the absorber reduces the heat loss of the collector, and in cloudy weather, the heat is taken from the atmosphere. In this case, the flat elements of the absorber-evaporator can be painted not only with black paint having the highest degree of blackness, 0.9, but also with other colors, such as green, blue, brown, violet (blackness > 0.8), red (0.44–0.57), and other colors, making such designs architecturally attractive. Evaporator dimensions which are not painted in black need to be increased slightly to compensate for the lower heat absorptivity.

The installation of such collectors is possible on the eastern, southern, western, southeastern, and southwestern facades, and on the roof at different angles. When installed on the walls it is necessary to provide an air gap with the wall of not less than 50 mm. For areas with winds above 10 m/s, appropriate mounting should be installed. Pipelines in the heat pump carrying cold coolant do not require thermal insulation. It is recommended that the piping manifold, which supplies the refrigerant to separate collectors, be placed vertically. To increase efficiency, the hot coolant from the condenser should have a temperature below 50 °C. The circulation of the refrigerant does not require an additional pump.

According to the results obtained in practice, at an ambient temperature of −10 °C and no sun, the COP is 2; for solar radiation of 300 W/m^2, it increases to 3, and at 600 W/m^2, it is 4. For 0 °C ambient temperature, the respective factors are 2.7, 3.6, and 4.5. At 10 °C ambient temperature and solar radiation of 600 W/m^2, the COP reaches 6.

The energy coefficient, *e*, which takes into account cloud cover, is calculated from:

$$e = \frac{E_a}{E_m} \tag{8.1}$$

where E_a is the actual amount of energy; E_m is the maximum energy without clouds for angles of inclination to the horizon of S = 60 degrees and S = 90 degrees, taking into account scattered radiation and cloudiness. These angles were chosen because in winter more energy can be obtained from vertical surfaces both in combination with heat pump systems and in passive solar heating systems, and the orientation of the collector at an angle of 60 degrees makes it possible to produce more energy in the autumn and spring months when the sun is lower. At the inclinations of the collector close to the optimal, e.g., 30–35 degrees for a city located at 48 degrees north latitude, in the summer months, a surplus of hot water is produced by the HWS, the consumption of which in the summer is 20% below the annual average.

8.4 Combined geothermal and heat pump heating systems

8.4.1 Schemes for geothermal heating systems

In general, systems in which geothermal fluid is sent straight to the end users are called "direct" systems. Those systems in which the geofluid is used to heat fresh water which in turn is then sent to the end users are called "indirect" systems. The former are used when the quality of the geofluid is of potable standards, which is rarely the case for high-temperature resources but may occur for low-temperature ones.

Furthermore, "open" systems refer to systems in which the fluid providing the final hot water service, either domestic hot water or space heating, is discharged to a drain or sewer. "Closed" systems are ones in which that fluid is recirculated after cooling to be reheated by the geofluid. Since the geofluid is not reused, it always flows in a once-through manner from the production wells to reinjection wells, or in rare cases, to a surface body of water such as a river or a bay. In the sections that follow all types of systems are illustrated.

Figure 8.13 shows a two-pipe closed geothermal hot water circuit. Geothermal fluid from wells 1 is fed via a storage tank 2 into the heat exchanger 5, and then discharged

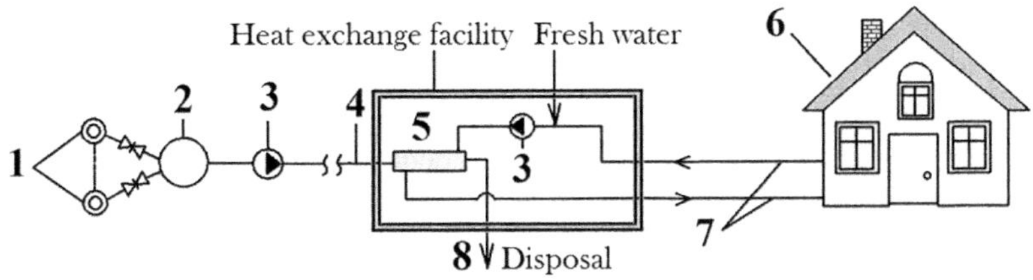

Fig. 8.13 Two-pipe closed geothermal hot water distribution system. *1*, geothermal wells; *2*, geothermal hot water storage tank; *3*, pumps; *4*, single-pipe transit heating mains; *5*, heat exchanger; *6*, domestic hot water supplies; *7*, two-pipe distribution heating system; *8*, cool geofluid network.

8. Fresh water that circulates through the two-pipe distribution network is heated in the heat exchanger of the central heating station and supplied to consumers. Fresh make-up water is supplied from the district water supply system as needed.

Figure 8.14 shows a closed geothermal heating system that provides both heating and domestic hot water via a four-pipe distribution network. The geothermal fluid is transported to a central substation via a single-pipe network. After the central heating plant, the geothermal water is either discharged or pumped back into the reservoir. Hot geofluid is divided and used in two parallel heat exchangers, one for producing domestic hot water 5 and one for space heating 6.

The specific arrangement of the system and the design of the equipment are determined by a number of General design considerations:

- distance of the central heat plant from the geothermal source
- thermal power requirements and nature of the heat consumer
- extent of geothermal water reserves
- temperature, chemical, and gas composition of the geothermal fluid
- conditions and limits for discharge of the spent geothermal fluid

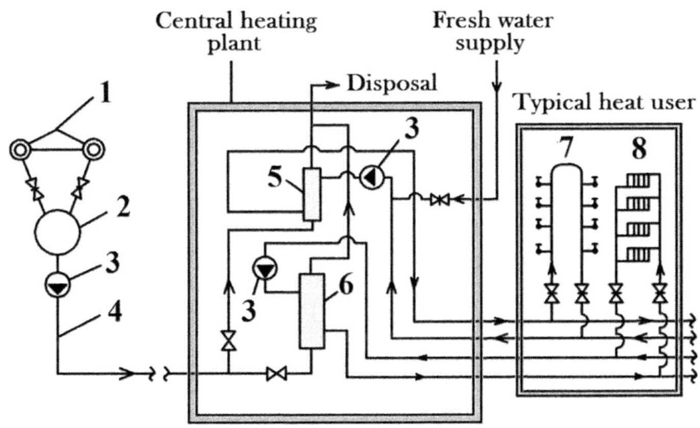

Fig. 8.14 Closed, four-pipe geothermal heating system with independent domestic hot water 7 and space heating 8 subsystems.

- availability of a source of fresh water
- distances between the geothermal fluid source, discharge point, and source of fresh water.

Whether the spent geothermal fluid is discharged near the geothermal intake (source) or is reinjected into the reservoir, the central heat plant is located near the geothermal intake. In the absence of a reinjection, the central heating unit may be sited near the heat consumer. In any case, the heat network of the geothermal fluid will consist of a one-pipe arrangement to the central heat supply station, whereas the local heat distribution network may be done by means of a two- or four-pipe network.

In all systems using geothermal heat supply, the thermal potential of the geothermal fluid should be put to maximum use, while the flow rate of the geofluid must be reliably on hand throughout the year. Also it is necessary to ensure that the maximum value of the COP is provided for all operating conditions, along with the minimum specific consumption of geothermal fluid per unit of calculated thermal load.

Figure 8.15 depicts an open geothermal heat supply system. The geothermal fluid is pumped from well 1, cooled in the heating system 2, and fed to the domestic hot water system 3. If it is necessary to boost the geofluid temperature, a boiler 4 or a heat pump 5 with evaporators 9 and condensers 8 can be used. The evaporators cool the waste geofluid prior to disposal, whereas the condensers can supply heat for the lower-temperature domestic hot water uses 3, such as tap water and bath/showers. Depending on the temperature of the geothermal heat carrier, the peak boiler operates either year-round or part of the heating season. The peak boiler works as long as the amount of heat supplied by the geothermal fluid is less than the required heat output of the heating system.

The heating system in Fig. 8.16 is closed; the geothermal fluid produced from wells 1 is fed to consumers via a pumping station 3. The geothermal fluid flow control is provided by the flow control valves 11 and regulator 12. Part of the geothermal fluid is used for heating greenhouses 4. The remaining part is used to heat fresh water in a 2-pipe heating system using a heat exchanger 7. If necessary, the supplied mains water is heated in condensers 8 of the heat pump unit HPU. After cooling, the geothermal fluid

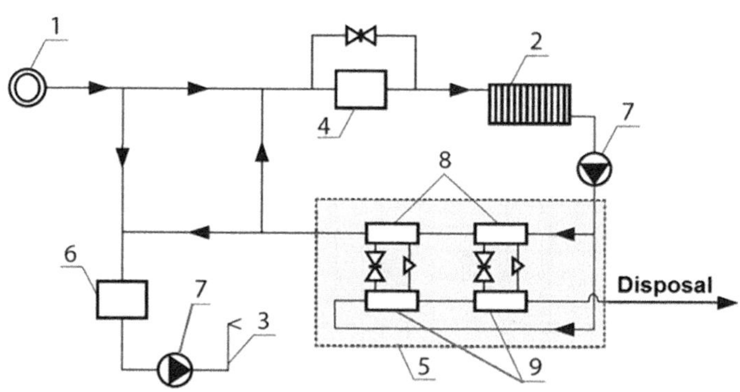

Fig. 8.15 Typical one-pipe district heating system with a peak boiler and heat pumps. *1*, production well; *2*, heating system; *3*, system hot water supply; *4*, peak boiler; *5*, heat pump installation; *6*, tank/accumulator; *7*, pump; *8*, condensers; *9*, evaporators.

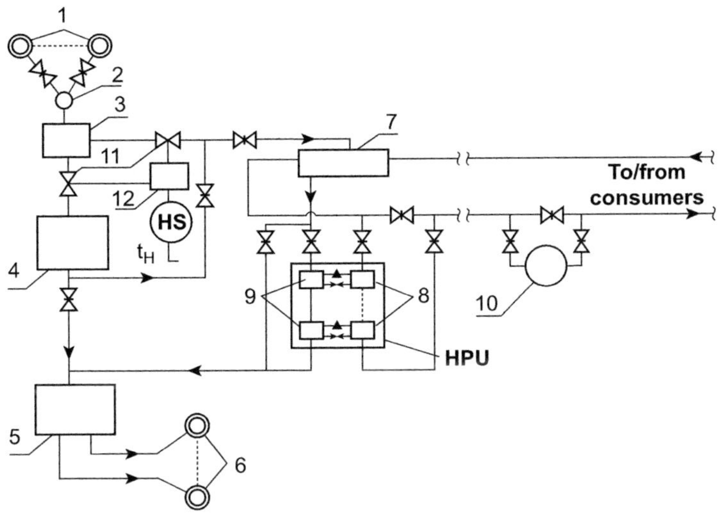

Fig. 8.16 Schematic of a two-pipe geothermal heat supply system with heat pump units (HPUs). *1,* geothermal wells; *2,* water collection tank; *3,* pumping station; *4,* greenhouse heating; *5,* reinjection pumping station; *6,* reinjection wells; *7,* network heat exchanger; *8,* condensers; *9,* evaporators; *10,* network storage; *11,* flow control valves; *12,* regulator.

is pumped into wells 6, rendering the geofluid circuit closed because all of the produced geofluid is reinjected and none is disposed to the surface.

The efficiency of heat pump installations increases with the use of low-temperature heating systems, or with sequential throughput from several units. One can improve the technical and economic performance of a geothermal heat supply system by integrating several geothermal systems, the heat consumers of which may be space heating systems of public, commercial and industrial buildings, industrial systems providing the technological needs of production, as well as greenhouses, car washes, laundries, hospitals, sanatoriums, and swimming pools.

Thermoeconomic analysis of the effectiveness of various geothermal schemes the heat supply shows that the efficiency of the systems depends essentially on the ratio of the domestic hot water supply load and space heating load.

8.4.2 Cogeneration and heat pumps

Often called "distributed energy systems," mini-thermal cogeneration power plants reduce the cost of energy consumed, provide a significant economic effect for final consumption, and solve the problem of peak loads, shortcomings of monopoly systems, and thus provide the consumer with high-quality, uninterrupted power supply.

At present, there are many arguments for choosing distributed cogeneration technologies: low cost of heat and electricity; proximity to the consumer; no need for expensive substations; environmental safety; mobility; ease of installation; and many other factors. An important factor in choosing cogenerators is their environmental safety, since the plants have low emissions of toxic substances into the atmosphere.

Power units of either gas reciprocating or gas turbine engines for gas-fired thermal electric stations are readily available commercially. Types of gas fuel used for thermal power plants include natural gas, propane, butane, flare gas, biogas, coal mine gas, coke oven gas, etc. The thermal power ratings of gas cogeneration plants range from 0.02 to 6 MWt, while the electric power ratings range from 0.5 to 8 MWe. Also available are micro-turbine generators capable of producing both electrical power and heat. One manufacturer, Elliott, manufacturers a micro-turbine with an electrical power of 100 kW and a thermal power of 172 kW that uses water as the heat transfer fluid. Heat transfer with secondary water produces water at 40 °C for hot water supply and at 70 °C for heating (Fig. 8.17).

In certain geothermal reservoirs called geopressured systems, dissolved combustible gases are contained in geofluids and can be used as fuel in cogeneration plants. The amount of dissolved gas increases with increasing depth. Such reservoirs are found in the waters of oil and gas basins where the volume of dissolved gas can reach 10 m^3 or more per 1 m^3 of formation water, after the gas is separated and reduced to atmospheric pressure. Although the geochemical composition of the dissolved gases varies widely, hydrocarbon gases predominate at 60–95%, sometimes reaching 98%, with the main share being methane. The concentration of heavy fractions (propane, butane, etc.) amounts to about 2–10%. The saturation pressure of dissolved gases in formation waters varies from 8 to 40 MPa and higher in rare cases. These gases are coproduced with the high-pressure, moderate-temperature geothermal fluids without the need for well pumps, i.e., by artesian flow.

The use of gas-containing thermal waters comes with certain difficulties. Hydrogen sulfide dissolved in water and free carbon dioxide cause intense metal

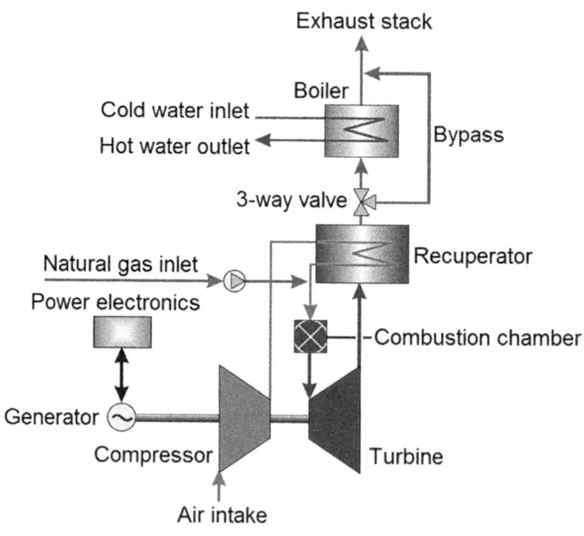

Fig. 8.17 Flow diagram for an Elliott TA 100 R CHP micro-turbine cogeneration unit [14].

corrosion. Degassing the thermal waters as soon as possible close to the well improves plant operation by removing the causes of scaling and active corrosion. However, the presence of gases in thermal water significantly increases their energy potential. With a gas factor of about 10 m^3 gas/m^3geofluid, the energy equivalent of the gas can be equal to the potential of thermal waters. By itself, the extraction of hydrocarbon gases from deposits of thermal waters is economically unfeasible because of the low concentration of hydrocarbons and methane in the water as a result of their poor solubility. Thus, separating the gases for separate sale and distribution is economically prohibitive. However, on-site use in cogeneration plants may be worthwhile.

The first cogeneration geothermal plant in Ukraine was developed and implemented in Medvedivka, Crimea (Fig. 8.18). This prototype plant provides 717 kWt of thermal power from a gas piston unit and 60 kWe of electric power [15].

Hypothetically, if a heat pump were to be added to the Medvedivka system, the heat output would increase from 717 to 1000 kWt, and the temperature of the water in the heating system would rise from 50 to 75 °C (Fig. 8.19). Various technological schemes are possible: gas piston engine; gas turbine engine; combined hydropower turbines. Thus, the combination of an autonomous cogeneration power plant with a geothermal heat pump installation provides highly efficient heat and electricity to consumers.

8.5 Solar and geothermal energy hybrid power systems

The general theme of this book is the enhancement of low-temperature energy resources. Geothermal energy is primarily in that category. Although there are

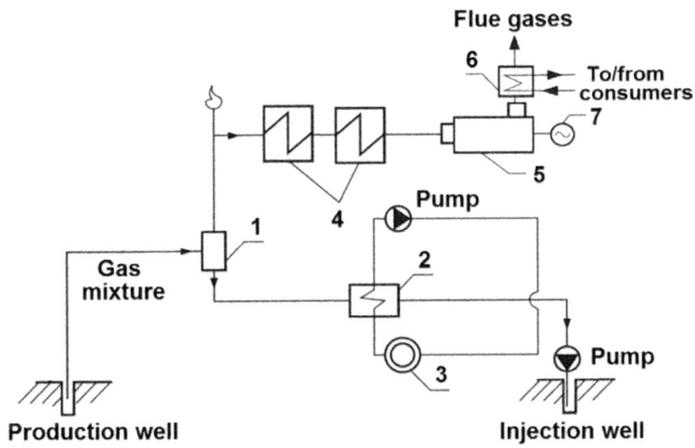

Fig. 8.18 Diagram of a cogeneration geopressured geothermal installation. *1*, separator; *2*, heat exchanger of the heating system; *3*, consumer of heat; *4*, gas preparation system; *5*, gas engine; *6*, waste heat recovery heat exchanger; *7*, generator.

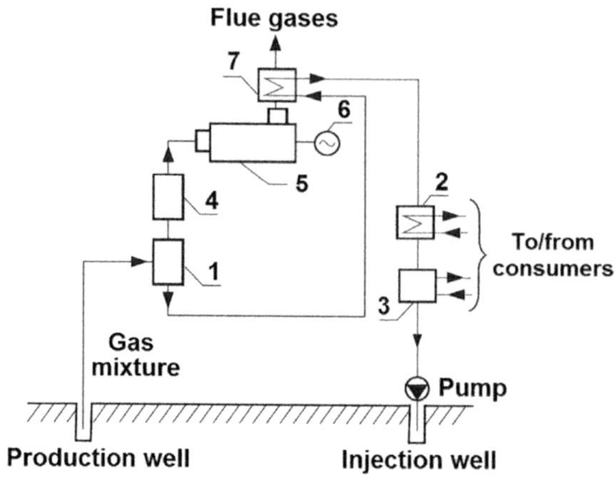

Fig. 8.19 Diagram of a geothermal installation with gas piston engines. *1,* gas separator; *2,* heat exchanger; *3,* heat pump installation; *4,* gas treatment system; *5,* gas piston engine; *6,* electric generator; *7,* flue gas heat exchanger.

high-temperature geothermal systems associated with volcanic systems, the vast majority of geothermal resources are at the lower end of the temperature spectrum. When used to generate power, these resources are exploited most often in conjunction with organic Rankine cycles (ORCs). However, with only a small temperature difference available between the heat source and the heat sink (i.e., ambient conditions), only very low thermal efficiencies can be achieved, despite many clever schemes designed to improve their performance.

In this section we consider combining two kinds of energy resources in a hybrid arrangement aimed at producing high efficiency and high sustainability. One energy source will be geothermal while the other will be selected from several options to achieve this goal.

8.5.1 Geothermal and non-renewable energy sources

From the earliest days of geothermal power plants, the notion of combining fossil fuel with geothermal was put forth. In 1924 Caufourier proposed enhancing a low-temperature geothermal flash plant by superheating the geothermal steam in a fossil-fueled boiler [16]. The plant idea was quite advanced for its time and incorporated a 4-stage flash process on water obtained from a hot spring [17]. It is not known whether the concept was ever implemented, but it seems unlikely.

In 1961 Hansen proposed using hot geothermal water in a conventional fossil-fired power plant to replace the low-temperature feedwater heaters [18]. This idea was adopted and expanded by the research group at Brown University (Providence, RI, US), and an engineering feasibility study was carried out for a large-scale coal plant in Utah. Ultimately, a large coal plant was built without the geothermal augmentation mainly because of difficulties in obtaining permits from the three states involved, California, Utah and Nevada, in the US [19,20].

In the early 1980s, ideas were generated to use fossil fuels to enhance geothermal power plants. In California, at The Geysers dry steam field, the first power plant was installed in 1960, and in 1984 a study was conducted to assess the feasibility of using natural gas to superheat the dry, slightly superheated steam from the field [21]. Although this study did not lead to a hybrid plant, later a kind of hybrid system was tried out wherein the hydrogen sulfide that was collected in the noncondensable gases was burned to provide some measure of superheat. All of this work was done before the world became invested in reducing carbon dioxide emissions. Nowadays, none of these proposals, regardless of their technical or economic advantages, would be viewed favorably owing to the negative environmental consequences.

8.5.2 Geothermal and other renewable energy sources

The possibility of combining geothermal energy, particularly low-temperature resources, with other renewable energy sources holds great promise. If synergistic applications can be devised, such systems would provide power at good efficiency, on a sustainable basis, and without damage to the environment. However, there are challenges to carrying this out owing to the intermittent nature of one of the most compatible renewable resources for use with geothermal, i.e., solar. Geothermal plants operate best as pure base-load plants, i.e., on a 24/7/365 basis. Solar is available only when the sun is shining, on average half of the time. In the following subsections, we will examine how solar and geothermal may be combined effectively.

Wind is rapidly growing worldwide as a renewable resource for electric power, but it is a non-thermal source and also intermittent, making it problematic as a partner with geothermal except as a stand-alone supplementary power generator in geothermal fields with high wind potential. Hydroelectric plants are renewable but also non-thermal, and therefore difficult to combine with geothermal. Biomass holds more hope as a hybrid partner with geothermal, and will be discussed in Section 8.6.

8.5.3 Solar-geothermal binary hybrid plants

In this and the next section we consider how solar, both photovoltaic (PV) and concentrated solar power (CSP), may be combined with geothermal power plants. In this section only geothermal binary cycles (ORCs) will be discussed.

On the one hand, since PV is not a thermal process but the direct conversion of sunlight into electricity, there is no possibility of combining this thermodynamically with an ORC. However, PV systems can be installed next to ORC plants to provide power for auxiliary equipment such as pumps and compressors, thereby allowing a higher net power from the ORC. The deciding factor on the feasibility of such a scheme is the relative cost of the PV power versus the value of ORC power. Most often such a system is retrofitted onto an existing ORC so that the full cost of the PV system must be considered in establishing the cost of power from the PV. At the present time, unless there are special circumstances prevailing regarding the ORC, such as resource limitations or lower output than contractual obligations, the economics do not favor this arrangement.

On the other hand, CSP does lend itself to thermodynamic integration with ORC plants. The collection of solar thermal energy is accomplished usually by means of parabolic trough collectors (PTCs) fitted with a fluid-filled tube running along the axis of the trough and located at the focal line. A suitable heat transfer fluid is pumped and circulated through the tube, extracting solar energy and delivering it to, say, the ORC working fluid by means of a heat exchanger. Alternatively, the heat could be used to raise the temperature of the geofluid prior to its use in the ORC heat exchangers. Furthermore, it conceivably might be used to reheat the geofluid leaving the ORC, allowing it or a portion of it to be recycled through the ORC. Figures 8.20−8.22 illustrate these concepts in highly simplified flow diagrams.

In all cases, when the sun is not shining the power plant reverts back to its original ORC design, absent some type of thermal storage. For a new integrated plant, a choice must be made regarding the cycle specifications so that equipment can be manufactured to meet the design conditions. This includes sizing of the heat exchangers and turbine blading and flow path. If the design is for "no sun" then

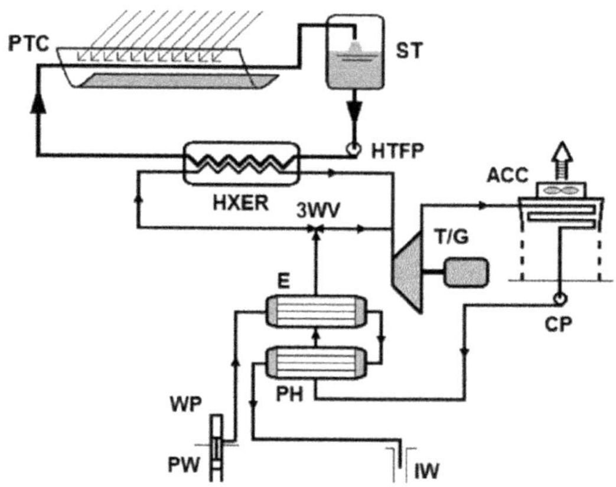

Fig. 8.20 Solar-augmented ORC: working fluid superheat [22].

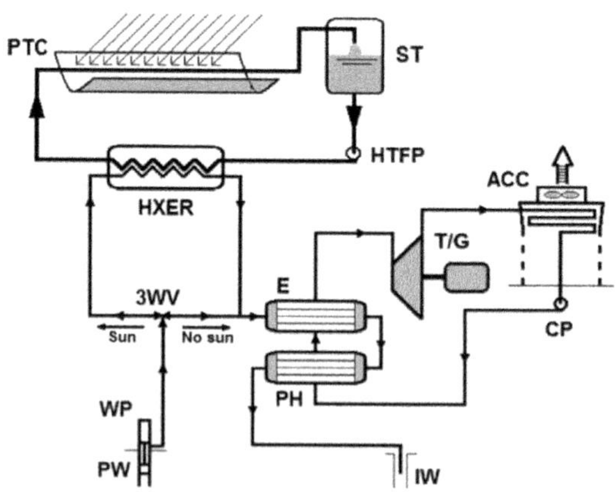

Fig. 8.21 Solar-augmented ORC: brine preheat.

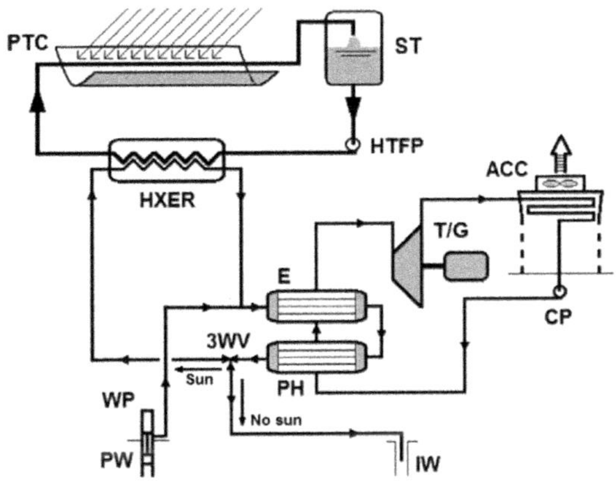

Fig. 8.22 Solar-augmented ORC: brine postheat.

when the sun is shining, the plant will produce more power than design but not the optimum amount. If the design is for "sun" then the plant will perform less well when sun is not available. Some compromise must be decided upon based on the long-term electrical generation and the cost-effectiveness of the combination.

For a retrofit of a solar system to an existing ORC, it will turn out that the solar system will produce cycle conditions that are off-design relative to the original plant, and thus the maximum benefit of the solar energy will not be achieved. However, if the plant was not performing up to the design conditions for any reason (cooler geofluid, lower geofluid flow rate, etc.), then the solar addition can be designed to return the plant to its design conditions, or perhaps improve on them.

A notable example of a hybrid geothermal-solar power plant is the Stillwater plant in Nevada, US [22]. This plant added both a parabolic CSP and a PV system to the original ORC plant with the goal of mitigating a loss of output caused mainly by the drop in geothermal fluid temperature since the plant began operating in 2009. The CSP system is similar to the schematic shown in Fig. 8.21. Given that the Stillwater site enjoys abundant sunlight year-round, the solar augmentation has proved to be effective. Furthermore, the plant has an air-cooled condenser that suffers dramatic declines in performance during hot summer days; the solar system counteracts this loss, as Fig. 8.23 shows for a typical spring day [23]. The PV portion is a stand-alone system that produces electricity for sale independently of the CSP-geothermal hybrid plant.

8.5.4 Solar-geothermal flash-steam systems

Whereas it is conceivable that solar energy could be added to geothermal flash-steam plants, no such plant has been built to date. One reason that this is less attractive than the ORC case described above is that flash plants do not employ air-cooled condensers and thus are not subject to the magnitude of power loss experienced by ORC in hot

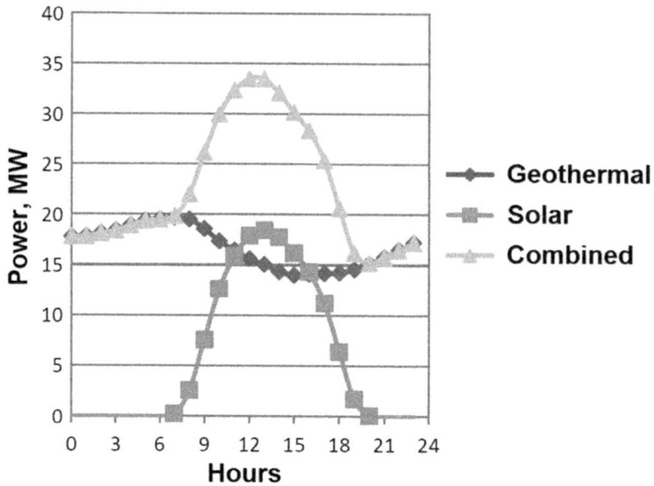

Fig. 8.23 Power output for Stillwater geothermal-solar power plant [23].

weather. Nevertheless, tests of hybrid solar-geothermal flash plants have been carried out [24–27].

Figure 8.24 shows four different ways solar energy might be used in a double-flash plant. The parabolic trough collector (PTC) heats the heat transfer fluid (HTF) for use in a heat exchanger (HXER). In the view presented, the HXER is placed between the flash vessel (F) and the turbine-generator (T/G). Thus, the low-pressure steam is boosted in temperature and superheated prior to entering the mixing chamber of the dual-admission turbine. The 3-way valve (3WV) controls the percentage of the flash steam that is sent to the HXER. Obviously, all the steam goes directly to the turbine when the sun is not shining, absent any thermal storage.

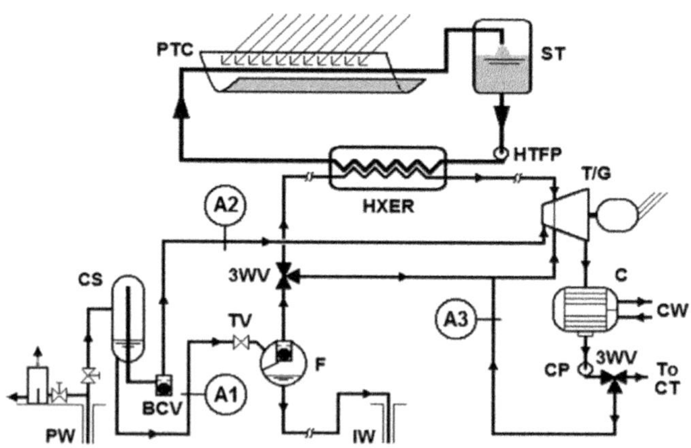

Fig. 8.24 Systems for incorporating solar energy with a double-flash geothermal plant [22].

The locations A1, A2 and A3 are alternative places for the HXER. A1 heats the brine coming from the main separators prior to being flashed. This results in more steam upon flashing and generates more power. A2 superheats the saturated steam coming from the main separator which allows for higher turbine efficiency owing to the absence of liquid during the high-pressure stages of the expansion. A3 can be designed to boil a portion of the steam condensate that can be added to the low-pressure steam going to the pass-in chamber of the turbine. Since there is always excess water at the cooling tower (CT), not all of the condensate needs to be sent to the CT. It is possible to design a system with any or all of these alternatives, although the complexity and cost of using multiple heat exchangers may work against this idea. These options can be analyzed and optimized for whatever conditions are found at a particular site, and if found economically viable, could be constructed to gain synergy between geothermal and solar energy.

8.6 Biomass and geothermal energy hybrid power systems

Biomass and geothermal are both considered renewable sources of energy. Combining them in a hybrid power plant furthers the policy of making the electric supply system sustainable. This section discusses several possible configurations of plant and the conditions under which such hybrid systems make technical and economic sense.

Biomass encompasses a wide variety of grown products, byproducts, and waste material from agricultural activity. As an example, the logging industry produces a high volume of waste in the form of bark, branches, and other tree material that cannot be cut into lumber. Furthermore some fast-growing species may be planted and harvested specifically for use as fuel in combustion boilers, if the economics are favorable. Another example is the sugar cane industry that produces large amounts of waste after the canes have been processed for their sugar. At sugar plantations it is normal to burn the wastes to raise steam for an on-site power plant.

There is one essential consideration when one contemplates designing a hybrid biomass-geothermal plant, namely, the co-location of the biomass source and the geothermal field. Since the geothermal energy, particularly in the form of steam or steam-water mixtures, cannot be shipped over long distances without heavy thermodynamic losses, the biomass must be available within a reasonable distance from the geothermal site. Fortunately there are many places around the world where this condition is met. One of the most important is New Zealand's North Island which is home to numerous geothermal power plants and a major logging industry [28]. Several plants are close to, if not surrounded by, tree plantations. Figure 8.25 shows an aerial image of the Ngatamariki geothermal power plant and its surroundings. Note the proximity of the Tahorakuri Forest and the numerous tree-harvesting sites in the area (small, light-colored clearings).

There are forests in the neighborhoods of nearly all geothermal plants in the region north of Lake Taupo. Figure 8.26 shows nine plants and the vast Kaingaroa Forest;

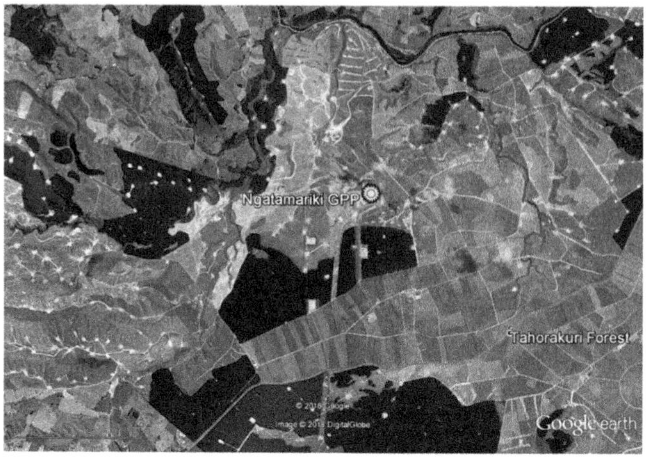

Fig. 8.25 Ngatamariki geothermal plant and surrounding forests; Google Earth image, January 10, 2015.

numerous smaller forests are not shown. At 188,000 ha areal extent, Kaingaroa is the largest plantation forest in the world. Sustainable harvesting is required for all logging in New Zealand.

Another example of a hybrid biomass-geothermal plant is at Larderello, Italy, the oldest geothermal field generating electricity, dating back to 1904. One of the two 20 MW units at Cornia has been converted from simple dry steam operation to hybrid operation with a biomass-fueled superheater. Since the geothermal steam supply in recent years had been insufficient to maintain full capacity of both units, one unit was converted to serve as the biomass boiler while the second continued to generate power, augmented by the superheater [29]. Figure 8.27 shows an aerial view of the power station and an adjacent PV plant.

Thain and DiPippo [28] presented several schemes for combining forest waste with geothermal power plants in New Zealand. Figure 8.28 shows how a biomass furnace might be used at the Rotokawa steam-binary combined cycle power station to superheat the saturated steam passing from the cyclone separator before it enters the steam turbine. The process state diagram in temperature-entropy coordinates is given in Fig. 8.29.

The net power generated by the basic system is 26,860 kW, whereas with the biomass boost the new power rises to 35,400 kW, an increase of 32%. The incremental power output relative to the heat supplied by the biomass combustion amounts to an equivalent thermal efficiency of 44%, much higher than can be obtained in a simple Rankine power cycle using the biomass heat. The example is theoretical owing to the fact that the basic plant exists and would need extensive modifications to the steam turbine and other components to accommodate the biomass superheater. However, the exercise demonstrates the advantage that can be gained by combining the two energy sources in this way, and could be used in new plants under consideration.

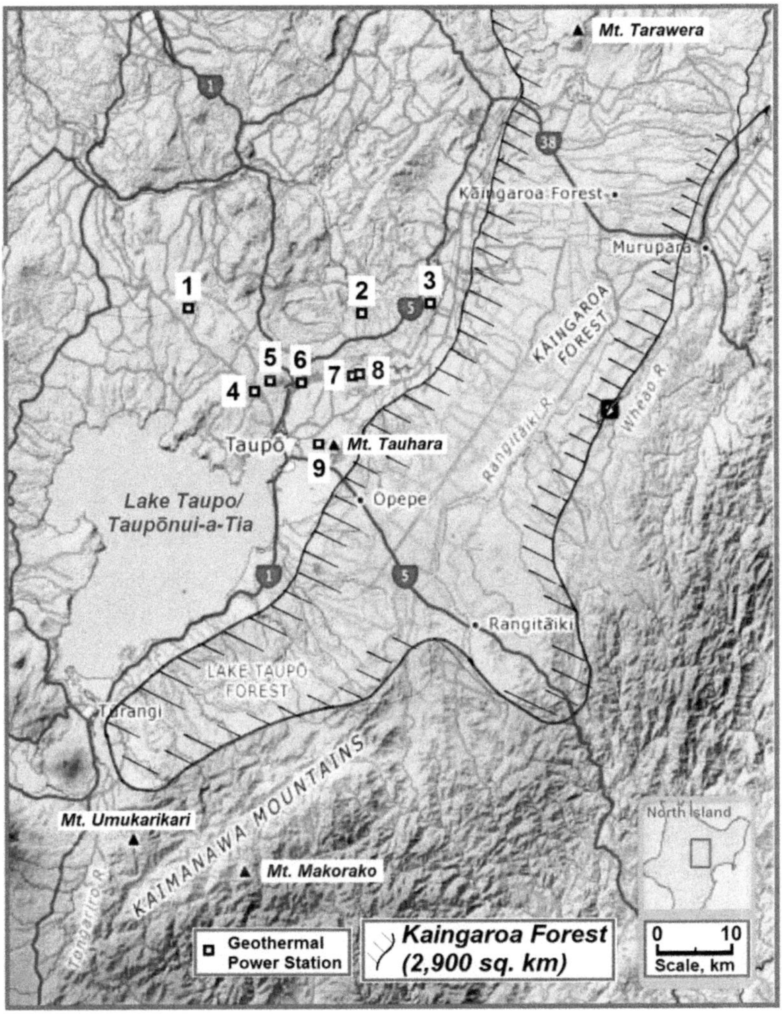

Fig. 8.26 Nine power plants in close proximity to the Kaingaroa Forest [28]. Key to power plants. *1,* Mokai I & II, 95 MW; *2,* Ngatamariki, 80 MW; *3,* Ohaaki, 65 MW; *4,* Poihipi, 55 MW; *5,* Te Mihi, 159 MW; *6,* Wairakei, 132 MW; *7,* Nga Awa Purua, 140 MW; *8,* Rotokawa, 32 MW; *9,* Te Huka, 23 MW.

8.7 Optimum operating conditions for hybrid fuel-geothermal heat pump

8.7.1 General considerations

The constant temperature of the geothermal heat-carrier at the production well outlet causes a number of difficulties in the operation of a geothermal heat supply system.

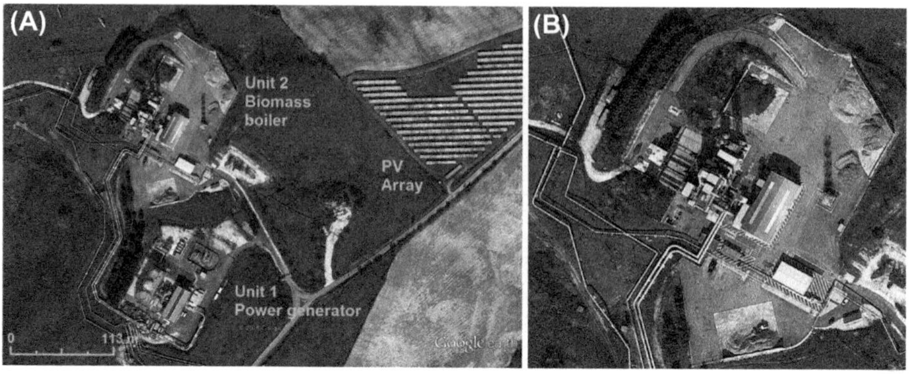

Fig. 8.27 (A) Cornia geothermal-biomass hybrid power station, Larderello, Italy; (B) Detail of Unit 2 used as the biomass steam generator that augments the geothermal steam Unit 1; modified Google Earth images, March 8, 2017.

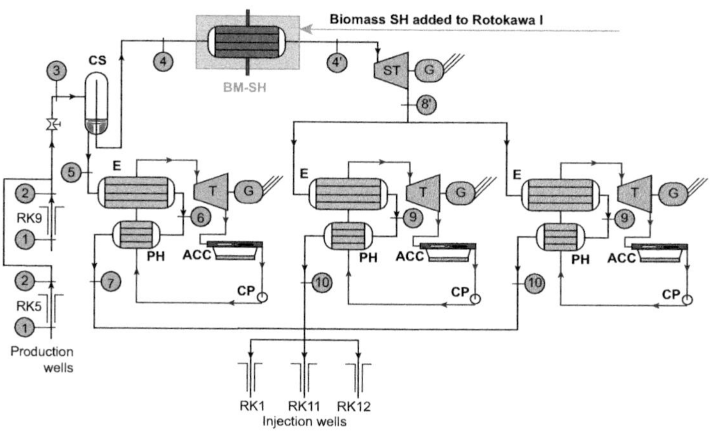

Fig. 8.28 Rotokawa I geothermal power station with proposed biomass augmentation [28].

With constant flow of a producing well during the heating season, as well as the non-heating season, the temperature potential of spent geothermal water that is returned to the formation can remain quite high. In this case, the system will have a low heat utilization factor, which leads to a low thermodynamic efficiency of the geothermal heat supply system. Thus, it is very difficult technically to be able to comply with the requirement for heat generation which depends on the temperature of ambient air by quantitative regulation in geothermal heat supply systems.

In the case where there is a high level of mineralization of the geothermal waters and the possible presence of dissolved aggressive gases, then it is not feasible to use the geothermal heat-carrier directly in heating and hot water supply systems. By using heat exchangers, the geothermal fluid transfers heat to clean water raising its temperature to the required level for the application, while reducing the temperature of the

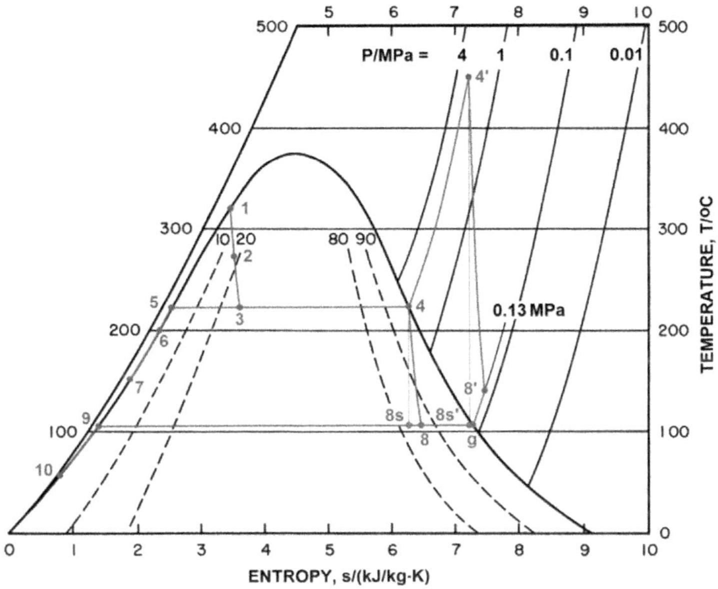

Fig. 8.29 Temperature-entropy diagram for hybrid system shown in Fig. 8.28 [28].

geofluid to about 5−10 °C above the saturation point for precipitation of the dissolved minerals.

Under certain extreme ambient conditions where the temperature of the thermal water is lower than required to meet the heating needs in the supply pipeline, then it is necessary to design a peak heating system such as an electric or gas-fired heat generator.

When combining two sources of thermal energy at a geothermal heat station, i.e., geothermal energy and, for example, natural gas, the distribution of heat generation from each source will depend on the adopted temperature chart of the heat supply system, i.e., on the heat required as a function of the ambient temperature.

8.7.2 Systems analysis

Figure 8.30 shows the scheme of a hybrid fuel and geothermal heat station (HFGHS) that provides thermal energy for heating and hot water supply systems and includes an additional peak heater, e.g., a heat generating plant operating on natural gas. Here it is assumed that the thermal load of the hot water supply system Q_{hws} is completely covered by geothermal energy, and Q_h heating systems are from two sources of thermal energy. The periods of operation of a geothermal source and a heat-generating plant for a heating system, both individually and jointly, are determined by the adopted temperature chart, the thermal power of the heating system, the flow rate and temperature of the geothermal fluid at the producing well outlet, and the climatic conditions of the surroundings.

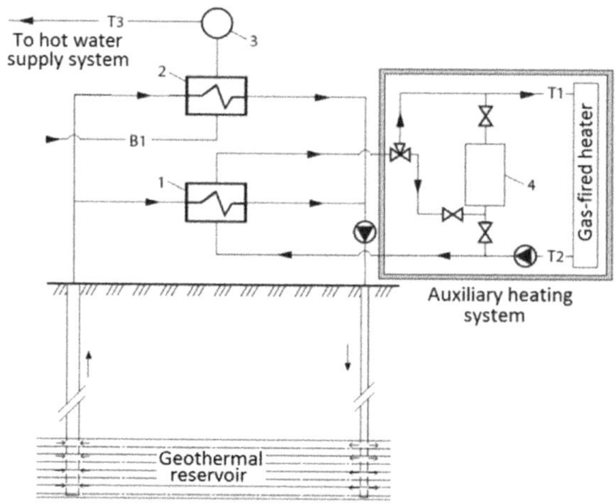

Fig. 8.30 Schematic of a hybrid fuel and geothermal heat station. *1,* heat exchanger of a space heating system; *2,* heat exchanger of a hot water supply system; *3,* tank/accumulator of a hot water supply system; *4,* heat storage tank.

It should be noted that the regulation of heat generation directly from the geothermal heat-carrier to the fresh water heat-carrier of the heating or hot water supply system is a quantitative one. The required mass flow rate of a geothermal fluid for a heating system with a counterflow heat exchanger and a minimum allowable temperature difference (pinch-point) of 10 °C between the heat-carriers, i.e., the one that provides the heat (geofluid) and the one that is heated (heat transfer fluid), will be determined by the following equations:

$$\dot{m}_g(t) = \frac{Q_0(t) \cdot 10^3}{c\left[t_{g1} - t_{g2}(t)\right]} \tag{8.2}$$

If $\tau_1(t) \geq t_{g1} - 10$, then the heat load of the heating system is provided exclusively by geothermal energy;

$$\dot{m}_g(t) = \frac{Q_{0\max}\left[t_{g1} - 10 - \tau_2(t)\right] \cdot 10^3}{c(\tau_1 - \tau_2)\left[t_{g1} - t_{g2}(t)\right]} \tag{8.3}$$

If $\tau_1(t) \geq t_{g1} - 10$, $\tau_2(t) \geq t_{g1} - 10$, then the coverage of the heating system load is ensured by the combined operation of the GCS and an auxiliary heat generating plant;

$$\dot{m}_g(t) = 0 \tag{8.4}$$

If $\tau_2(t) \geq t_{g1} - 10$, then the thermal load of the heating system is provided exclusively by the auxiliary heat generating plant.

The terms in these equations have the following meanings:

t = ambient air temperature, °C
t_{g1} = geofluid temperature at outlet of the producing well, °C
t_{g2} = geofluid temperature at outlet of the heating system heat exchanger, °C
τ_1, τ_2 = maximum temperature of the heat transfer fluid in the heating system supply and return pipelines, respectively, in accordance with the temperature chart, °C
$\tau_1(t)$, $\tau_2(t)$ = temperature of the heat transfer fluid in the heating system supply and return pipelines, respectively, depending on the temperature of ambient air
$\dot{m}$ = mass flow rate of geofluid, kg/s
Q_{0max} = maximum thermal power of the heating system, MWt.

Heat balances of the hot water supply and heating system heat exchangers as well as Eqs. (8.2)–(8.4) can be used to predict the heat generation modes of the hybrid fuel and geothermal heat station with different variants of the heating system temperature chart and heat power of the heat supply system.

8.7.3 Case study: 1 MWt system in Dzhankoy, Crimea

As an example of this analysis, a case study was made of the modes of heat generation from a hybrid fuel-geothermal heat station with a maximum heat power of 1 MWt: space heating load being 0.75 MW; hot water supply load being 0.25 MW; plant site was the town of Dzhankoy in Crimea; and the temperature of the geothermal fluid at the well outlet was 64 °C.

As can be seen from Fig. 8.31, auxiliary heating becomes necessary when the ambient temperature falls below +2 °C, for a 90–70 °C temperature chart of the heating system Case (a), and below −5 °C for a 70–50 °C temperature chart Case (b). Also, as the water temperature the supply pipe decreases, the required output power of the peak heater and its operating time likewise decrease. Note: the notation "T_1–T_2 °C temperature chart" is the heating requirement as a function of outdoor temperature when T_1 and T_2 are the temperatures in the supply pipe and return pipe, respectively.

The maximum heat output of the auxiliary peak heater in Case (a) was 750 kW, in Case (b) it was 648 kW. Taking into account that the average ambient temperature for the heating period is 1.5 °C for the town of Dzhankoy, the duration of the peak heater operation with a temperature chart of 70–50 °C will be less than with a temperature chart of 90–70 °C.

The heat output of a geothermal source to cover the thermal load on the heating system is presented in Fig. 8.32.

It is interesting to note that for Case (a), the geothermal source can no longer cover the thermal load of this system once the ambient temperature falls to −4 °C and lower. The temperature of the geothermal heat-carrier is not high enough to carry out the heat exchange process with the heat-carrier of the heating system in accordance with the temperature chart. In Case (b) the geothermal source completely or partially covers the heat load of the heating system during the entire heating period.

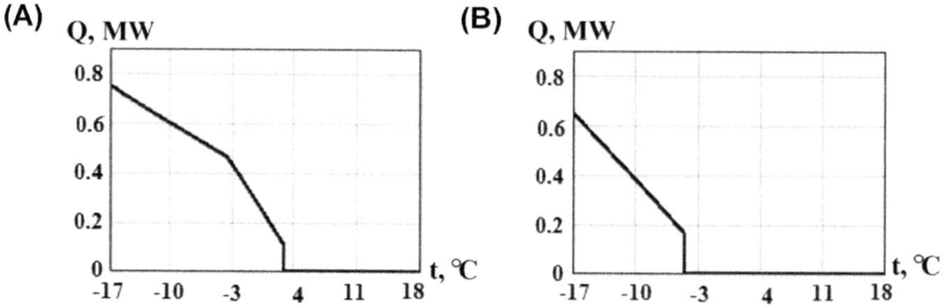

Fig. 8.31 Graphs showing the amount of auxiliary peak thermal power needed as a function of ambient temperature to satisfy the heating demands as determined by temperature charts for (A) 90–70°C; (B) 70–50°C.

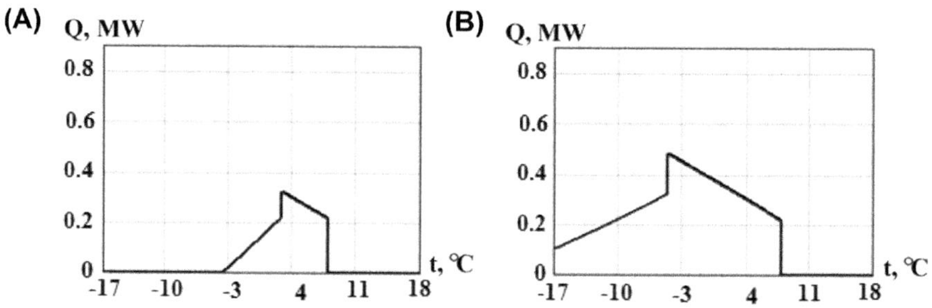

Fig. 8.32 Graphs showing the amount of geothermal heat needed as a function of ambient temperature to satisfy the heating demands as determined by temperature charts for (A) 90–70°C; (B) 70–50°C.

Figures 8.33 and 8.34 present the variation in the temperature of the geofluid leaving the heat exchanger of the heating system and at the inlet of the reinjection well, respectively, as functions of ambient temperature.

For Case (a), higher temperatures of the discharged geofluid are typical, which causes a relatively lower efficiency of using geothermal energy than for Case (b). The residual heat of the geofluid after the heating system heat exchanger can be utilized in preheaters for heating the hot water supply system or using heat pumps as an additional source of heat for the heating system.

According to the graphs of heat consumption for the HFGHS (Fig. 8.35), during the heating period, geothermal energy supplied 76.5% of the total heat load of the heat supply system for Case (a); for Case (b) that percentage rose to 97.5%. In other words, for a 70–50 °C temperature chart, the consumption of natural gas for the generation of thermal energy will be lower by 21% compared with a 90–70 °C temperature chart. The analysis presented shows the significance of the adopted temperature chart for the heat generation modes from the HFGHS.

Two versions of the heating system are considered in the temperature charts of (A) 90–70 °C and (B) 70–50 °C (Fig. 8.35), where the temperature range for each case indicates the water temperature in the supply pipe and the water temperature in the return pipe. The left part of each graph shows the dependence of the heat load on the

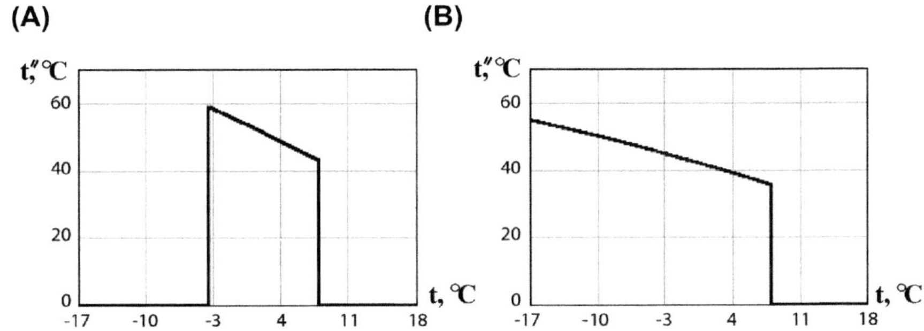

Fig. 8.33 Graphs showing the variation in the temperature of the geofluid leaving the heat exchanger of the heating system as a function of ambient temperature: (A) with a 90−70°C temperature chart; (B) with a 70−50°C temperature chart.

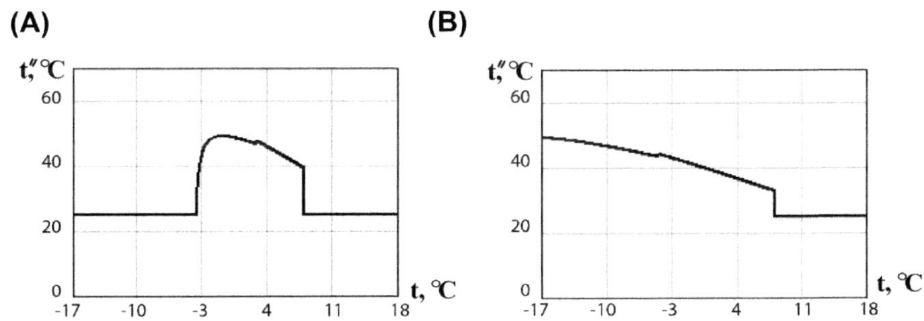

Fig. 8.34 Graphs showing the variation in the temperature of the geofluid entering the reinjection well as a function of ambient temperature: (A) with a 90−70°C temperature chart; (B) with a 70−50°C temperature chart.

outside temperature. These are called the heating curves, in that the power of the peak boiler and the power of the geothermal heating system are determined as a function of the outdoor temperature. The duration of the constant outdoor temperature is determined by climatic data. The right part of the graph shows the duration of heat loads, given as the number of hours of the heating period per year, out of a total of 8760 h (non-Leap Year).

To select the optimal modes of heat generation for a HFGHS, a comparison was made on the basis of exergy efficiency and energy costs (electric energy and natural gas) in the production of 1 GJ of thermal energy for the heating system at temperature charts: 90−70 °C, 70−50 °C, 50−35 °C and distribution of thermal loads: 75% - space heating, 25% - hot water supply.

Figure 8.36 shows the results, namely, the exergy efficiency of a HFGHS as a function of ambient temperature for various temperature chart scenarios. The results of the study show that the lower temperature charts correspond to optimal heat generation modes. These charts provide maximum values of exergy efficiency and minimum energy costs since a significant part of the thermal load or total load is covered by geothermal energy.

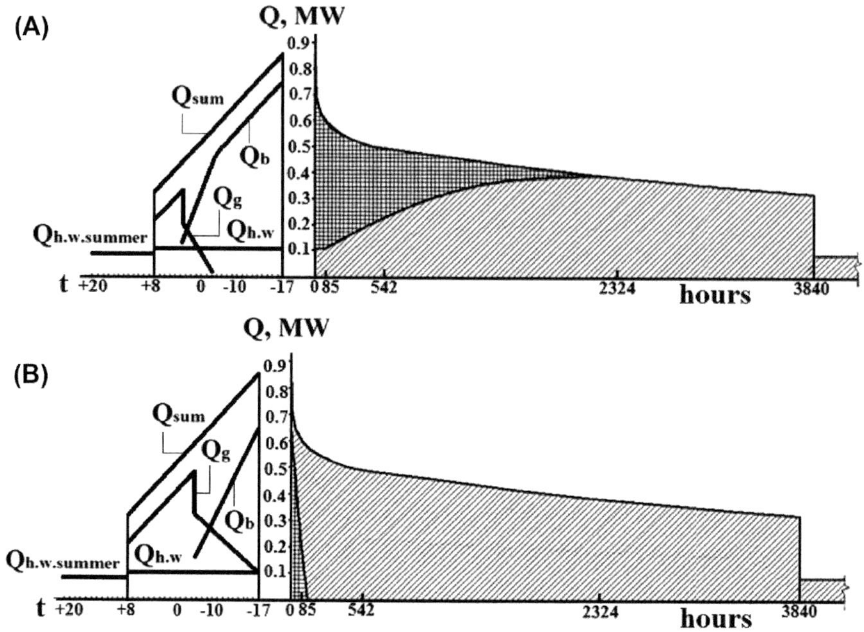

Fig. 8.35 Graphs of HFGHS heat consumption: (A) for a 90–70°C temperature chart; (B) for a 70–50°C temperature chart; square hatching is the part of the heat covered with a peak heater (gas boiler) and sloped hatching is the part covered with geothermal energy.

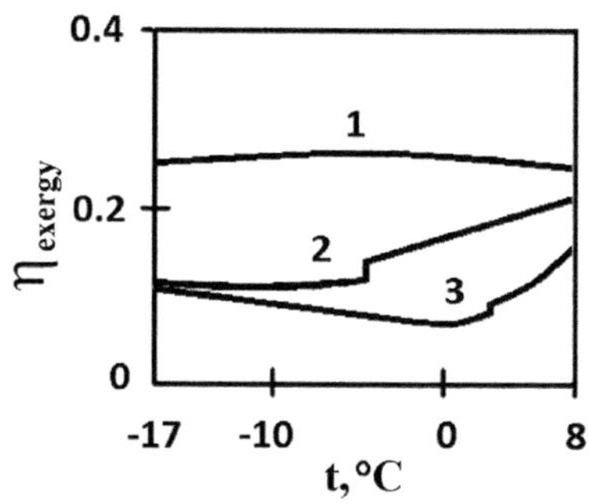

Fig. 8.36 Variation of exergy efficiency a HFGHS as functions of ambient temperature and temperature chart: 1. 50/35°C; 2. 70/50°C; 3. 90/70°C.

The exergy efficiency for systems with a temperature chart of 50–35 °C (line 1) exhibits a maximum at a value of 0.27 at an ambient temperature of −5 °C. For high-temperature heat supply graphs of 90–70 °C (line 3), the exergy efficiency is always lower than the other cases, and shows a minimum at about 0 °C. Thus, lower-temperature heating systems are more efficient, as we wrote earlier in Chapter 3, Section 3.5. If the outdoor air temperature stays at +8 °C for 5 days, then the heating season is over, and the system only delivers domestic hot water.

8.8 Summary

Synergistic benefits can be achieved by combining different kinds of energy resources in both thermal heating and electric power generation. Careful consideration must be paid to several important factors in deciding on the feasibility of a particular arrangement. This is especially critical when combining renewable energy resources because some of these can only be used close to their origin, such as geothermal and waste energy. Although biomass in principle can be shipped from its point of origin, economics often dictates that the energy be used near the biomass source. Therefore, co-location of the resources plays a vital role in the feasibility of a hybrid system. This chapter demonstrates that renewable and sustainable energy sources of geothermal, solar, hydro and biomass can be combined in various hybrid schemes to achieve higher utilization efficiencies than each of them used separately in individual plants. However, the practical and economic constraints placed on their use must be taken into account during feasibility studies prior to committing to a development.

Nomenclature

c	heat capacity
COP	coefficient of performance
e	effectiveness
J	heat supply coefficient
$\dot{m}$	mass flow rate
P	pressure
Q	heat power
t	temperature
V	volumetric flow rate

Greek letters

τ	temperature

Subscripts

C	compressor
CSP	concentrating solar parabolic trough systems
CHI	global horizontal irradiance
CHP	central heating plant
CST	concentrated solar thermal
CS	solar collector
COP	coefficient of performance
Cond	condenser
DNI	direct normal irradiation
E	evaporator
ET	expansion tank
GHE	ground heat exchanger
HE	heat exchanger
HP	heat pump

HPU	heat pump unit
HR	heated room
HWS	hot water supply
HFGHS	hybrid fuel and geothermal heat station
L	work
P	pump
PV	photovoltaic
ST	storage tank
TA	tank/accumulator

Review questions

1. Define a hybrid energy heating and cooling system.
2. What is the motivation to use hybrid heating systems?
3. Explain the use of waste energy resources in heating and cooling systems.
4. What combinations of low-energy or waste energy resources (sun, wind, geothermal, biomass, hydro) are used in hybrid energy systems?
5. What determines the duration of use of the peak boiler (heat pump) in heating systems during the year?
6. Explain why co-location of renewable or sustainable energy resources is important.
7. Discuss the pros and cons of using a non-renewable fuel such as natural gas in conjunction with renewable energy resource such as solar or geothermal.

Example

The heat supply arrangement shown in Fig. 8.E1 allows for greenhouse heating and domestic hot water for residential buildings. The water supply is the fresh drinking

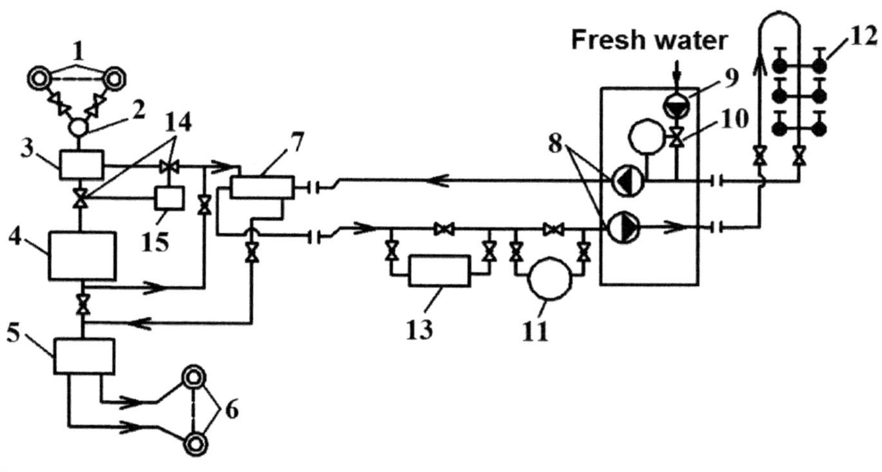

Fig. 8.E1 Integrated two-pipe geothermal heating system with a peak boiler. *1,* geothermal wells; *2,* intermediate collection tank; *3,* pumping station; *4,* heated greenhouse complex; *5,* reinjection pumping station;*6,* reinjection wells; *7,* network heat exchanger; *8,* network pumps; *9,* fresh water pump; *10,* flow regulator; *11,* network storage tank; *12,* domestic hot water users; *13,* peak boiler; *14,* control valves; *15,* regulator.

water for the settlement. Let us assume that the temperature of the water source is 15 °C in summer and 5 °C in winter. An organic-fueled boiler (biomass) located near the hot water consumers in the residential buildings serves to meet the peak demand for heat.

Given the quantitative regulation of the heating load during the year that is achieved by varying the thermal water flow rate, the modes of operation are as follows:

1) Summer period: A constant flow of thermal water with a temperature of 90 °C, providing heat to supply hot water at a temperature of 62 °C.
2) Beginning of the heating season when outdoor temperature falls to 8 °C and before firing up the peak boiler 2.9 °C: Thermal water flow is regulated depending on the heating load and geothermal energy provides the full heating needs of the greenhouse and hot water supply.
3) When the outdoor temperature falls below 2.9 °C: Thermal water flow changes depending on the outdoor temperature and provides the load for the greenhouse heating up to a maximum of 500 t/h. However, geothermal energy alone is unable to meet the needs of hot water demand, i.e., $Q_g < Q_{hw}$. Therefore, the hot water load is provided by the peak boiler. It receives the geothermal water after it leaves the greenhouses having a temperature of 50 °C. The requirements of the heat load are met by changing the heat output of the peak heat source.

Given data: Note: Climate data are for Kiev, Ukraine.

1) Number of degree-days = 3626
2) Duration of the heating period = 176 days
3) The duration of the period with average daily air temperature ≤ 0 °C = 103 days
4) Geofluid temperature, $t_g = 90$ °C
5) Return water temperature from the greenhouse heating system, $t_r = 50$ °C
6) Ambient air temperature, $t_a = -22$ °C
7) Fresh water temperature: summer, $t_w = 15$ °C; winter, $t_w = 5$ °C
8) Greenhouse internal air temperature, $t_i = 18$ °C
9) Geofluid mass flow rate, $\dot{m}_{gw} = 140$ kg/s (504 t/h)
10) Hot water temperature leaving the central heating plant $t_2 = 62$ °C
11) Average mass flow rate of domestic hot water: summer, $\dot{m}_{hw} = 75$ kg/s (270 t/h); winter, $\dot{m}_{hw} = 82$ kg/s (295 t/h)
12) Specific heat of water $c = 4.19$ kJ/kg·K

Analysis:

1. Temperature of the fresh water after the heat exchanger: $t_{hw} = t_r - \Delta t_h = 50-5 = 45$ °C
2. Effectiveness of hot water heat exchanger: $e = \frac{t_{hw}-t_w}{t_r-t_w}$

$$e = \frac{45 - 15}{50 - 15} = 0.857\ (summer)$$

$$e = \frac{45 - 5}{50 - 5} = 0.889\ (winter)$$

3. Heat exchanger $K \cdot F$ value:

$$KF = \frac{c\dot{m}_{hw}}{\dot{m}_{hw}/\dot{m}_{gw} - 1} \ln \frac{1 - e}{1 - e \cdot \dot{m}_{hw}/\dot{m}_{gw}}$$

$$KF = \frac{4.19 \times 75}{75/140 - 1} \ln \frac{1 - 0.857}{1 - 0.857 \times 75/140} = 0.900\, MW/^{\circ}C\ (summer)$$

$$KF = \frac{4.19 \times 82}{82/140 - 1} \ln \frac{1 - 0.889}{1 - 0.889 \times 82/140} = 1.213\, MW/^{\circ}C\ (winter)$$

4. Thermal power of the peak heat boiler: $Q_p = c\dot{m}_{hw}(t_2 - t_1)$where
$t_1 = t_0 - \Delta t_h$; $\Delta t_h = 10$ K— design return water temperature after network heat exchanger.
t_2 = fresh water temperature after peak heat generator, °C.
t_l = temperature of the return water after the greenhouse, °C.
$t_1 = t_0 - \Delta t_h$; $\Delta t_h = 10$ K— design return water temperature after network heat exchanger.

$$Q_p = 4.19 \times 82 \times (62 - 50) = 4.12\, MW$$

$$Q_{gw} = 4.19 \times 140 \times (90 - 50) = 23.5\, MW$$

$$Q_{hw} = 4.19 \times 75 \times (62 - 15) = 14.8\, MW\ (summer)$$

$$Q_{hw} = 4.19 \times 82 \times (62 - 5) = 19.6\, MW\ (winter)$$

5. Heat supply coefficient, corresponds to the on (off) peak heat boiler:

$$J_{heat} = 1 - \frac{Q_p}{ec\dot{m}_{hw}(t_g - t_r)} = 1 - \frac{4.12 \times 10^3}{0.889 \times 4.19 \times 82 \times (90 - 50)} = 0.66$$

6. The ambient temperature at which the peak heat generator is turned on:

$$t_{pb} = t_a + \frac{Q_p(t_i - t_a)}{ec\dot{m}_{hw}(t_g - t_r)} =$$

$$= -22 + \frac{4.12 \times 10^3 \times [18 - (-22)]}{0.889 \times 4.19 \times 82 \times (90 - 50)} = -8.5^{\circ}C$$

7. Schedule of regulation of the duration of geothermal fluid flow; see Fig. 8.E2:
Process 0–1 – heating of hot water in summer (16–62 °C) for $t_a = 8$ °C.
Process 1–2 – heating of hot water in winter in the heating period (5–62 °C) for $t_a = 8$ °C.
Process 2–3 – heating of hot water supply for $t_a = -9$ °C.
Process 3–4 – heating the greenhouse during the heating period (90–50 °C).

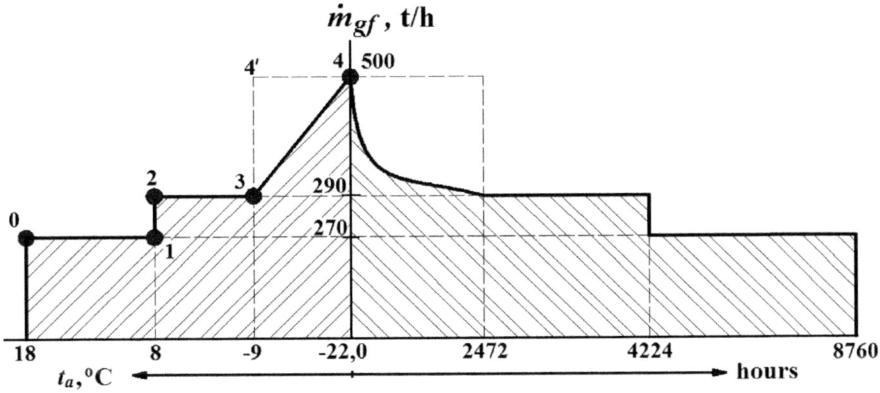

Fig. 8.E2 Regulation of the duration of the flow of geothermal fluid.

8. Heat-load/duration curve of the peak boiler; see Fig. 8.E3:

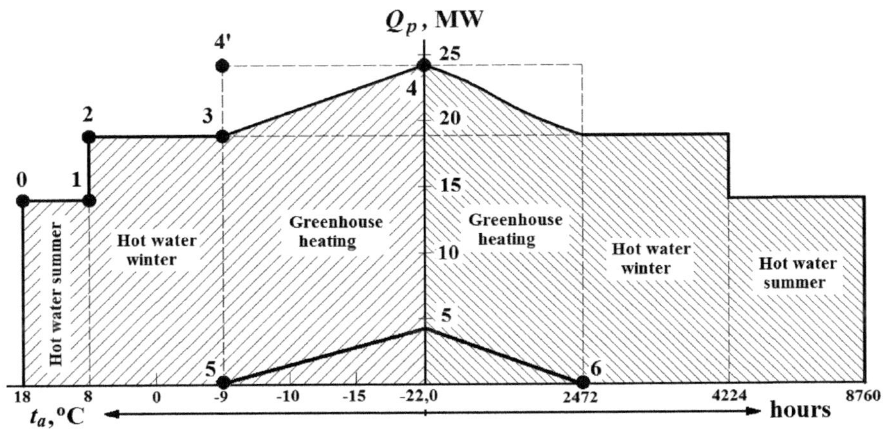

Fig. 8.E3 Heating curve and heating load duration curve; Process 5–6 is the peak boiler heat load.

An alternative means of regulating the heat load is also possible by supplying the maximum flow of geothermal water once the ambient temperature falls to −9 °C, and using it for the hot water system and the greenhouse heating system. Note: the dashed boxes in Figs. 8.E2 and 8.E3 represent the case when the maximum geothermal load is applied at an ambient temperature of −9 °C, and remains constant as the temperature drops to −22 °C.

As the air temperature drops, the thermal loads are redistributed. The share of greenhouse heating increases, and the peak boiler is turned on to ensure meeting the load of the domestic hot water system. Reducing the number of hours of use of the peak heater will improve the economic efficiency of the hybrid heating system. The current greenhouse heat load depends on the outdoor temperature and is found using:

$$Q = Q_{\max}\left[\frac{t_i - t_{a,current}}{t_i - t_{a,calc}^{t=-22^\circ C}}\right]$$

The duration of the heating period and the schedule of negative outdoor temperatures is determined by the HDD data which was explained in Chapter 3, Section 3.4, Table 3.2 and Ref. [20] in that chapter.

Exercises

1. Create a process flow diagram (system schematic) for a hybrid heating system using a geothermal heat source and a heat pump.
2. Create a process flow diagram (system schematic) for a hybrid heating system using a heat pump (heat source is sea water) and a solar collector.
3. Create a process flow diagram (system schematic) for a hybrid heating system using a cooling system from an absorption heat pump and a solar collector.
4. Create a process flow diagram (system schematic) for a hybrid heating system using a biofuel boiler and an absorption heat pump.
5. Create a process flow diagram (system schematic) for a hybrid heating system using a heat pump, a solar collector and a biofuel boiler.
6. Refer to Figs. 8.2 and 8.6 to solve this exercise. For the Kagoshima Nanatsujima Mega-Solar Power Plant on Kyushu, Japan, calculate the following: (A) Estimated annual solar energy input (GWh) to the PV system using the GHI basis; (B) Efficiency of the PV plant relative to the solar input energy.
7. Refer to Figs. 8.2 and 8.10 to solve this exercise. For the Silkeborg, Denmark solar thermal heating plant, calculate the following: (A) Estimated annual solar energy input (GWh) to the collector system using the GHI basis; (B) Efficiency of the heating plant relative to the solar input energy.
8. The first commercial geothermal power plant in the U.S. was the PG&E Unit 1 at The Geysers in California in 1960. Consider a proposal to superheat the natural steam using a natural gas-fired furnace. For purposes of this exercise, assume that the following specifications pertain:

   Natural steam conditions at the plant inlet: pressure $P_1 = 647.6$ kPa, temperature $T_1 = 175.6$ °C, mass flow rate $\dot{m} = 30.2$ kg/s; Condenser pressure $P_c = 13.8$ kPa; Steam temperature leaving the gas heater $T_{sh} = 350$ °C. Assuming that the condenser pressure is the same for both the basic geothermal plant and the hybrid plant, calculate the following:

   Part A. The gross electrical power output of the basic plant, assuming a turbine isentropic efficiency of 75% and a generator efficiency of 95%.

   Part B. The gross power output of the hybrid plant, assuming a turbine isentropic efficiency of 85% and a generator efficiency of 95% and allowing for a 5% pressure drop through the heater.

   Part C. The ratio of waste heat discharged from the condenser to the power output for both plants. (Ans: 5.00 vs. 3.92)

   Part D. The marginal thermal efficiency of the heat added in the hybrid plant, based on the incremental power output per unit of heat added in the gas heater. (Ans: 44.7%)

References

[1] Solargis s.r.o. Download solar resource maps and GIS data for 180+ countries. The World Bank, solar resource data - Solargis, Bratislava, Slovakia; 2017. https://solargis.com/maps-and-gis-data/download/world/.

[2] Perovo Solar Park. https://en.wikipedia.org/wiki/Perovo_Solar_Park.

[3] Osborne M. China completes largest floating solar power plant. PV Tech; May 18, 2017. https://www.pv-tech.org/news/china-completes-largest-floating-solar-power-plant.

[4] Yirka B. 70MW: Kyocera and partners to build largest photovoltaic power plant in Japan. Phys.Org; April 12, 2012. https://phys.org/news/2012-04-70mw-kyocera-partners-largest-photovoltaic.html.

[5] Kagoshima Nanatsujima Mega Solar Power Plant. Power Technology. https://www.power-technology.com/projects/kagoshima-nanatsujima-mega-solar-power-plant/.

[6] Kyocera starts operation of 70 MW solar power plant, the largest in Japan. News Release; November 5, 2013. https://global.kyocera.com/news/2013/1101_nnms.html.

[7] Kato S. 5 MW solar plant built on 26-degree sloping dam wall. Solar Power Plant Business; June 21, 2016. https://tech.nikkeibp.co.jp/dm/atclen/news_en/15mk/062000651/?ST=msbe?ST=msbe&P=1.

[8] Weiss W, Spörk-Dür M. Solar heat worldwide, 2018 edition. IEA solar heating & cooling programme, Fig. 2. May 2018. https://www.iea-shc.org/Data/Sites/1/publications/Solar-Heat-Worldwide-2018.pdf.

[9] Silkeborg forsyning relies on the sun. Skørping, Denmark: Arcon-Sunmark; 2018. http://arcon-sunmark.com/cases/fjernvarme-silkeborg-danmark.

[10] Solar thermal energy. Wikipedia; September 24, 2018. https://en.wikipedia.org/wiki/Solar_thermal_energy.

[11] Bezrodny MK, Zagoruchenko MV, Redko AF, Ratushnyak GS, Khmelnuk MI. Low potential power engineering. Kharkiv, Ukraine: Ukraine Education and Science Ministry; 2016. 412 pp. [in Ukrainian].

[12] Bezrodny MK, Pritula NO. Energetic efficiency of heat pump schemes heat supply. NTUU "KPI"; 2012. 208 pp. [in Russian].

[13] Bernier J. La pompe de chaleur. Mode d'emploi, Paris, Edition, 553 p.

[14] Elliott Energy Systems, Inc. TA 100 R CHP: 100 KW micro-turbine unit for combined production of electricity and heat – description and Specifications. On-line brochure [in Russian].

[15] Basok BI, Kolomeyko DA. Analysis of cogeneration plants. Part II. Energy efficiency analysis. Prom. Heat Engineering 2006;28(4):79–83 [in Russian].

[16] Anon. Recent developments in the utilization of the earth's heat. Mechanical Engineering 1924;46(8):448–9.

[17] DiPippo R. An analysis of an early hybrid fossil-geothermal power plant proposal. Geothermal Energy Magazine 1978;6(3):31–6.

[18] Hansen A. Thermal cycles for geothermal sites and turbine installation at the Geysers power plant, California. In: Paper G/41, Proc. conf. on new sources of energy, Rome, United Nations. 3; 1961. p. 365–79.

[19] Anon. System design verification of a hybrid geothermal/coal fired power plant. The Ralph M. Parsons Company; 1978. Project 5905.

[20] Khalifa HE, DiPippo R, Kestin J. Geothermal preheating in fossil-fired steam power plants. In: Proc. 13th intersociety energy conversion engineering conf., vol. 2; 1978. p. 1068–73.

[21] Janes J. Evaluation of a superheater enhanced geothermal steam power plant in the Geysers area. Rep. 700-84-003, California Energy Commission, Siting and Environmental Div. 1984.

[22] DiPippo R. Geothermal power plants: principles, applications, case studies and environmental impact. 4th ed. Oxford, England: Butterworth-Heinemann: Elsevier; 2016.

[23] Di Marzio G, Angelini L, Price W, Chin C, Harris S. The stillwater triple hybrid power plant: integrating geothermal, solar photovoltaic and solar thermal power generation. In: Proc. World geothermal congress 2015, Melbourne, Australia, April 19–25, 2015.

[24] Hiriart G,L, Gutiérrez-Negrin LC. Calor del subsuelo para generar electricidad: combinacion solar-geotermia (Subsoil heat to generate electricity: solar-geothermal combination). Ingenieria Civil May 1995;313:13–22 [in Spanish].

[25] Lentz A, Almanza R. Parabolic troughs to increase the geothermal wells flow enthalpy. Solar Energy 2006;80:1290–5.

[26] Handal S, Alvarenga Y, Recinos M. Geothermal steam production by solar energy. Geothermal Resources Council - Transactions 2007;31:503–10.

[27] Alvarenga Y, Handal S, Recinos M. Solar steam booster in the Ahuachapán geothermal field. Geothermal Resources Council - Transactions 2008;32:395–9.

[28] Thain IA, DiPippo R. Hybrid geothermal-biomass power plants: applications, designs and performance analysis. In: Proc. World geothermal congress 2015, Melbourne, Australia, April 19–25, 2015.

[29] DiPippo R. Combined and hybrid geothermal power systems [chapter 14]. In: DiPippo R, editor. Geothermal power generation: developments and innovation. 1st ed. Oxford, England: Woodhead Publishing: Elsevier; 2016.

Industrial waste heat resources

9

9.1 Overview of industrial waste energy utilization

The use of waste heat is one of the most important ways of conserving energy resources. Waste heat (WH) is the energy potential of products, waste, by-products and intermediate products formed during technological processes in facilities and plants that is not used in the production process itself, but may be partially or completely used for saving energy in other processes [1].

Industrial waste energy may be divided into three main groups: combustible (fuel), thermal (heat), and excess pressure; hybrid systems that use combinations of these main groups may be considered a fourth type. General methods of using low-potential waste heat are as follows:

1. Fuel – Used directly as combustion fuel. WH fuel comprises combustible gases and waste products that can be used as fuel, such as, blast furnace and coke oven gases in metallurgy, coal mine methane, biogas in the processing of biomass, and purge gas for ammonia production.
2. Thermal – Used directly as heat or heat generation in appropriate plants. WH thermal is the heat of waste (or off) gases of process units and various processes of fuel combustion.
3. Pressure – Used to produce electric or mechanical energy in appropriate plants. Waste overpressure is the potential energy of gases, steam or water with elevated pressure that can be used to obtain mechanical or electrical energy; for example, high pressure steam in the steam supply systems of a plant or natural gas pressure let-down stations that feed gas to consumers.
4. Hybrid – Thermal and electric or mechanical energy, which are simultaneously produced from waste energy products in appropriate plants.

The waste energy potential from technological processes constitutes a secondary energy resource that can be used for the generation of electric energy and heat in cogeneration units [2]. In some cases it may be possible to reorganize existing heat generating facilities into combined heat and power generation plants in order to increase the efficiency of fuel usage and environmental safety.

At the end of 2016, there were 376 MW of installed electric power capacity based on waste heat recovery in the world, with 39 MW of new capacity under construction in 16 projects [3]. Figure 9.1 shows the percentage of the installed capacity by source of the waste heat.

Waste heat is most widely used in internal combustion engines (ICEs) of both the diesel and gasoline types, together with gas turbines. In the first case, the energy comes both from exhaust gases and jacket cooling water; in the latter case it is only from the

Low-Temperature Energy Systems with Applications of Renewable Energy
https://doi.org/10.1016/B978-0-12-816249-1.00009-1

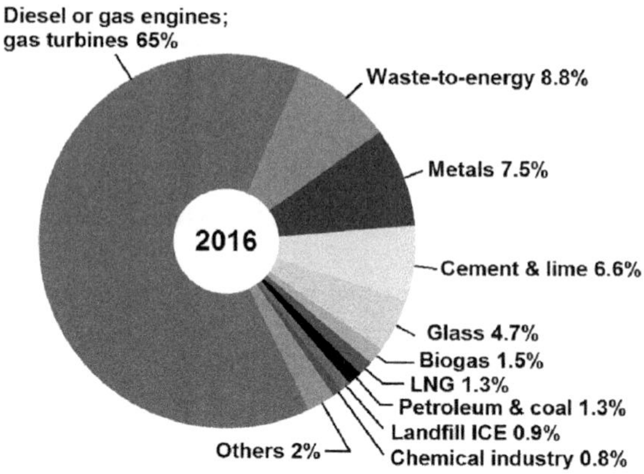

Fig. 9.1 Share of 376 MW installed electric capacity derived from various waste heat sources [4].

hot exhaust gases. Waste-to-energy plants, ferrous and nonferrous metallurgy, and the cement/lime industries are the next largest sources of waste heat for power production. The use of waste heat ensures a reduction in the gross energy consumption of a product.

Waste thermal energy may conveniently be classified as high-, medium- or low-temperature. High-temperature waste heat constitutes energy sources with a temperature of at least 600°C. Low-temperature waste heat involves liquids with a temperature of less than 100 °C and gases with a temperature below 300 °C.

When using high-temperature waste heat, heat recovery boilers and combined heat-power plants are normally employed. Utilization of low potential waste heat, for example, heat of polluted effluent, aggressive fluids, ventilation waste, flue gases, and waste flue gases requires the use of specialized equipment. In assessing the feasibility of using waste heat, a company needs to: (a) determine their demands in steam and hot water for technical and sanitary needs, (b) justify the choice of principles and methods of construction and regulation of heat supply systems, and (c) choose the most effective means to exploit the waste heat.

Currently, less than half of the primary energy used in the transport, industrial and residential sectors is used in the application processes; the remainder is wasted. It is possible to significantly reduce the consumption of primary energy resources of fuel, electricity and heat by developing an efficient energy technology production scheme, taking into account the relationship between technological and energy processes. It is logical that the most promising system is one in which the primary energy system is directly connected with the secondary or waste heat recovery equipment, creating a unified system. For example, global manufacturers of absorption refrigeration technologies are developing innovative technologies that reduce, recycle and reuse waste energy [5].

9.2 Quantitative measures of waste heat outputs

Figure 9.2 is a simple schematic to illustrate the general system of heat recovery. The waste heat from a main technological unit is partly captured by the WH recovery plant. The potential amount of WH, Q_T, is easy to calculate once the parameters of the main plant are given; see Eq. (9.1). The total annual WH energy available is also easy to find if the hours of operation are known.

For flue gases, steam or hot water, the annual potential for thermal WH generation, kJ/y, is determined by the formula:

$$Q = (H_1 - H_2)\dot{V}\beta\, l \tag{9.1}$$

where H_1 is the volumetric enthalpy of the hot exit fluids from the technological unit (industrial furnace, etc.) and H_2 is the volumetric enthalpy of those fluids after the WH utilization plant or heat exchanger, kJ/m^3; $\dot{V}$ is volume flow rate, m^3/y; β is a coefficient that takes into account the nominal load reduction and the number of operating hours of the WH utilization plant and the technological unit; l is a coefficient that takes into account losses of heat from the WH utilization plant into the environment ($l \cong 0.8-0.9$).

9.3 Waste energy recovery methods

There are many ways to improve the utilization of systems burning fuel gas. For example, heat exchangers may be used for preheating the air supplied to the burners and even the actual fuel. These techniques increase the efficiency of fuel-using heat generators by about 1–2%. Furthermore, it is possible to capture the exhaust heat for supplying district hot water heating systems.

Currently, modern absorption chillers and heat pumps, steam turbine and gas turbine plants, organic Rankine cycle plants (ORCs), and cogeneration plants are being manufactured which allow for more complete conversion of waste heat recovery. An interesting example is the technology of trigeneration wherein heat, electricity, and cold (i.e., refrigeration) are simultaneously produced.

Energy audits aim to develop measures for transformation of secondary energy flows outside the main fuel-fired unit. Possible approaches include deep cooling of

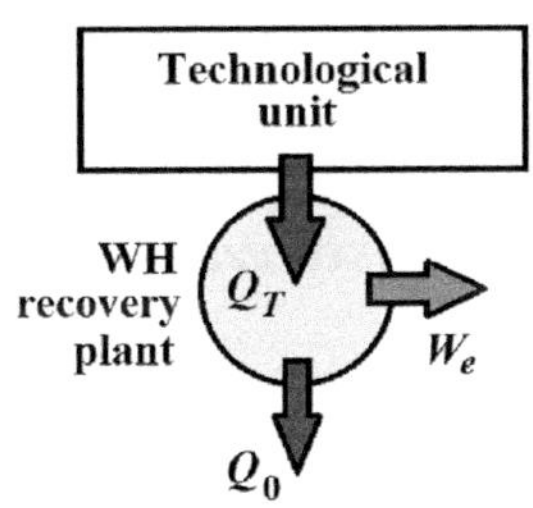

Fig. 9.2 Schematic of waste heat recovery. Note: If only heating applications are used, W_e is zero.

flue gases resulting in water vapor condensation (so-called condensing boilers), capturing waste heat to supply greenhouses, extraction of gases from combustion products, for example, CO_2, which may improve the efficiency of the fuel-fired heat generator by 8–12%. Therefore, attention is now being paid to develop technologies that reduce, recycle, and reuse waste energy, or the Reduce-Recycle-Reuse initiative.

9.3.1 Heat recovery of flue gases

Source of gaseous waste heat include: heating furnaces of metallurgical workshops with gas temperatures of 1000–1100 °C, furnaces of thermal and heat-pipe shops and sections with gas temperatures of 600–800°C, melting units in foundry production with gas temperatures of 800–900°C, various driers and stands for drying ladles in foundries with temperatures of 300–400 °C, and industrial boiler houses with exhaust gas temperatures of 150–200 °C.

Techniques for using the waste heat of combustion products may be divided into two groups: 1) Closed systems for internally preheating the air supplied for gaseous combustion in a fuel-fired unit; and 2) Open systems that use the exhaust heat outside the main unit to generate steam, hot water, heated air, and so on. In the latter process, it is usual to:

- Send the combustion products to a heat recovery boiler to produce a saturated vapor; or
- Send the combustion products to a recuperative or tubular air heater for heating the inflow air of heating, ventilation, and air conditioning systems.

In cases where both air and water heating is desired, one usually installs the air heater and contact water heater sequentially to make most effective use of the exhaust gas flow.

9.3.2 Flue gas heat recovery circuits

There are several variants of connecting engineering systems to WH recovery boilers: single, parallel, sequential, and parallel-sequential ones; see Fig. 9.3.

The first four systems are single-use waste heat recovery where the following end uses are achieved: (A) space heating; (B) hot water supply; (C) refrigeration by means of an absorption machine; and (D) electric power generation by means of an organic Rankine cycle (ORC). The other four systems are trigeneration waste heat systems where (E) and (G) are parallel arrangements and (F) and (H) are series arrangements that achieve the end uses just described in different combinations. System (D) is elaborated in Section 9.4.4.

9.3.3 Heat recovery from waste water and evaporative cooling systems

Discharge cooling systems for equipment and technological processes (e.g., industries in mechanical engineering, metallurgy, chemistry, petrochemicals) are sources of a significant amount of waste heat. Often this WH is discharged to the surroundings

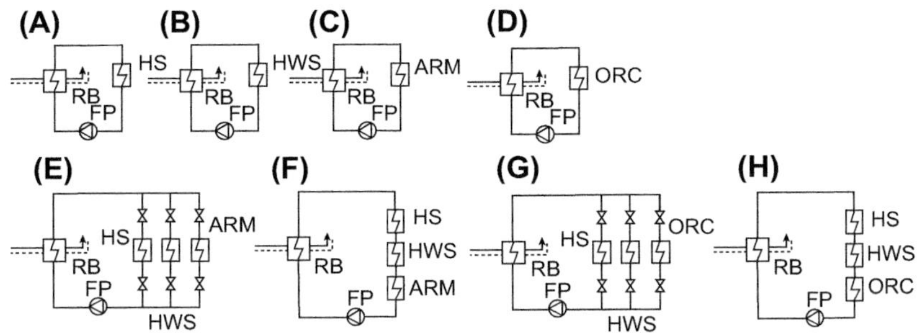

Fig. 9.3 Arrangements of WH recovery boiler equipment. *FP*, feed pump; *RB*, heat recovery boiler. (A, B, C, D) single connection. *A*, heating and ventilation system (HS); *B*, hot water supply system (HWS); *C*, absorption refrigeration machine (ARM); *D*, electricity generation (ORC); *E*, systems (A, B, C) connected in parallel with RB; *F*, systems (A, B, C) connected in series with RB; *G*, systems (A, B, D) connected in parallel with RB; *H*, systems (A, B, D) connected in series with RB.

via cooling towers that involve evaporation. The main deterrents for using such WH are low temperature (25–35 °C) and the presence of waste water pollutants. However, the heat from these systems may be used by means of lithium-bromide absorption heat pumps and vapor-compression refrigeration machines.

Some examples of waste heat will illustrate the magnitude of the wasted energy. In evaporative cooling units for metallurgical furnaces (Fig. 9.4), steam output is 0.2 tons of steam per ton of iron; 0.25 tons of steam per ton of open-hearth steel; 0.3 tons of steam per ton of products in heating furnaces. Vapor pressure ranges from 0.8 to 4.7 MPa.

Heat losses by blast furnaces that are cooled with water are from 3 to 5% or 15–25 MJ/h per furnace. Heat losses from heating furnaces cooled with water are 15–25%. The cooling of furnaces consumes the largest portion of water used by the entire facility. The heat balance for a furnace shows that the heat of cooling the furnace is 10–30% of the heat which is fed into the furnace.

With single-phase liquid water cooling, up to but not more than 40–45 kJ/kg of heat may be removed provided the water is not heated to more than 60–65 °C. With evaporative cooling, 2500 kJ/kg of heat may be removed which leads to a significant reduction in the water flow rate. The spent vapor has low pressure and variable

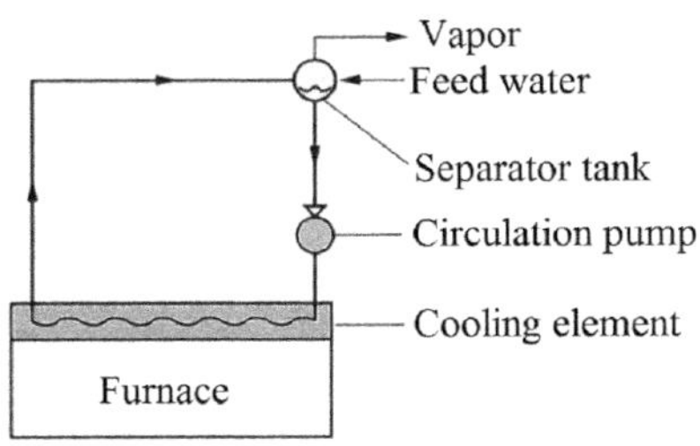

Fig. 9.4 Evaporative cooling system with forced liquid circulation.

flow rate and may be contaminated with chemical and mechanical impurities, thereby limiting it further use. The main possible uses of the vapor are heat supply, electricity generation, or a combination thereof.

9.3.4 Heat recovery of ventilation air

Health regulations require that the air in a building must be refreshed via a continuous replenishment of conditioned air with outside air. Since the air being exhausted carries thermal energy, it is advisable to recover that energy to the extent feasible. Modern designs employ supply and exhaust ducts combined into one unit, allowing easy interaction between the two streams. However, in existing, older structures, the main obstacle to heat recovery of ventilation emissions lies in inconvenient placement of the supply and exhaust ducts.

Energy recovery and consequent savings can be provided by highly efficient heat exchangers such as plate-and-frame units for supplying fresh ambient and recirculation air. Also common are supply-and-exhaust units with recuperative rotary heat exchangers and heat pump units (HPUs), which ensure high efficiency heat recovery. When direct heat transfer is not advisable, air heat recovery units using an intermediate heat-transfer fluid are used.

The choice of type of recovery unit is determined by technical and hygienic requirements. These include the structural configuration of the supply and exhaust ventilation systems, the inadmissibility of the transfer of odors and harmful substances, and the ratio of volumes of supply air and exhaust air. When air ducts are located far from each other, heat recovery units with an intermediate heat-transfer fluid must be used. In cases where air ducts are adjacent, but there are hygienic restrictions on the contact of intake air and exhaust air, it is recommended to use heat recovery units and heat exchangers with an intermediate heat transfer fluid, as well as heat exchangers with heat pipes as heat utilizers. Rotary and alternating regenerators as well as direct-contact heat exchangers with an intermediate fluid may be recommended for side-by-side systems, as long as there are no hygienic restrictions on the contact of supply and exhaust air.

The choice of the type of heat recovery unit is driven by energy costs or savings relative to the equipment costs. One of the most effective solutions of using low-potential heat is the use of vapor-compression refrigeration (VCR) equipment of air-conditioning systems operating in the heat pump mode. The thermal power of the HPU is determined by the required cooling performance.

During the summer season, the reverse water supply of the air conditioning system which is cooled in the evaporators of the VCR is used as a source of low-temperature heat, and subsequently used as the cooling agent. In the heating season, the circulating water from the HPU is used to heat the circuit water of the heat supply system. The amount of heat used in the total thermal balance of a typical enterprise is about 30%.

9.3.5 Heat recovery from industrial compressor installations

Compressor technology is widely used in modern production of various enterprises (e.g., chemical and petroleum industries), as well as in pneumatic systems. The capacity of compressors may reach 27,000 m^3/min; the power of the electric drive is up to 40,000 kW. The minimum work for a given compression job is achieved thermodynamically by an isothermal process, but this requires continuously cooling the gas while it is being compressed, a daunting engineering challenge. Real compression processes are closer to adiabatic, but are often modeled as polytropic. Thus, actual compression of gases results in an increase in temperature that is often unwanted. Thus, cooling is required before the gas is delivered to the end use. The amount of heat removed by the cooling system is typically about 80–90% of the electric drive power consumed. The amount of heat removed from the compressor units also depends on the capacity of the compressor and the gas pressure at the outlet of the compressor. When the compressor pressure ratio, $P_{out}/P_{in} = 200$, the temperature of the compressed gas rises from ambient to about 200–250 °C.

Various types of compressors (e.g., piston, rotary, screw, vane, and centrifugal) can be used. Compressor installations differ in efficiency and cooling system, so that the amount of heat removed varies with the cooling system. In real conditions for, say, screw compressors, one of the most common, the compression process is polytropic, i.e., $PV^n = C$, where $1 < n < k$, and k is the isentropic index (1.4 for air). Thus, the outlet gas temperature is higher than during an isentropic compression. The waste heat from compressor installations may be used in hot water supply and heating systems, in steam supply systems, and for electric power generation using ORC technologies. We will elaborate on the ORC application in Section 9.4.

9.3.6 Energy recovery from natural gas pressure let-down stations

Natural gas is sent from gas fields to consumers via long-distance, high-pressure pipelines. The pressure is maintained by compressor stations located along the pipeline; see Section 9.4.2. Just prior to entering local gas distribution networks, the gas is sent through pressure-reducing valves to allow safe usage by the consumers. The pipeline pressure prior to the let-down station may be anywhere from 1.5 to 10 MPa. Simply using a throttle valve wastes significant thermodynamic potential because this is a completely irreversible process with no benefit. If a power-producing machine were to replace the throttle valve, some of that potential would be recovered.

One method involves using a positive displacement machine such as a helical screw expander (HSE); see Fig. 9.5. Often these devices are used as compressors and are driven by a motor. However, they can be used as power generators if high-pressure fluid is available to drive the rotors in the opposite direction. They are rugged machines that can withstand considerable pressure. They have been used with geothermal resources and have achieved commercial success.

The following description shows a theoretical application of a HSE. Fig. 9.6 shows two configurations: (A) normal let-down station and (B) let-down station supplemented with a HSE. The HSE is installed in parallel with the pressure let-down valves (PLDs) to allow flexibility of operation. When the main valves are both closed, all of the gas flow is directed to the HSE (as shown), or the gas may be apportioned as

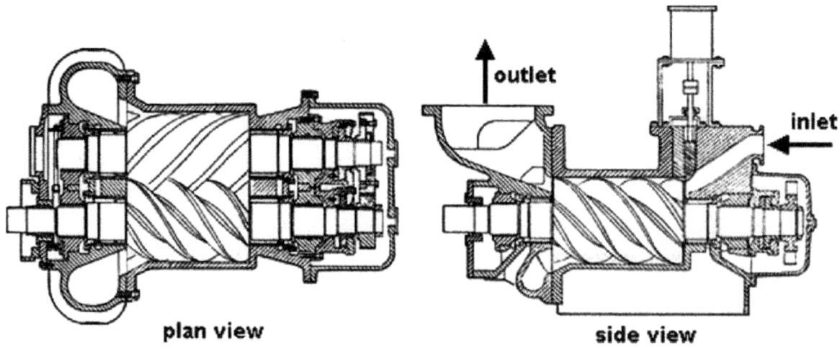

Fig. 9.5 Helical screw expander that can replace pressure let-down valves [6].

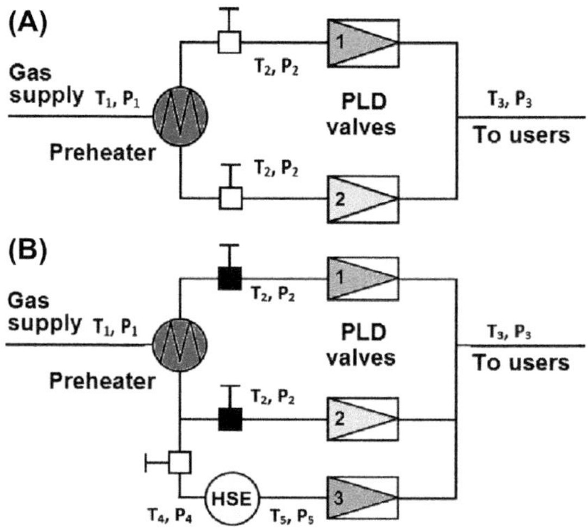

Fig. 9.6 Natural gas PLD station configurations, without (A) and with (B) energy recovery.

needed. Owing to the drastic gas temperature drop created either throttling or expansion in the HSE, it is usually necessary to preheat the gas before it enters the PLD station. This is especially important when the HSE is used because the temperature drop is larger than that caused by throttling owing to the removal of additional energy in the form of work. The valve 3 allows for trimming the HSE exhaust pressure to match the required distribution line pressure.

The analysis of the system is based on the configuration shown in Fig. 9.7. This greatly simplified schematic shows a small side stream of gas 5 being diverted from the gas pipeline to provide the preheating needed to raise the temperature incoming gas to the appropriate level to keep the HSE exhaust temperature at the level demanded by the application, typically not less than 5 °C.

A study of typical gas conditions was carried out, assuming 0.65 isentropic efficiency for the HSE, with the following results. For a particular composition of natural gas, an inlet pressure of 2.23 MPa, temperature of 25 °C, and volumetric flow rate of 3800 m^3/h (0.8 kg/s), the HSE would produce about 230 kWe. Preheating to 150 °C

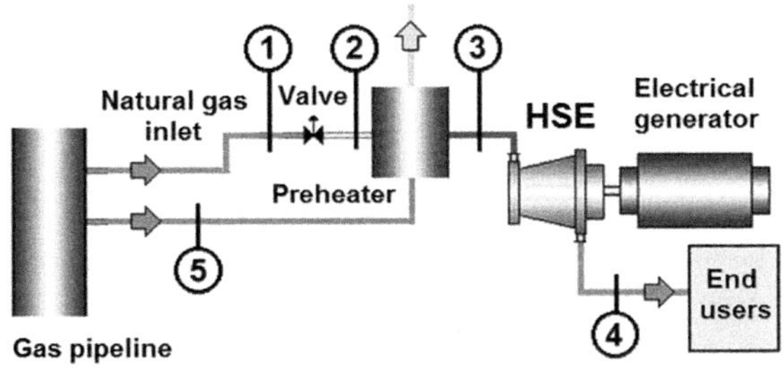

Fig. 9.7 Simplified flow diagram of HSE energy recovery with gas preheating.

would be needed to maintain a gas outlet temperature of 17 °C from the HSE. The amount of preheat (in kWt) is about the same as the amount of power generated (kWe), but the value of the power is several times greater than the cost of fuel for the preheat, leading to positive cash flow during operation. The amount of natural gas burned for preheat is about 0.5% of the main gas flow into the PLD station.

9.3.7 Other sources of low-potential heat

Here are several other waste heat sources that may be of value:

- Sewage of residential buildings, hotels, medical-preventive, and other institutions (savings being about 35% of consumption for heating hot water)
- Sewage treatment plants of various types
- Mine water and ventilation air from mines
- Waste water from industrial enterprises, heat generating stations, swimming pools
- Chemical processes in industry.

As an example of the last case, consider the production of sulfuric acid, a highly corrosive substance. Waste heat is generated in the form of the heat of reaction. In the process of obtaining the acid, it must be cooled from about 80 to 140 °C down to 40–60 °C. This means that producing 1000 kg of sulfuric acid gives about 3.35 kJ of heat. The cooling is carried out using water and an intermediate heat transfer fluid, as shown in Fig. 9.8. Owing to the corrosive nature of the acid, it is impossible to prevent acid from entering the cooling medium if surface heat exchangers are used. Therefore, devices with an intermediate heat transfer fluid are used (Fig. 9.8). The inter-tubular space is evacuated and filled with a suitable fluid (ammonia, special cooling agent, etc.). Acid is fed into the lower tubular surface, and cooling water is supplied into the upper one. The acid heats the intermediate fluid, which boils, sending its vapor upwards where it condenses on the upper bundles of tubes, heating the cooling water, while the condensate returns to the bottom to be reboiled.

9.3.8 Equipment used in waste heat recovery

This section lists and describes briefly some of the main pieces of equipment that are involved in waste heat recovery systems.

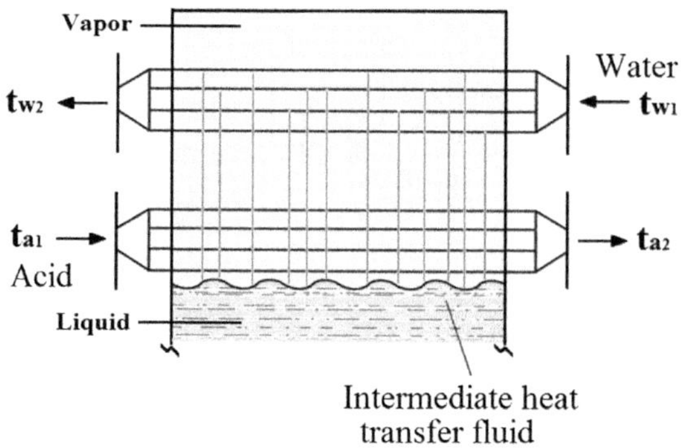

Fig. 9.8 Diagram of an intermediate heat transfer system for a corrosive product.

- Waste heat recovery steam generator. Generally used with gas turbines where the hot exhaust heats and boils water to make steam for a bottoming steam turbine cycle.
- Radiation coolers of converter gas in steel production. Converter gas is created during the production process in which pig iron is converted to steel.
- Air heaters. May be made of cast iron, sectional, recuperative-type with fin heating surfaces on the gas and air sides.
- Direct-contact devices with an active nozzle. Flue gases are cooled and transfer heat to a heated medium using a moving dispersed nozzle. The fluids are contained in separate tubular sections and do not mix.
- Condensers, heat pipes. Given the low temperatures in condensers, waste heat may be suitable for district heating, soil heating, or greenhouses. Heat pipes convey heat from one system to another by means of a phase-change fluid.
- Rotary heat exchangers ("thermal wheel"). Consists of a circular honeycomb matrix of heat absorbing material, which is slowly rotated within the supply and exhaust air streams of an air handling system.
- Economizer: In the case of process boilers, waste heat in the exhaust gas is passed along a recuperator where it heats the incoming fluid for the boiler and thus decreases the heat requirement from the primary fuel.

9.3.9 *Intermediate heat transfer fluids*

The need for intermediate heat transfer fluids (HTFs) stems from the fact that gases have low density and low heat capacity (about 1.5 kJ/(kg K)) and require duct pipes with significantly large diameters. It is difficult to provide thermal insulation of such ducts, and insufficient insulation at high gas temperatures leads to high heat losses and rapid gas cooling before it can be captured in the waste heat recovery unit. One way to cut down on these losses is to place the WHU close to the heat source, but this is often impractical or unfeasible. Furthermore, the low specific heat capacity means that large volume flow rates must pass through the heat exchanger of the WHU, which also results in large dimensions and costly equipment.

The heat exchangers tend to be very large because of the low coefficient of heat transfer from the gases to the tube wall (less than 50 W/(m^2 K)). Other difficulties arise from uneven heating and the complexity of regulating the process due to significant changes in the temperature of flue gases during the heat exchange process and the possibility of condensation of water vapor. Heat transfer is diminished due to contamination of the heat transfer surface with soot and products of combustion. Corrosion may occur on the walls of the heat recovery unit because of the content of corrosion-active substances such as water, oxygen, sulfur and nitrogen oxides.

Therefore, rather than send the exhaust gases through very large pipes where these problems would be encountered, high-temperature liquid heat transfer fluids are widely used to capture the waste heat at the main plant and transport the energy to the WHU energy converter. For example, such a fluid can carry thermal energy from a gas turbine exhaust stack to an ORC power plant.

We will discuss three categories of HTF: (1) mineral, (2) synthetic, and (3) silicone. Their characteristics involve advantages and disadvantages as well as cost considerations that will determine which HTF is best for any given application. The main criteria for choosing a heat carrier are: thermophysical properties; operating temperature range; degree of fire hazard; toxicity; environmental characteristics; resource indicators; and cost.

(1) Mineral These HTF (lubricants) are produced on the basis of extracts of phenolic purification of distillates of sulfuric oils by successive de-paraffinization and purification. Mineral HTF are least resistant to overheating (local and short-term) since they are degraded at high temperature with the formation of high viscosity gummy substances resulting in an increase in viscosity and fouling of heat exchanger surfaces. The operating temperature range is from 20°C to 300°C. The primary disadvantage of these HTF is that of thermal decomposition mentioned above. The specific heat capacity of mineral lubricants is 1.88–2.1 kJ/(kg K). Mineral HTF are the least expensive to purchase, but have the highest operating costs. Some commercial examples include *AMT-3001, Mobilterm, MultiTherPG-1*, and others.

(2) Synthetic These HTF (lubricants) are made on the basis of polyalkybenzenes, diphenyloxide mixtures and terphenyls. These HTF may be used from 40 °C to 340 °C. Synthetic HTF are more resistant to overheating than mineral ones, but the effects are similar. The specific heat capacity of synthetic HTF at t = 200 °C is 2.34–2.85 kJ/(kg K). These fluids are more expensive than mineral HTF. Some commercial examples include *T1V-330, ALOTERM, Thermolan, Terminol-72*, and others.

(3) Silicone These HTF (diathermal oil lubricants) are based on thermo-stabilized silicon-organic liquid. The silicone fluids are characterized by a low temperature of solidification (from −60 °C to −140°C), high thermal conductivity, low viscosity, and are odorless, nonflammable, noncorrosive, environmentally safe, and non-toxic ... significant advantages over the other two categories of HTF. Silicone HTF have the widest temperature operating range from -60 °C to 400–450 °C. Silicone HTF when overheated behave differently than the other HTF; they are subjected to regrouping with the formation of cyclic products, and there is no thermal decomposition, so that no high-viscosity products are formed. They also have low cost and long life (several decades). These are the most expensive HTF, but the high initial cost can be offset by the lower cost of operation and the elimination of costs for ensuring fire safety. Silicone HTF are used in solar collectors, boiler plants, and heat recovery units. Some commercial examples include *Syltherm-800, DurathermS, Sovexil TSZH-v*, and others.

9.4 Power generation technologies using waste heat

9.4.1 Rankine cycle cogeneration

When it is desired to generate both electrical power and thermal energy for processes or direct heating, there are two possibilities: (1) build two separate facilities dedicated to each output or (2) build a combined system that produces both outputs. According to the laws of thermodynamics, combined energy conversion or cogeneration is more efficient than running separate plants.

Using typical thermal efficiencies, Fig. 9.9 illustrates this fact. The combined plant coefficient of performance COP is 55% higher than that of separate plants, i.e., 90% versus 58%. This translates into reduced primary energy (fuel) requirements and reduces carbon dioxide and other atmospheric emissions.

A cogeneration plant is a thermal power cycle that utilizes the output of a single boiler to produce more than one useful form of energy, usually electricity and heat. A cogeneration cycle is shown in Fig. 9.10. The steam from the boiler is delivered to the steam turbine that puts out mechanical power $\dot{W}_T$ to drive the generator which makes electricity. An extraction (state 4) from the turbine removes steam and sends it

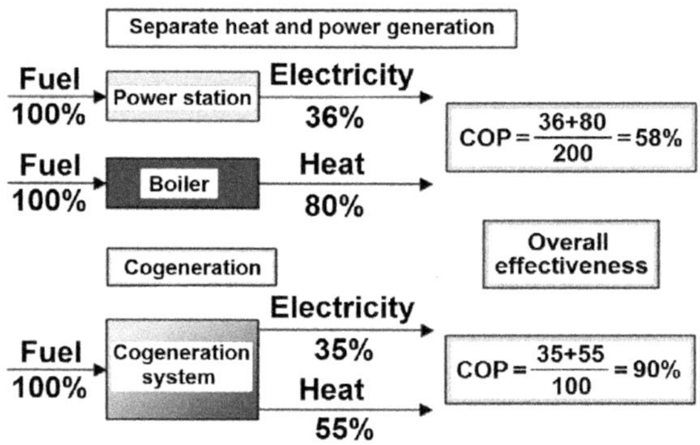

Fig. 9.9 Comparison of typical separate and combined power and heat plants, illustrating the advantage of cogeneration over separate plants.

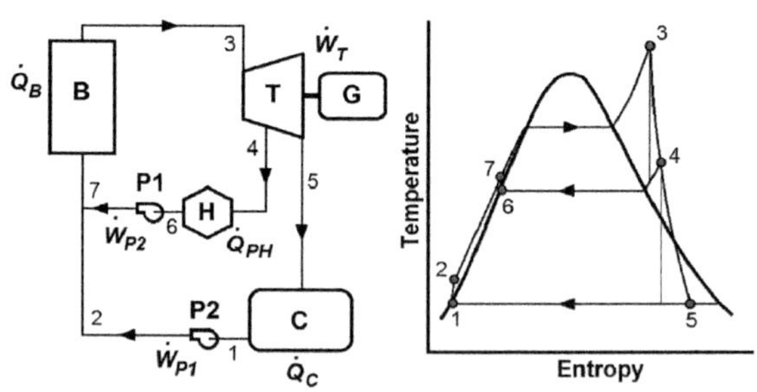

Fig. 9.10 A simple cogeneration cycle and its T-s process/state diagram. *B*, boiler; *T*, turbine; *G*, generator; *H*, heater; *P1, P2*, feed pumps; *C*, condenser.

to a heat exchanger or other element where it delivers process heat $\dot{Q}_{PH}$. The condenser C removes waste heat, reducing the steam to liquid condensate which is pumped back to the boiler (P2). Since the heater H generally condenses the process steam, the condensate may be returned to the boiler by another feed pump P1 for reboiling.

The cycle thermal efficiency, the turbine power, and the boiler heat requirement are found from the following equations:

$$\eta = \frac{\dot{W}_T}{\dot{Q}_B} \tag{9.2}$$

$$\dot{W}_T = \dot{m}_3(h_3 - h_4) + \dot{m}_5(h_4 - h_5) \tag{9.3}$$

$$\dot{Q}_B = \dot{m}_1(h_3 - h_2) + \dot{m}_6(h_3 - h_7) \tag{9.4}$$

Considering that the power to run the feed pumps is small compared to the power of the turbines, the following approximate equation may be used for the boiler heat input:

$$\dot{Q}_B \approx \dot{m}_1(h_3 - h_1) + \dot{m}_6(h_3 - h_6) \tag{9.5}$$

A coefficient of performance or utilization factor ξ may be defined to assess the effectiveness of this plant. The factor ξ is the ratio of the sum of all the output energy to the energy input to the boiler. If we assume that some fraction α of the condenser heat is put to use in direct heating of buildings, then the utilization factor may be written as:

$$\xi = \frac{\dot{W}_T + \dot{Q}_{PH} + \alpha\dot{Q}_C}{Q_B} \tag{9.6}$$

where the process heat is

$$\dot{Q}_{PH} = \dot{m}_6(h_4 - h_6) \tag{9.7}$$

and the condenser disposes of the following heat rate to cooling water

$$\dot{Q}_c = \dot{m}_5(h_5 - h_1) \tag{9.8}$$

If there is no direct heating usage of the condenser heat, then $\alpha = 0$.

9.4.2 Combined Brayton and Rankine cycles – steam and organic

Gas turbines are used for many applications such as jet aircraft propulsion and electric power generation. In the latter case, the engine is stationary and housed in a power plant building where the hot exhaust gases are available for waste heat recovery. The following technologies may be considered for converting the heat of exhaust gases into useful output: (a) WH recovery boiler, (b) steam turbine plant with a WH boiler, (c) ORC plant with a waste heat exchanger. Combined gas turbine-steam turbine systems are commonly used in industry and demonstrate very high thermal efficiencies, greater than 60%.

One of the best opportunities for WH recovery from gas turbines (GTs) involves natural gas compressor stations which serve to boost the gas pressure along pipelines [7]. For example, across Canada, the United States and Europe there are numerous GT installations that burn some of the natural gas passing through the pipeline and drive compressors, with the exhaust from the GT being available for energy recovery. Figure 9.11 is a map of the gas pipelines and the locations of the gas compression stations across the contiguous U.S. The lines that intersect the northern border extend into Canada where similar stations are found.

A typical gas compression station may have several GT-compressors as shown in the schematic Fig. 9.12. Each GT unit is housed in a building with separation along the pipeline between stations being on the order of 150 km. More than 70% of the heat generated by burning the natural gas is carried off with the exhaust gases. Furthermore, the temperature of exhaust reaches 450−500 °C, while the flow rates range from 60 to 120 kg/s, which means that the waste heat amounts to 20−46 MWt. Clearly, the hot exhaust gases are a suitable target for use in a waste heat recovery unit, assuming favorable economic conditions.

Europe has numerous WH recovery stations arrayed along natural gases pipelines in several countries, as can be seen from Fig. 9.13. Pressure boosting is done by centrifugal superchargers driven by gas turbines. The efficiency of older gas turbine plants

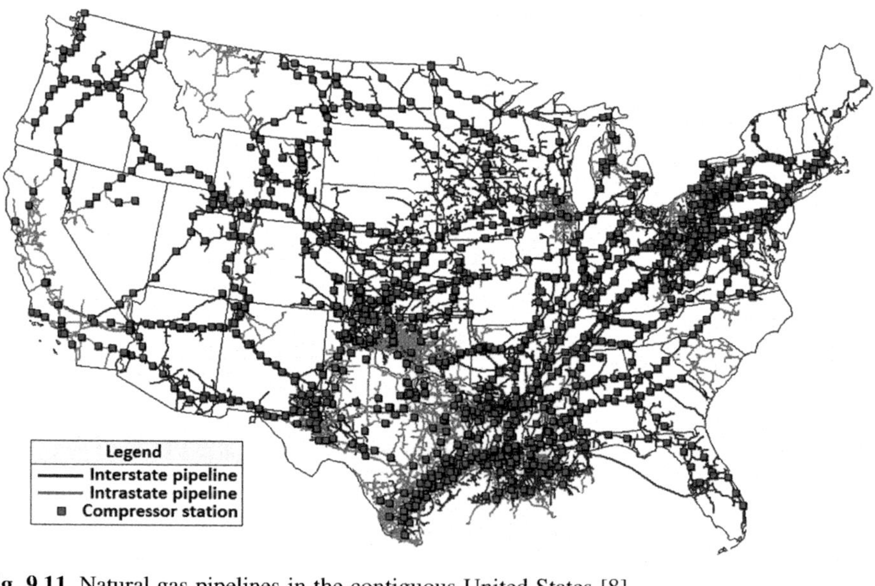

Fig. 9.11 Natural gas pipelines in the contiguous United States [8].

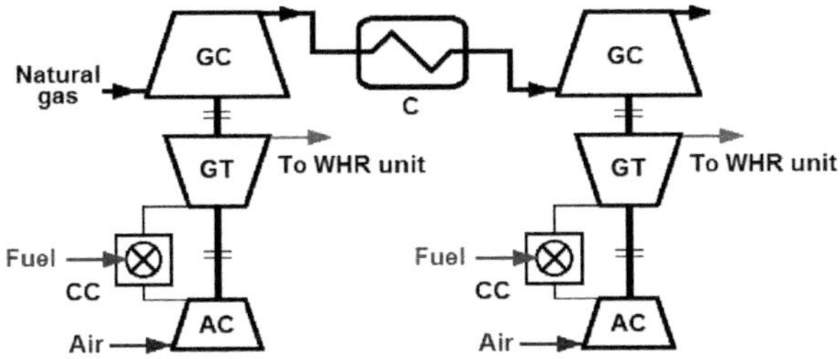

Fig. 9.12 Typical arrangement of natural gas compressor stations along gas pipelines. *AC*, air compressor; *C*, cooler; *CC*, combustion chamber; *GC*, gas compressor; *GT*, gas turbine.

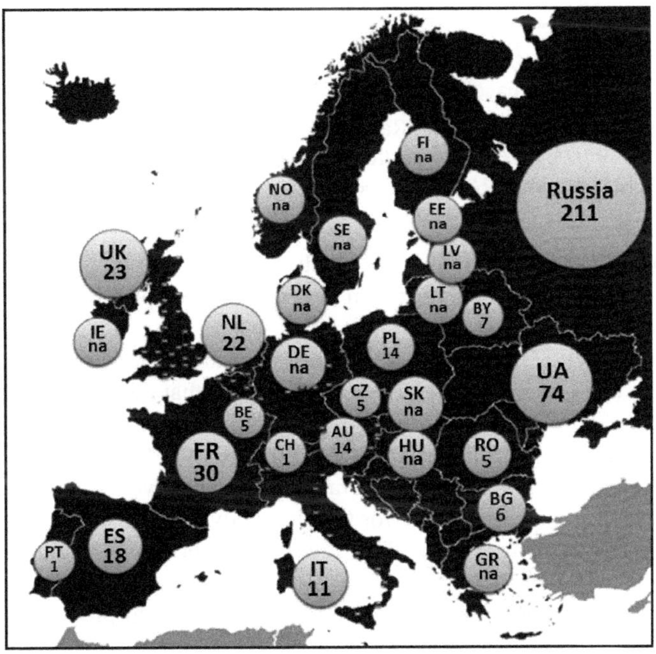

Fig. 9.13 European gas compressor stations by country and number of stations [9].

which have been in operation for many years is 23−28%, but modern units are higher. Waste stack gases with temperatures starting from only 150 °C can be economically used to generate power from 500 kW up to several tens of MW.

Figure 9.14 shows a schematic of a typical steam power plant with steam raised in the waste heat boiler/superheater system. This would be used when the exhaust gases are of high temperature and steady in time. For variable conditions that might correspond to low or fluctuating gas volumes caused by variations in demand, exhaust temperatures will be lower. Then, rather than deploying a steam plant, it might be more appropriate to use an organic Rankine cycle (ORC) power plant. In such cases, the vapor-generating system would be simpler (only a heat exchanger would be needed) and the condenser could be air-cooled. ORC units are smaller in output than steam

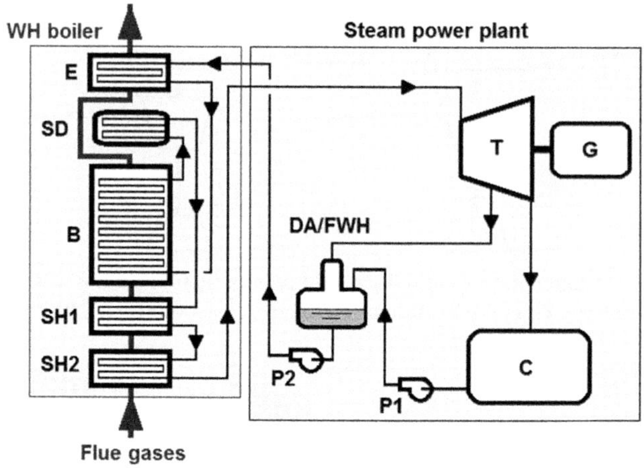

Fig. 9.14 Waste heat boiler and steam Rankine cycle with a single, open feedwater heater. *B*, boiler; *C*, condenser; *DA/FWH*, deaerator/feed water heater; *E*, economizer; *G*, generator; *P1*, *P2*, pumps; *SD*, steam drum; *SH1*, *SH2*-superheaters; *T*, turbine.

units and typically generate about 5 MW each. For high-temperature WH, thermal efficiencies of 40−50%, or even higher, can be obtained by sophisticated cycles but with high associated capital costs.

Calculations of ORC WHR power modules with different working fluids (R600a, R601a, R602a, and others) were carried out and show that the specific electric power generated at an exhaust gas temperature of 450 °C and an turbine inlet vapor temperature of 350 °C is 100−115 kWe/(kg/s). For a range of GT exhaust mass flow rates from 60 to 120 kg/s, the potential for generated electrical power is 7−12 MWe [10,11].

In any case, there must be a buyer for the electricity, typically a local utility willing to enter into a long-term agreement. The price paid to the plant developer is a negotiated value and will determine the economic feasibility of the plant. The electricity would serve to offset fuel that would otherwise be needed to produce that power in a conventional power plant. That translates into an important environmental advantage in that the produced electricity comes without any additional emissions of potentially harmful gases.

An example of a waste heat recovery unit at a natural gas compressor station is shown in Fig. 9.15. The organic Rankine cycle technology was developed by Ormat and GE. There are 18 similar facilities across Canada and the northern U.S. installed by Ormat recovering waste heat and generating about 90 MW of power without emitting any new greenhouse gases or other emissions into the environment [12]. The first plant was installed in 1999. Furthermore, as they are air-cooled, they do not use water for the operation of the power plant.

The schematic flow diagram and heat/mass balance for a typical 6.3 MW Ormat plant is given in Fig. 9.16 [12]. The working fluid in the Ormat Energy Converter (OEC) cycle is n-pentane. The cycle net thermal efficiency of the OEC is 16.3%;

Fig. 9.15 The 4 MW OREG 4 power unit installed at the Peetz compressor station on the Trailblazer Pipeline in Colorado, US: compressor station (right) and ORC with air-cooled condenser (left); Therminol pipelines run from exhaust stacks to the ORC heat exchanger. Photograph © Ormat Technologies, Inc.; used with permission.

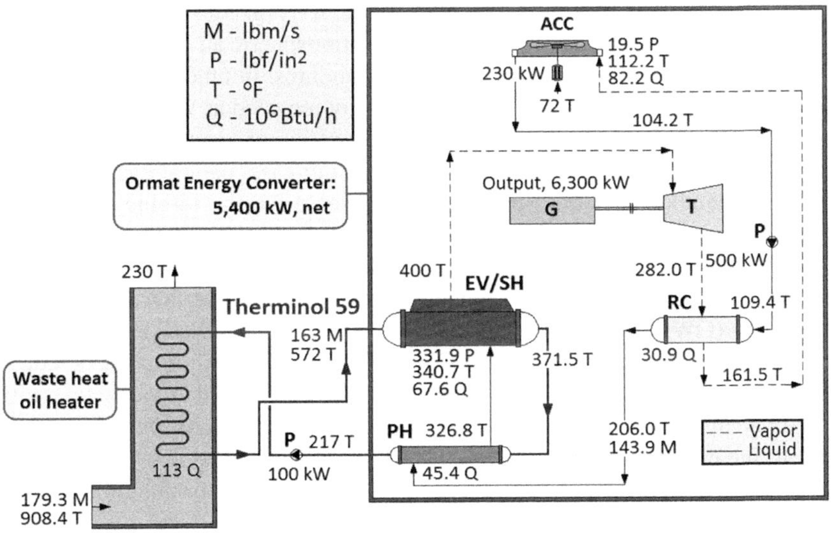

Fig. 9.16 Heat balance diagram for typical Ormat WHR ORC power plant, modified from [13]. *ACC*, air-cooled condenser; *EV/SH*, evaporator/superheater; *G*, generator; *P*, pump; *PH*, preheater; *RC*, recuperator; *T*, turbine.

the gross efficiency is 19.0%. The difference is caused by parasitic power requirements for the air-cooled condenser, two circulating pumps, and other small loads (not shown) which in total consume 900 kWe.

These WH resources can also theoretically be used for heat supply, heating and ventilation, mechanical energy, and air conditioning (cold) applications. However, since many compressor stations are remotely located, such uses are seldom adopted.

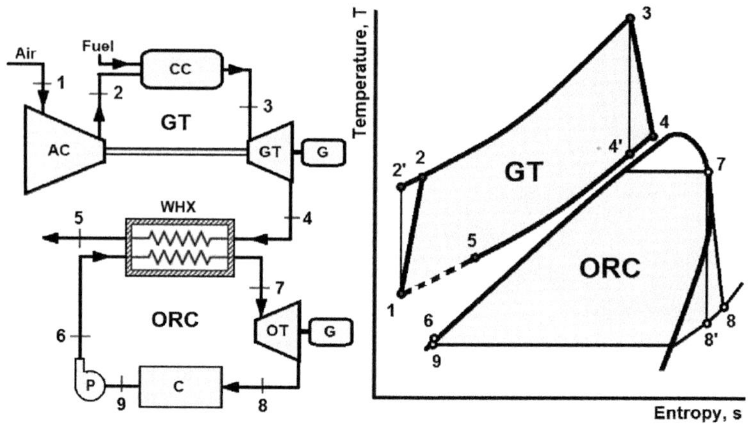

Fig. 9.17 The combined Brayton-organic Rankine cycle with waste heat recovery.

Gas turbines are modeled by the Brayton cycle; see Fig. 9.17. Generally, GT operate on an "open" cycle whereby the inlet fluid is atmospheric air and the exhaust products of combustion are discharged into the surroundings. In other words, the working fluid passes continuously through the GT and is not recycled as would be the case in a closed cycle. Typical applications include jet propulsion and stationary electric power plants. The air is compressed, mixed with fuel, in our case, natural gas, and fired in a combustion chamber. The hot, high-pressure gases drive the turbine which spins a generator to make electricity.

Figure 9.17 depicts a combined Brayton-ORC installation together with the *T-s* diagram showing the processes undergone by the air, combustion gases, and the organic working fluid (OWF). The exhaust gases transfer heat to the OWF to heat, vaporize, and possibly superheat (not shown) the OWF. Many OWF's exhibit a retrograde saturation curve (as shown) that allows expansion from the saturated vapor line without causing condensation in the turbine, as is the case with water/steam. This results in higher turbine efficiencies than with wet expansions. Combined Brayton/Rankine cycles can be designed to produce power outputs in the range of tens or hundreds of MW, with high efficiency.

These types of combined plants find application in several areas: industries that have high consumption of hot water and electricity such as cement, glass, drying, pulp and paper mills, oil and gas refineries, and chemicals production; banks; office buildings; shopping centers; hospitals; hotels; airports; and railway stations. In industrial installations for the production of cement, glass, ceramics, and metals, the temperature of the flue gases is 500–800 °C. More than 60% of the heat of flue gases is generally not used; see Sections 9.4.5 and 9.4.6. ORCs are capable of working with heat sources lower than 350 °C. Modern steam and ORC turbine and heat pump technologies can utilize and transform low-potential energy for higher effectiveness, while simultaneously lowering the temperature of the waste heat source. This benefit comes at the expense of additional capital and operating costs, more so with the use of advanced and complex technology.

9.4.3 Supercritical CO_2 Brayton cycle

As discussed in the previous section, Brayton cycles are usually high-temperature power plants from which significant waste heat is available. In this section we examine a Brayton cycle that can be operated using relatively low-to-moderate temperature heat sources. This comes about owing to the use of carbon dioxide, CO_2, as the working fluid in the cycle which for this application is closed. The CO_2 circulates through the plant, being heated, expanded, cooled, and pressurized. The main features that makes this suitable for lower temperature heat sources is that the CO_2 pressure is always maintained above the critical pressure of 7.4 MPa, and the temperature remains above the critical temperature of 31.06 °C. Having so low a critical temperature means that many waste heat sources can be used to supply heat to this cycle. Furthermore, the efficiency of the supercritical CO_2 cycle (sCO_2) may be enhanced by the use of an internal heat recuperator that captures much of the heat that would otherwise be rejected to the atmosphere. The following selected references have been used as sources of information on sCO_2 Brayton cycles: [14–19].

The sCO_2 cycle was known long ago but only recently have advances in technology made it practical. A wide variety of heat sources can be combined with the sCO_2 cycle including concentrated solar thermal, waste heat from many industries, fossil fuel-fired boilers, refuse incinerators, geothermal resources, and even nuclear power plants.

The advantages of the sCO_2 cycle include:

- Single-phase operation that eliminates all 2-phase equipment
- High density fluid similar to liquid water throughout the cycle
- Small size equipment owing to high density throughout the cycle
- Lower capital cost due to small sizes
- High heat capacity
- Liquid-like compression, thereby reducing the work of compression compared to a conventional gas turbine cycle
- Reduced compressibility factor near the critical point
- Non-toxic, inexpensive, nonflammable
- Effectively sequesters an abundant and important greenhouse gas within the plant
- Dry (air) cooling possible in arid climates
- Reduced water consumption.

The simplest sCO_2 cycle is shown in Fig. 9.18. The closed cycle consists of six thermodynamic state points: states 1, 5 and 6 are low pressure, chosen here as 12.4 MPa,

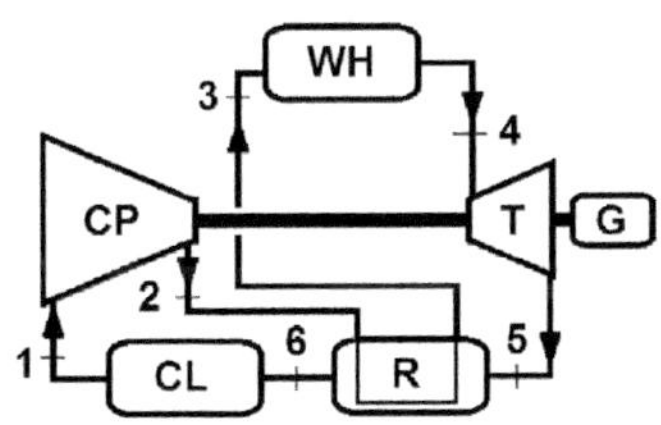

Fig. 9.18 Simplest closed Brayton cycle using supercritical CO_2 as working fluid. *CL,* cooler; *CP,* compressor; *G,* generator; *R,* heat recuperator; *T,* turbine; *WH,* waste heat heater.

while states 2, 3 and 4 are high pressure, 20 MPa. As an example, the temperature at the inlet to the compressor is 35 °C, so that all states in the cycle are greater than the critical temperature. Depending on the temperature of the waste heat exhaust gases, more complex systems may be appropriate, such as multiple turbines with interstage reheat.

The recuperator is an internal heat exchanger then heats the pressurized sCO_2 by heat exchange with low-pressure, post-expansion sCO_2. After exiting the recuperator, the sCO_2 is heated in a second heat exchanger (WH) by some waste heat source, raising the temperature to its ultimate value of approximately 200 °C. The heated and pressurized sCO_2 is then expanded through the turbine that drives the generator to make electricity. The hot exhaust fluid gives up much of its thermal energy in the recuperator, leaving a small amount to be discharged in the cooler.

Table 9.1 lists some thermodynamic properties at the six state points; Figs. 9.19 and 9.20 show the cycle processes in temperature-entropy and pressure-enthalpy diagrams, respectively. The following assumptions have been made to arrive at these results:

- No pressure losses in pipelines and heat exchangers
- Compressor efficiency = 0.80
- Turbine efficiency = 0.85
- Recuperator effectiveness = 0.60
- Thermodynamic properties of sCO_2 taken from Refprop V. 10 [20].

The broken lines in the diagrams indicate where heat is exchanged in the recuperator. Heat transfer with the surroundings occurs during processes 3-4 (into cycle from waste heat source) and 6-1 (rejected to air or cooling water); work transfer occurs during processes 1-2 (input to drive the compressor) and 4-5 (output from turbine).

Table 9.2 summarizes the heat and work terms and thermal efficiency. Although the efficiency of 10.7% may seem low, the input heat is low-temperature waste heat that would otherwise not be put to use. In other words, once the plant is built and operating, there is no fuel cost.

Table 9.1 State-point properties for sCO_2 Brayton cycle.

State	Temperature, °C	Pressure, MPa	Density, kg/m^3	Enthalpy, kJ/kg	Entropy, kJ/kg.K
1	35	12.4	775.0	278.69	1.2350
2	46.0	20	807.4	290.64	1.2425
3	104.5	20	459.9	436.49	1.6622
4	200	20	258.8	597.79	2.0465
5	156.5	12.4	181.5	568.56	2.0586
6	72.4	12.4	351.7	422.71	1.6756

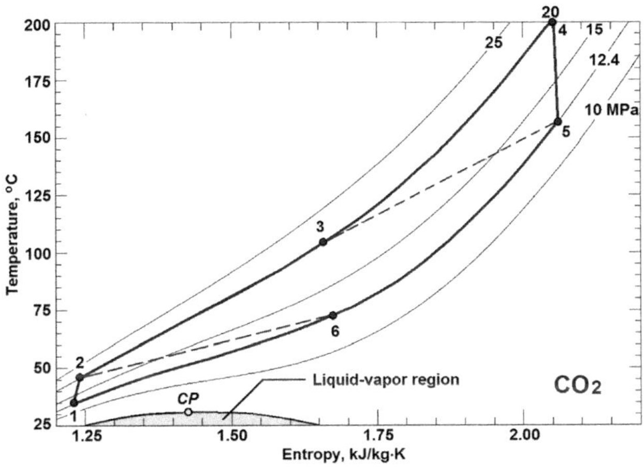

Fig. 9.19 Temperature-entropy process diagram for sCO_2 cycle: *CP*, critical point.

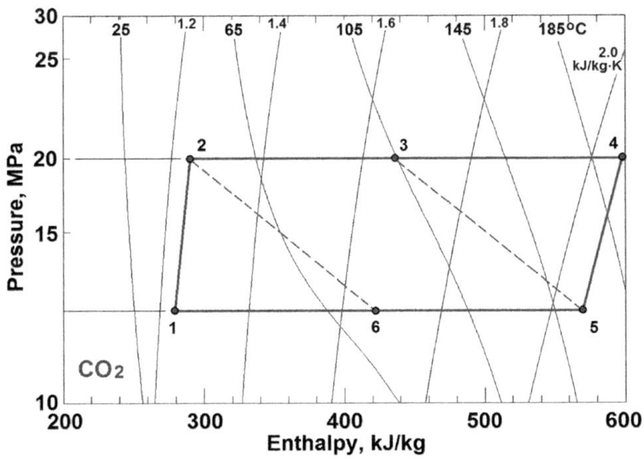

Fig. 9.20 Pressure-enthalpy process diagram for sCO_2 cycle.

Table 9.2 Summary of heat, work and efficiency for sCO_2 cycle.

Assumed mass flow of CO_2	**100 kg/s**
Work of compression	1195 kW
Waste heat input	16,130 kWt
Work output from turbine	2923 kW
Heat rejected from cooler	14,402 kWt
Heat exchanged in recuperator	14,585 kWt
Net power output	1728 kW
Thermal efficiency	0.107

9.4.4 Boiler plant heat recovery with ORC power cycles

Industrial furnaces are widely used in many industries (metallurgy, mechanical engineering, chemicals, ceramic and building materials). Industrial furnaces using fossil fuels are characterized by heat losses from flue gases with a temperature of 500 °C and above. Recuperators and regenerators are used to partially cool the flue gases and preheat the air supplied to the furnace burners. However, the temperature of the flue gases after cooling remains high and reaches 250–350 °C.

The same characteristics apply to heating boiler plants of large technological facilities and municipalities. These plants consume a significant amount of energy, usually as natural gas, whose price fluctuates depending on supply, demand, and world economic and political conditions. The efficiency of boiler units nowadays is about 0.85–0.90, with the loss of heat via waste gases being about 6–7% or more. The internal use of heat from exhaust gases of boiler units is used for preheating the combustion air supplied to the burner and heating the return circuit water of the heat supply system, but significant heat is still lost to the surroundings.

Since the purpose of boilers is to raise steam for various uses, these boilers do not produce electricity, which means a low energy conversion factor for the fuel. Furthermore, in the event of an emergency power outage and shutdown of their operation, heating mains may be damaged and consumers will lose their steam supply unless an alternative power supply in available on site. One way to use waste gas heat is to generate electricity to meet the boiler house needs such as motor drive of fans, smoke exhaust fans, circulating pumps, etc. A system for a heat recovery power plant using an organic Rankine cycle (ORC) is shown in Fig. 9.21.

The waste gas heat recovery after the boiler A and the economizer B is carried out in the recovery evaporator C. The ORC working fluid vapor is fed into the turbine D, where expansion occurs performing work to drive the electric generator E. Then, the vapor condenses in the condenser F and the condensate is fed back to the heat recovery system by the pump G. The thermodynamic efficiency of the cycle will not be high but the heat supply is effectively free as waste heat.

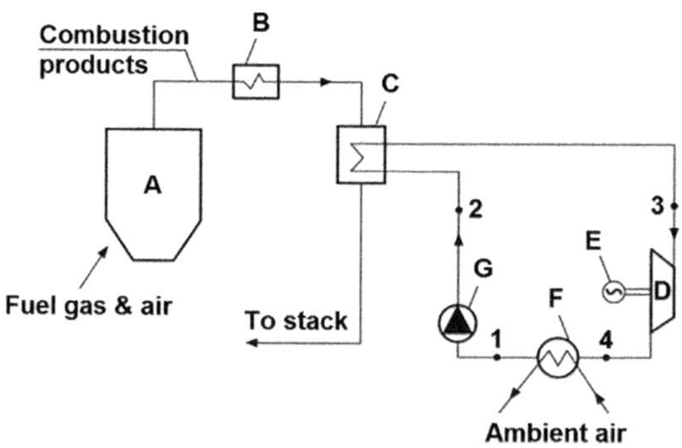

Fig. 9.21 Flow diagram of a boiler unit with heat recovery cogeneration power plant. *A*, boiler; *B*, economizer; *C*, heat recovery unit/evaporator; *D*, turbine; *E*, electric generator; *F*, condenser; *G*, pump.

Calculations were performed on an ORC power unit with different working substances (R600; R600a; R601a, R602, and others). For boilers with a heat output of 60–120 MWt, a mass flow rate of combustion products of 25–40 kg/s, a gas temperature of 280–300 °C, and a turbine inlet temperature in the ORC of 200 °C, the specific electric power generated is from 35 to 50 kWe/(kg/s), and the potential total electrical power for the WHR unit is 875–2200 kWe.

9.4.5 Heat recovery from cement production with ORC power cycles

In this and the next section, we will describe waste energy potential in the glass and cement industries, both of which are highly energy intensive. Significant amounts of WH are available at such enterprises. The main problem with attempting to capture these waste heat quantities is the lack of consumers of secondary thermal energy resources at the facilities themselves. Therefore, the waste heat can only be effectively utilized for heating purposes by being transferred to the ultimate end users, which can include city district heating systems. However, if that waste heat is converted into electricity at the plant site, then the electricity can be delivered to distant end users via transmission lines.

Some data on worldwide cement production will set the framework for this discussion. Of all the energy expended in the non-metallic mineral sector (~9% of total global energy use), manufacture of cement accounts for 70–80%. The weighted average among cement-producing countries for specific energy consumption comes to 4.4 GJ per tonne of product. China produces nearly one-half of all cement in the world. With so much energy being expended, there is a comparable high potential for energy savings: 2.5 to 3 EJ per year may be saved (28–33% of all energy consumed in this sector) by various means, including waste energy recovery. Such savings in primary energy would have corresponding reductions in greenhouse gas emissions, particularly CO_2 [21].

The process of producing cement and its follow-on product concrete are shown schematically in Fig. 9.22 [22]. The raw material, mainly limestone, is crushed in ball mills, passed through an electrostatic precipitator, stored, preheated, and reacted in a high-temperature rotary kiln which yields clinker. To make cement, the clinker must first be cooled. Prior to being crushed in the cement mill, the gypsum produced in the kiln is separated from the main product stream. The output from the cement mill may be blended with other constituents to meet certain specifications depending on the end use. The packaged product is then shipped to the consumers. Electricity is one of the main energy inputs; worldwide, the electricity intensity of cement production is about 91 kWh per tonne of cement. An international goal has been established to reduce this to 87 kWh/t by 2030 [23].

The main energy consumption (in 10^9 kJ) are for: raw grinding = 8.346 (1.88%), kiln heating (fuel combustion) = 410.464 (92.68%), and finish milling = 24.057 (5.43%) [22]. Although the firing of the kiln consumes the bulk of the energy, there are other places along the production line where waste energy can be recovered. Figure 9.23 [24] focuses on the preheater, kiln, and clinker cooler, showing the

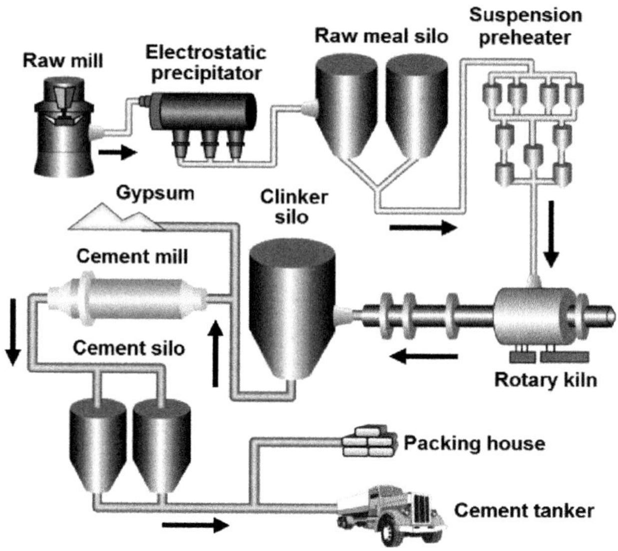

Fig. 9.22 Schematic flow diagram for cement production [22].

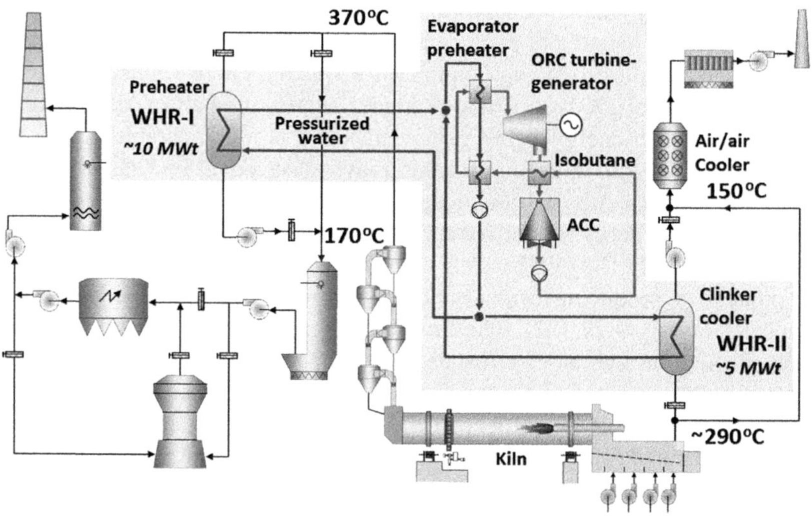

Fig. 9.23 Waste heat opportunities in cement manufacturing processes, modified from [24]. Schematic and data are for the Untervaz (Switzerland) plant. Total electrical power is 2.6 MW [25].

primary waste heat sources (WHR-I and WHR-II); secondary waste heat may be recovered at the shells of the preheater and the kiln. The primary ones are suitable for power generation while the secondary ones may be appropriate for direct heat applications using hot water [26]. The most commonly used WHR power technologies are steam Rankine cycle with various enhancements and ORC (shown in Fig. 9.23), including Kalina, and supercritical CO_2 Brayton cycles.

One of the first commercial waste heat power generation plants using ORC technology was implemented by Turboden using the exhaust gas from a cupola furnace in Torbole, Italy [27]. Around the same time, another plant came on-line at Heidelberg Cement in Lengfurt, Germany (1998) by Ormat Technologies.

Figure 9.24 shows the heat balance for a dry cement kiln for the following conditions: exhaust temperature = 290–390 °C; cooler exit temperature = 250–350 °C. Approximately 35% of the total energy involved can be used for drying the product and for WHR power generation.

Table 9.3 provides some information for selected examples of the recovery of waste heat from cement production facilities [21]. The Ait-Baha plant is shown in Fig. 9.25 [30]. This plant began with an annual production capacity of 2.2 million tonnes of cement, but currently puts out about 3.6 million tonnes of clinker and 4.9 million tonnes of cement [29].

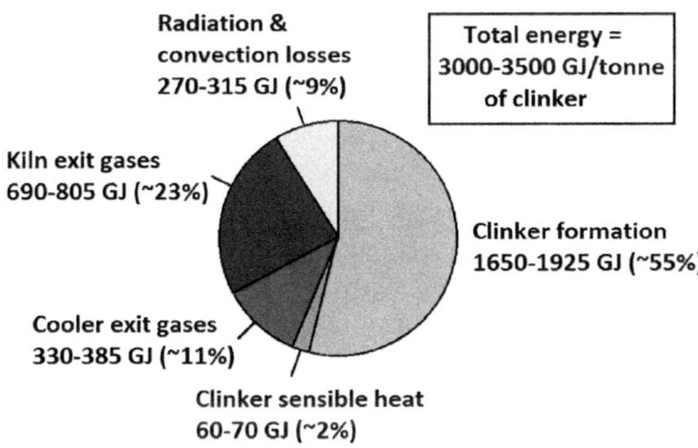

Fig. 9.24 Heat balance of a dry cement kiln, modified from [24].

Table 9.3 Selected examples of WH power generation at cement production plants [28].

Location	Power, MWe	Year of start-up	Plant owner
Ait Baha, Morocco[a]	2	2010	Ciments Du Maroc (Heidelberg Group)
Aleşd, Romania	4	2012	Holcim SA
Rohožník, Slovakia	5	2014	CRH
Fieni, Dâmboviţa, Romania	3.8	2015	S.C. Carpatcement Holding S.A. (Heidelberg Group)

[a]Hybrid plant, includes a concentrating solar power unit installed in 2015.

Fig. 9.25 Ciments Du Maroc production facility at Ait-Baha, Morocco. Elevated structure houses the conveyor from the mine which is about 2 km away [29].

9.4.6 Heat recovery from glass factories with ORC power cycles

Glass manufacture also is a heavily energy intensive industry. This includes the production of container glass (bottles), sheet or plate glass (via the float process), fiberglass, specialty glass, and secondary products made from glass. Data from Europe show that there has been a dramatic improvement in energy efficiency. Whereas in 1960 the energy intensity in the glass industry was 35 GJ/tonne of product, the most recent data show 8 GJ/tonne, a reduction of 77% [28,31,32]. Nevertheless, there is still room for improvement through waste heat recovery (WHR).

WHR options suitable for the glass industry include electricity generation, steam or hot water production, recuperation, natural gas fuel preheating, and batch/cullet preheating. It has been established in practice that ORC systems are a good option with important opportunities in the glass industry [28].

The primary feedstock for glass making today, as it has been for 7000 years, is sand. Figure 9.26 shows a schematic flow diagram for the entire glass production line, including a waste heat recovery system. An excellent video of the making of flat glass via the float method (or float glass), invented by Sir A. Pilkington, can be found at Ref. [34].

Melting furnaces are the major energy users, accounting for more than 75% of all energy needed. The melting is done via combustion of fuel, usually natural gas, electric heating, or a combination of the two. The availability of waste heat is evident considering that the melting furnaces have of 50–60% efficiency; heat losses through the walls and flue gas losses represent 20–25% and 25–35% of losses, respectively [28]. It is estimated that internal modifications to the system components can lead to energy savings of 20–25% overall, mostly coming from improvements in the melting and refining areas [35].

As a means of reducing the specific energy consumption and consequently the CO_2 emissions during glass production, a widely proven method taps the waste heat of exhaust gases, which typically have a heat content of about 25–30% of the furnace

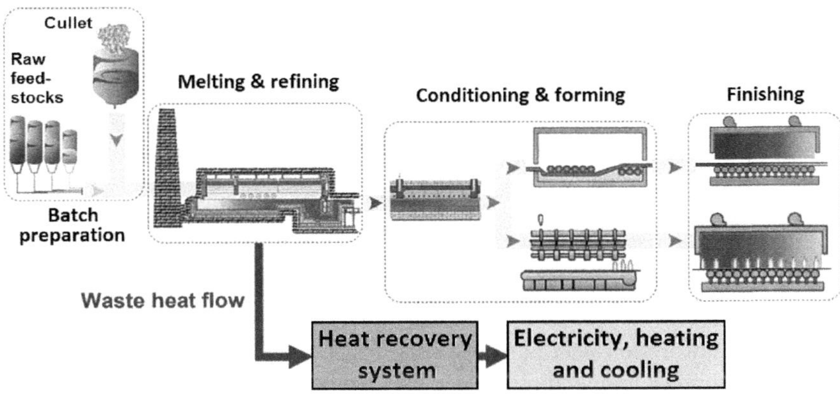

Fig. 9.26 Glass production processes. Raw feedstocks include sand (primarily), soda ash, and limestone; cullet is recycled glass from previously used material and broken glass in the production process. Melting and refining are the most energy intensive phases of production, modified from [33].

energy input. The exhaust gas temperature downstream of the air regenerators is 400–500 °C, and can reach 700 °C or higher with recuperators of air-fired furnaces, whereas at oxygen-fired glass melting furnaces temperatures higher than 1100 °C can be seen. For WH temperatures of 450–500 °C, ORC systems and steam Rankine cycle systems can show cycle thermal efficiencies of 15–19%. As a rough estimate, a 500 tonne/day glass manufacturing plant will have potential for a 1 MW ORC WH power plant. Table 9.4 gives some information on two WH glass plants in Italy [28].

The following additional sources of information on glass making and its energy and environmental implications are available to interested readers: [24,36–38].

9.4.7 Waste-heat-driven absorption chillers and heat pumps

Heat-driven absorption chillers may be used in heat recovery systems. Absorption chiller machines (ABCMs) can be driven by waste heat (WH) from various sources such as exhaust gases, steam, and hot water including geothermal, which can amount to a savings of 35% of the initial primary thermal energy source. ABCM using WH can produce chilled water with a temperature of about 5 °C [39].

Table 9.4 Examples of WH power plants at glass production plants in Italy [28].

Location	Power, MWe	Year of start-up	Plant owner	Glass type
Cuneo	1.3	2012	Gea Bischoff/ AGC	Float glass
Villotta di Chions (PN)	0.5	2015	BDF Industries	Container glass

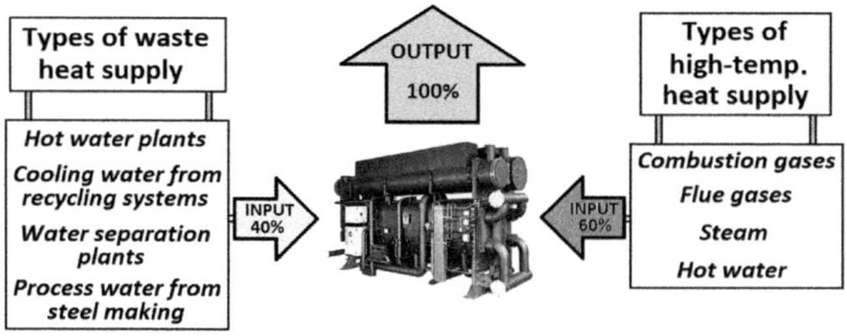

Fig. 9.27 Schematic operation of a typical ABHP, modified from [39].

Heat-driven absorption heat pumps (ABHPs) are used as heating systems to produce heat of a higher temperature than a low-temperature source via water heating (40 °C–175 °C) or steam production with pressures up to 0.6 MPa.

ABHP may be designed to deliver heat either at the high-temperature end or the intermediate-temperature side, depending on the requirements of the end user. An example of the latter application is depicted schematically in Fig. 9.27. Here the system uses a high-temperature heat resource in combination with heat from a low-temperature source to obtain an intermediate-temperature heat stream. The ABHP is capable of increasing the temperature of the heated medium by 30–60 °C above the temperature of the low-temperature source. The quantity of heat at the exit of ABHP consists of the heat received from the low-temperature WH source (~40%) and the heat received from the high-temperature source (~60%).

A specific case example is shown in Fig. 9.28 which is for the district heating system in Thisted, Denmark [40]. The arrangement is actually two heat pumps connected at the mid-level temperature where the district heating return water is reheated to its specified temperature. The high-temperature heat comes from a straw-fired boiler, a

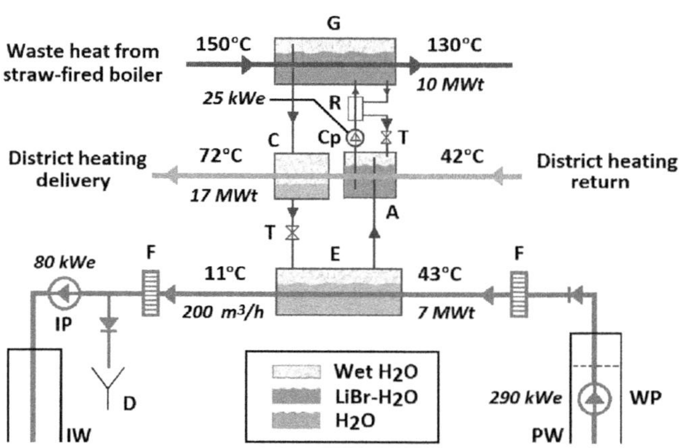

Fig. 9.28 Geothermal plant in Thisted, Denmark, redrawn from [40]. *A*, absorber; *C*, condenser; *Cp*, compressor; *E*, evaporator; *F*, filter; *G*, generator; *IP*, injection pump; *IW*, injection well; *PW*, production well; *R*, recuperator; *T*, throttle valve; *WP*, well pump.

renewable biomass energy source. The heating fluid enters the generator of the ABHP at 150 °C and leaves at 130 °C; there is a 10 MWt heat transfer. The return water of the heating system at 42 °C enters the absorber, and then the condenser, where it heats up to 72 °C and sent back to the district heating users. For efficient operation of the heat pump in ABHP, water is pumped from a geothermal well with a temperature of 43 °C to the evaporator where it is cooled to 11 °C, and then disposed of via an injection pump in a disposal well.

The heat balance of Thisted ABHP is composed of the supplied heat of hot water, with a capacity of 10 MWt, and the heat of the supplied geothermal water, 7 MWt. The total heat generated in the ABHP and supplied to the heating system is, therefore, 17 MWt.

9.5 Summary

There is ample waste heat produced in industry for entrepreneurs to consider as sources of energy for a variety of applications. In this chapter we have examined many such cases where recovery of waste heat and other forms of energy are technically feasible. Since the decision to proceed with any such project depends on the economic feasibility which in turns is site- and time-specific, that aspect is left to those who know those conditions for their particular proposed project.

This chapter considers several diverse waste heat sources, but focuses on three industrial type applications: (1) natural gas distribution pipeline network, (2) cement fabrication processes, and (3) glass-making. The main objective in all cases is the generation of electricity, but direct heat applications are also discussed.

The requirement to maintain gas pressure in the pipelines that carry natural gas from the well fields to end users means that compressor stations are situated at regular intervals along the pipeline. Gas turbines are the natural choice to drive the compressors given that the product flowing in the pipe is a suitable combustible fuel. The high-temperature exhaust from the gas turbines is readily captured either by a bottoming steam Rankine cycle or by an organic Rankine cycle using one of any number of working fluids. Since the former type generally need to be complex systems in order to achieve high efficiency, the latter type are generally preferred owing to their simplicity. Thus, there are now hundreds of megawatts being generated along pipelines in Canada, the U.S., Europe, and Asia. Although the thermal efficiencies are low, ~15%, they are sized for the application, ~5 MW/unit, and can be operated and monitored remotely. One of the main advantages of these units is that they produce no new atmospheric emissions while generating power that can be sold to utilities.

Although they differ in detail, both cement manufacture and glass-making facilities offer numerous waste heat recovery opportunities, at different levels of temperature. Well-designed plants use some potential waste heat internally to lower the consumption of external heat supply from primary fuels, but there are processes where there is sufficient un-reused heat to justify a waste heat recovery power unit. The practical power in the form of electricity that can be recovered is generally less than that from natural gas compressor stations and typically is from 1 to 5 MWe.

In order for waste heat recovery to be considered, there must be a buyer for the product, namely, either electricity or heat. Furthermore, the contract for energy purchase must be adequate for the developer to earn a decent return of the investment. Variations and fluctuations in the price of both the primary product of the industry and of the fuels used to generate electricity in central stations make it difficult to decide on the feasibility of all waste heat recover options even for industries where ample waste heat is available, thus mandating thorough site-specific feasibility studies before committing to such a venture.

Nomenclature

Latin letters

l	coefficient taking into account heat loss
m	specific mass flow rate
h	enthalpy
H	enthalpy
Q	amount of heat
V	consumption
W	power

Greek letters

α	some fraction of the condenser heat
β	coefficient accounting for the mode of plant operation
η	efficiency
ξ	utilization factor

Subscripts

AC	air compressor
ABCM	absorption chiller machines
ABHP	absorption heat pump
ARM	absorption refrigeration machine
B	boiler
CC	combustion chamber
C	cooler
C	condenser
DA/FWH	deaerator/feed water heater
E	economizer
FP	feed pump
G	generator
GT	gas turbine
HS	heating system
HWS	hot water supply system

HSE	helical screw expander
HTF	heat transfer fluid
ICE	internal combustion engines
ORC	organic Rankine cycle
OWF	organic working fluid
P	pump
PLD	pressure let-down valves
RB	recovery boiler
SD	steam drum
SH	superheater
VCR	vapor-compression refrigeration
WH	waste heat
WHR	waste heat recovery
WHU	waste heat unit

Review questions

1. Explain the concept of "secondary energy resources.
2. To which groups do the secondary energy resources belong?
3. What are the sources of waste heat recovery of various industries?
4. What types of heat exchangers used to capture the heat of gases and liquids?
5. How can one make use of the heat of chemically aggressive gases or liquids?
6. Name some high-temperature heat-transfer media for heat transfer over long distances.
7. Cogeneration units: describe the principle of operation and thermodynamic efficiency.
8. Combined steam-gas turbine units: describe the technological scheme.
9. Open and closed gas turbine installations: what are the advantages and disadvantages?
10. Describe the open cycles of Brayton/Rankine.
11. Describe the closed cycle of Brayton/Rankine.
12. How can one combine simple Brayton/ORC cycles?
13. How can one combine Brayton/ORC cycles with a recuperator?
14. Describe the supercritical carbon dioxide cycle (sCO_2).
15. Gas turbines and boilers: describe the technological schemes.
16. List and compare several heat conversion technologies.
17. Make an ORC flowchart.
18. ORC Modules: list and describe various cycle types.
19. ORC modules: describe selection criteria for working fluids.
20. ORC modules: describe the basic equipment.
21. Give some examples of ORC technology used in WHR.
22. Describe how absorption refrigeration machines are used in WHR schemes.
23. Discuss the future of technologies in waste heat recovery.

Exercises

1. Steam is produced in the ideal cogeneration cycle, show in Fig. 9.10, at 10 MPa and 700 °C, at a mass flow 25 kg/s. Thirty percent of the steam is extracted from the turbine at 700 kPa and diverted to a process heater. This steam leaves the process heater at 700 kPa as a

saturated liquid. The remaining steam leaves the turbine at 20 kPa, but 30% of the condenser heat transfer is used for heating and cooling buildings. Sketch this process on a T-s diagram and calculate the utilization factor for the cycle.

2. Air enters the compressor of the Brayton cycle of Fig. 9.17 at 300 K and 100 kPa, where it is compressed to 700 kPa. It exits the combustor at 1200 K. The exhaust gases exit the heat exchanger at 350 K. The pressure in the Rankine cycle is raised by the pump from 20 kPa to 5 MPa. The heat exchanger provides the steam turbine with steam at 400 °C. Determine the efficiency of the combined cycle if the steam turbine power output is 10 MW.

References

[1] Jouhara H, Khordehgah N, et al. WHR technologies and applications. Thermal Science and Engineering Progress 2018;6:268–89.

[2] Rezaie B, Rosen MA, et al. District heating and cooling: review of technology and potential enhancements. Applied Energy 2012;93:2–10.

[3] Panayiotou GP, Bianchi G, et al. Preliminary assessment of waste heat potential in major European industries. In: 1st int. conf. on sustainable energy and resource use in food chains, ICSEF 2017, 19–20 April 2017, Berkshire, UK. Energy Procedia, vol. 123; 2017. p. 335–45.

[4] Tartière T, Astolfi M. A world overview of the organic Rankine cycle market. In: IV international seminar on ORC power systems, ORC2017, 13–15 September, 2017, Milano, Italy.

[5] Langan M, O'Toole K. A new technology for cost effective low grade waste heat recovery. In: 1st int. conf. on sustainable energy and resource use in food chains, ICSEF 2017, 19–20 April 2017, Berkshire, UK. Energy Procedia, vol. 123; 2017. p. 188–95.

[6] DiPippo R. Geothermal power plants: principles, applications, case studies and environmental impact. 4th ed. Oxford, England: Butterworth-Heinemann: Elsevier; 2016.

[7] Bianchi M, Branchini L, et al. Techno-economic analysis of ORC in gas compression stations taking into account actual operating conditions. In: IV int. Sem. on ORC power systems, ORC201713-15, September 2017, Milano, Italy. Energy Procedia, vol. 129; 2017. p. 543–50.

[8] Energy Information Administration. Office of Oil & Gas, Natural Gas Division, Natural Gas Transportation Information System.

[9] Campana F, Bianchi M, et al. ORC waste heat recovery in European energy intensive industries: energy and GHG savings. Energy Conversion and Management 2013;76: 244–52.

[10] Cleveland T, Mokhatab S. Capturing waste heat from gas compressor stations. Pipeline and Gas Journal September 20, 2017.

[11] Biščan D, Lončar D. Optimization of waste heat utilization in gas turbine pipeline compressor station. Strojarstvo (Engineering) 2010;52(4):475–87.

[12] Hedman BA. Waste energy recovery opportunities for interstate natural gas pipelines. Report for Interstate Natural Gas Association of America (INGAA). February 2008.

[13] Bronicki LY. Ormat Rankine cycle configurations for utilization of low temperature heat sources. In: Proc. 9th biennial ASME conf. On engineering systems design and analysis, ESDA2008-59126, July 7–9, 2008, Haifa, Israel.

[14] Dennis R. Overview of supercritical carbon dioxide based power cycles for stationary power generation. In: International seminar on organic Rankine cycle power systems, Politecnico di Milano, Milano, Italy, September 13, 2017.

[15] Anderson KR, Wells T, et al. Waste heat energy supercritical carbon dioxide recovery cycle analysis and design. Advances in Renewable Energy 2014;1(1).

[16] Vasquez R,P, Tooa YCS, et al. Exergetic analysis of supercritical CO_2 Brayton cycles integrated with solar central receivers. Applied Energy June 2015;113.

[17] Iverson BD, Conboy TM, et al. Supercritical CO2 Brayton cycles for solar-thermal energy. Applied Energy 2013;111:957–70.

[18] Pasch J, Conboy T, et al. Supercritical CO2 recompression Brayton cycle: completed assembly description. Sandia Report No. SAND2012-9546. October 2012.

[19] U.S. Dept of Energy. Supercritical carbon dioxide Brayton cycle. Quadrennial Technology Review 2015, Chapter 4 Technology Assessments.

[20] Thermodynamic and Transport Properties Database (REFPROP): version 10, National Institute of Standards and Technology (NIST), US Department of Commerce. https://www.nist.gov/srd/refprop.

[21] Tracking industrial energy efficiency and CO_2 emissions. International Energy Agency. https://www.iea.org/publications/freepublications/publication/tracking_emissions.pdf.

[22] Energy efficiency and saving in the cement industry. ClimateTechWiki – A Clean Technology Platform. https://www.climatetechwiki.org/technology/energy-saving-cement.

[23] Cement - tracking clean energy progress. OECD/IEA; January 25, 2019. https://www.iea.org/tcep/industry/cement/.

[24] Hepberger M. ORC Waste Heat Recovery system using kiln exit and cooler vent air. Düsseldorf: VDZ Jahrestagung; 2012.

[25] Holcim Untervaz plant improves energy-efficiency by 20% by using ABB low-temperature ORC technology. Global Cement June 14, 2011. http://www.globalcement.com/magazine/articles/371-holcim-untervaz-plant-improves-energy-efficiency-by-20-by-using-abb-low-temperature-orc-technology.

[26] Best energy consumption. ICR Research; February 16, 2015. https://www.cemnet.com/Articles/story/156121/best-energy-consumption.html.

[27] Waste Heat Recovery. Turboden S.p.A., Brescia, Italy. https://www.turboden.com/applications/1053/waste-heat-recovery.

[28] Hatzilau CS, Karellas S, et al. Energy saving incentives for the European glass industry in the frame of the EU Emissions Trading Scheme. In: ECEEE industrial summer study proceedings, Industrial efficiency. 2016. pp. 531–541.

[29] Morocco - reference projects in Morocco. Heidelbergcement Group. https://www.heidelbergcement.com/en/morocco.

[30] Usine d'Aït Baha (AIT Baha Plant). Ciments du Maroc, HeidelbergCement Group. https://www.cimentsdumaroc.com/fr/usine-dait-baha [in French].

[31] Karellas S, Zourou K, et al. Energetic and exergetic assessment of waste heat recovery systems in glass industry. In: ASME ORC2013, 2nd int. sem. on ORC power systems October 7–8, 2013, De Doelen, Rotterdam, The Netherlands.

[32] Karellas S, Giannakopoulos D, et al. The potential of WHR/batch and cullet preheating for energy efficiency in the EU ETS glass industry and the related energy Incentives. Energy Efficiency 2018;11:1161–75.

[33] Glass. Industrial efficiency technology database. The Institute for Industrial Productivity. http://ietd.iipnetwork.org/content/glass.

[34] Pilkington float glass process. NSG Group; October 18, 2013. https://www.youtube.com/watch?v=ig4G5WbOMLc.

[35] Glass manufacturing is an energy-intensive industry mainly fueled by natural gas. Today in Energy August 21, 2013. https://www.eia.gov/todayinenergy/detail.php?id=12631.

[36] Antonio MD, Hildt N. Energy efficiency opportunities in the glass manufacturing industry. https://aceee.org/files/proceedings/2003/data/papers/SS03_Panel4_Paper_08.pdf.
[37] Industrial decarbonisation & energy efficiency roadmaps to 2050. March 2015. https://assets.publishing.service.gov.uk/government/uploads/system/uploads/attachment_data/file/416675/Glass_Report.pdf.
[38] Scalet BM, Garcia Muñoz M, et al. Best available techniques (BAT) reference document for the manufacture of glass. Industrial Emissions Directive 2010/75/EU. 2013. http://eippcb.jrc.ec.europa.eu/reference/BREF/GLS_Adopted_03_2012.pdf.
[39] Абсорбционные бромистолитиевые теЦловые насосы (Absorption Lithium-Bromide Heat Pumps). Cold Energy Ltd., Official rep. of Thermax, St. Petersburg, Russia. http://abxm-thermax.ru/abxm/absorbcionnye-teplovye-nasosy/.
[40] Mahler A, Magtengaard J. Country update report for Denmark. In: Proc. World geothermal congress 2010, Bali, Indonesia, April 25–29, 2010. https://www.geothermal-energy.org/pdf/IGAstandard/WGC/2010/0131.pdf.

Index

'*Note*: Page numbers followed by "f" indicate figures and "t" indicates tables.'

I

K

L

M

9780128162491